Variational Principles and Distributed Circuits

OPTOELECTRONICS AND MICROWAVES SERIES

Series Editor: **Professor T Benson and Professor P Sewell**
The University of Nottingham, UK

* forthcoming

Variational Principles and Distributed Circuits

A. Vander Vorst
Université catholique de Louvain
and
I. Huynen
National Fund for Scientific Research,
Université catholique de Louvain

RESEARCH STUDIES PRESS LTD.
Baldock, Hertfordshire, England

RESEARCH STUDIES PRESS LTD.
16 Coach House Cloisters, 10 Hitchin Street, Baldock, Hertfordshire, England, SG7 6AE
and
325 Chestnut Street, Philadelphia, PA 19106, USA

Marketing:

Research Studies Press Ltd.
16 Coach House Cloisters, 10 Hitchin Street, Baldock, Hertfordshire, England, SG7 6AE

Distribution:

NORTH AMERICA
Taylor & Francis Inc.
International Thompson Publishing, Discovery Distribution Center, Receiving Dept.,
2360 Progress Drive Hebron, Ky. 41048

ASIA-PACIFIC
Hemisphere Publication Services
Golden Wheel Building # 04-03, 41 Kallang Pudding Road, Singapore 349316

UK & EUROPE
ATP Ltd.
27/29 Knowl Piece, Wilbury Way, Hitchin, Hertfordshire, England, SG4 0SX

Library of Congress Cataloguing-in-Publication Data

Vorst, A. vander (André), 1935-
    Variational principles and distributed circuits / A. Vander Vorst and I. Huynen.
      p. cm. – (Optoelectronic and microwaves series)
    Includes bibliographical references and index.
    ISBN 0-86380-256-7
     1 Microwave transmission lines—Mathematical models. 2.
Electromagnetism—Mathematics. 3. Variational principles. I. Huynen, I. (Isabelle),
1965-II. Title. III. Series.

TK7876 .V6324 2000
621.381'31—dc21   00-039022

British Library Cataloguing in Publication Data
A catalogue record for this book is available from the British Library.

ISBN 0 86380 256 7

Printed in Great Britain by SRP Ltd., Exeter

Editorial Foreword

Variational principles have a long history of providing insight and practical solutions to problems in a wide range of scientific and engineering disciplines. In particular, their substantial contribution toward the development of high frequency technologies during and after World War II, by underpinning the equivalent network approach for the design and analysis of microwave circuits, is clearly recognised. At a time before modern computers, variational principles provided solutions to immediate engineering problems with unparalleled accuracy. However, even in the modern era, as the power of computers has increased at an astonishing pace, variational methods have evolved and remain an invaluably elegant and powerful tool for researchers in a variety of fields. Therefore, we are very pleased to welcome this book, which provides both a comprehensive and in-depth treatment of all aspects of variational principles applied to microwave, millimetre-wave and optical circuits, to this series.

Starting with a thorough review of the field, the book clearly establishes the form and physical significance of variational principles as well as their relationship with other approaches. A detailed discussion of the application of variational principles to planar microwave geometries is followed by the establishment of a general variational principle for electromagnetics and its application to a wide variety of practically important configurations, including transmission lines, cavities, junction problems as well as gyrotropic devices.

Throughout, the mathematical basis and practical implementation of the variational approaches is clearly and elegantly presented and evidently benefits from the authors' extensive knowledge and original contributions to the field. Therefore, this book is likely to be of interest both to experienced researchers in the field as well as those wishing to explore the stimulating and practically valuable topic of variational principles for the first time.

TM Benson
P Sewell

University of Nottingham
July, 2001

Contents

Preface

When, in 1888, Hertz deduced the concept of propagation from Maxwell's theories and when, a few years later, in 1895, Marconi made his first propagation experiments, nobody would have expected the extraordinary development of radio that was to come. In those days, transmission lines were already present, such as one or two wires above the ground, transmitting telegraphy. Consideration at the time and distance dependence of the signal resulted into what was called the telegraphist's equations. It did not take long to establish the link between these two-variable equations and the four-variable Maxwell's equations. Simultaneously, circuit theory evolved from lumped circuits to distributed circuits. The two developments merged, and transmission line theory was born.

The first transmission lines were analyzed using a quasi-static approach. Homogeneously filled waveguides necessitated the use of transverse electric (TE) and transverse magnetic (TM) waves which were solutions developed in the 19th century. It is quite surprising to see how much scientific material was already available at that time, with solutions developed by Laplace, Poisson, Helmholtz, Bessel, Legendre, Lamé, Floquet, Hill, Mathieu, and others.

During World War II, microwave network theory became an recognized subject, mainly due to the efforts developed at the Massachusetts Institute of Technology (MIT) Radiation Laboratory, USA. Computers were not available, fortunately one might say, enabling efforts to be concentrated on theory. It is extremely interesting to observe that advances came out of the fruitful interaction of different scientific disciplines: methods characteristic of quantum mechanics and nuclear physics were focused on the application of electromagnetic theory to practical microwave radar problems. Out of these rather special circumstances emerged a strategic lesson. In seeking to apply a fundamental theory at the observational level, it was found very advantageous to construct an intermediate theoretical structure, a phenomenological theory, capable of organizing the body of experimental data into a set of relatively few numerical parameters, and using concepts that facilitate contact with the fundamental theory. The task of the latter then became the explanation of the parameters of the phenomenological theory, rather than the direct confrontation with raw data. Right after the war came the wonderful scattering matrix tool, simultaneously adapted to microwave circuits from scattering in physics, by Belevitch in Belgium and Carlin in the US.

Schwinger played a central role in the development of microwave network theory at the MIT Radiation Laboratory. He was interested in developing "a systematic analytical procedure for reducing complicated problems to the fundamental elements of which they are composed by the application of generalized transmission line theory, symmetry considerations, and Babinet's principle". He used essentially two methods. One is the Integral Equation Method, in which imposing the boundary conditions on a discontinuity in a waveguide leads directly to integral equations from which the field is determined. He replaced the discontinuity by a lumped parameter network in a set of transmission lines. The circuit parameters were expressed directly in terms of the fields, so that the solution of the equations leads to the elements of the equivalent circuit.

The other method he called the Variational Method, where the impedance elements are expressed in such a form that they are stationary with respect to arbitrary variations of the field about its true value. By judiciously choosing a trial field, he obtained remarkably accurate results. Moreover, these could be improved by a systematic process of improving the trial field, which, if carried far enough, will lead to a rigorous result. Furthermore, in solving a given problem, several methods can be used in conjunction. For example, one might solve an integral equation approximately by a static method and then use this approximate solution as a good trial field in the variational expression. These developments resulted in a rigorous and general theory of microwave structures in which conventional low-frequency electrical circuits appeared as a special case. Particular use was made of the distinction made between the theoretical evaluation of microwave network parameters, which entails the solution of three-dimensional boundary-value problems and belongs to electromagnetic theory, and, on the other hand, the network calculations of power distribution, frequency response, resonance properties, etc., characteristic of the "far-field" behavior in microwave structures, involving mostly algebraic problems and belonging to microwave network theory.

In the 1950s, there was a tremendous breakthrough in transmission line technology. It started with the request for higher bandwidth and hence higher frequencies. The stripline, and more obviously the microstrip line, opened the era of planar hybrid distributed electronics. They brought up a new problem, that of an inhomogeneous transverse section. Stripline and microstrip were rapidly followed by the slot-line, the electromagnetic complement of microstrip. Then came the coplanar waveguide.

At the lowest frequencies, a quasi-static approach was still usable although not at the higher frequencies, and new approaches and methods became necessary. Quite naturally, quasi-static methods led to full-wave dynamic methods. New approaches were developed, such as the moment method. Old strategies were improved, among which were the Rayleigh-Ritz and Galerkin procedures. With the increasing use and capabilities of the computer, discretized formulations were developed, such as finite-differences, finite-elements, and time-domain finite-differences.

Technology continuously evolved, not only in terms of topologies, but also toward smaller size and higher frequencies, with microwave integrated circuits, both hybrid and monolithic. Today the variety of structures is quite large: microstrip, microslot, and coplanar wave guide, on dielectric substrates possibly with ferromagnetic inserts, as well as on semiconducting substrates. The frequency range is enormous: it goes from the lowest range of microwaves, with mobile telephony circuits at 900 MHz as a main application, up to millimeter waves, with a view to automotive applications at 100 GHz. Communications with optical carriers, in particular for ultra-high bandwidth data transmission, require solutions for optomicrowave devices.

Today, the practicing engineer has at his disposal a number of methods for analyzing lines and resonators, as well as professional software. So, his question is: what method shall I use?

The purpose of this book explore the methods for calculating circuit parameters, complex propagation constants and resonant frequencies, for the whole variety of configurations and over the whole frequency range. We will show that variational methods can do that. They are analytical methods and we strongly believe in the value of pushing analytical methods as far as possible when calculating circuit parameters. The reason is simple. In general, we are not interested in calculating the parameters of just one configuration. What we are mostly interested in is circuit synthesis and, with this in mind, we are eager to know how changes in geometrical and physical parameters can affect the circuit's performance. This is precisely what analytical methods can do.

Perturbational methods have also been used for decades. The variational approach, however, can be applied to a wider variety of problems. Indeed it is not necessary, as with perturbation theory, that the departure from a simple known case is small. The variational approach can be used successfully even when the departure is large. As an example, it can be applied to deformed lines and resonators, possibly inhomogeneously loaded, and to singularities such as corners of re-entrant conducting surfaces or of blocks of dielectric or magnetic material.

So, the main subject of the present book is clearly that of microwave network theory. It is interesting to observe that the general variational principle developed in Chapter 4, for solving problems of shielded and open multi-layered lines with gyrotropic non-Hermitian lossy media, exactly follows the guidelines described by Schwinger: the proof of the principle is established; an equation is solved by a static method, using conformal mapping; the approximate solution is used as a good trial field in the variational expression; the trial field is improved by a systematic process, and very accurate results are obtained with a remarkably small number of iterations.

However, the field used to obtain results, may not be a good approximation of the field in the exact structure; in particular this is the case near abrupt discontinuities. It is however, important that the integral of the field is stationary. These field values are an excellent starting point for iterative procedures in which the field is the quantity of interest.

The reader will find in Chapter 1 a review of the areas in which variational methods have been used, in some cases for centuries. The importance of variational methods and principles for solving microwave engineering problems is also highlighted.

In Chapter 2, variational principles are introduced in general terms, for calculating impedances and propagation constants. They are first developed so that the value of the variational quantity has a direct physical significance, such as energy and eigenvalues. Then, by re-arranging Maxwell's equations, it is shown that explicit expressions can be obtained as well as implicit expressions. Variational principles are finally compared with the reaction concept, moment method, Galerkin procedure, Rayleigh-Ritz method, mode matching technique, and perturbational methods.

Quasi-static and full-wave dynamic methods for analyzing planar lines are reviewed in Chapter 3. The quasi-static methods usually provide a straightforward expression for a line parameter. On the other hand, full-wave methods do not provide a simple explicit description of the behaviour of a line parameter over a range of frequencies. The main part of the chapter is devoted to detailed variational formulations, applicable to a variety of configurations, including anisotropic media. Discretized formulations are compared with analytical formulations. The combination of variational principles with other methods is also discussed.

The efficiency of variational principles is illustrated in Chapter 4 by a general principle developed by the authors. It is applicable to planar multilayered lossy lines, including those involving gyrotropic lossy materials. In combination with conformal mapping, it drastically reduces the complexity and duration of numerical calculations. Both spatial and spectral domain approaches are derived. The mathematical and numerical efficiency are demonstrated, while the choice of the method for deriving trial quantities is discussed. Mathieu functions are shown to be very efficient expressions for trial fields in slots. The method is validated by measurements on topologies such as slot-lines, coupled slot-lines, finlines, and coplanar waveguides, with YIG-layers.

Finally, Chapter 5 is devoted to applications operating up to optical frequencies, including multilayered lines with dielectric and semiconducting layers, lines coupled to gyrotropic resonators, uniplanar and microstrip-to-slotline junctions, and optomicrowave devices. It describes an original measurement method, applicable to a variety of substances, including bioliquids such as blood and axoplasm, for investigating microwave bioelectromagnetics, and soil characteristics for demining operations. Combined with a variational principle, it is used for determining the complex permittivity of a planar substrate.

The first author started using variational principles in 1967, extending the variation-iteration method described in Chapter 2. This led to several doctoral theses investigating waveguides and cavities loaded with two- and three-dimensional dielectric and ferrimagnetic inserts. Later, during her doctoral thesis, the second author developed the general variational principle demonstrated in Chapter 4, which has since been used in a variety of configurations. This book, however, would not have been written without the stimulating environment offered by Microwaves UCL, Louvain-la-Neuve, Belgium. Discussions were held continuously with close colleagues, which were tremendously fruitful. We owe special thanks to Prof. D. Vanhoenacker-Janvier, who has been involved in all what happened in our Laboratory for fifteen years. She was the key person in setting up the high performance microwave measurement facilities and had a central role in developing microwave passive and active circuits on silicon-on-insulator technology. Our thanks also go to Dr. T. Kezai, who analyzed millimeter wave finlines using the Galerkin procedure, Prof. J.-P. Raskin and Dr. R. Gillon, who brought in new problems with their research on silicon-on-insulator technology, raising a need for analysis of planar lines on semiconducting substrates, and Dr. M. Serres, who presented us with the problem of analyzing optomicrowave devices.

Our most sincere thanks go to Dr. M. Serres, who kindly accepted to devote time in professionally editing our manuscript, as well as to M. Dehan, Ph.D. student, who also was of great assistance in this task. We are most grateful to both of them. We wish to thank the series editors, Prof. T. Benson and Dr. P. Sewell, for their encouragement and their very useful critical remarks. We also wish to thank the Publishing Editor of RSP who was very helpful in sorting out the questions about the presentation of the manuscript. Despite the help we have had, any mistakes or inaccuracies are, of course, the sole responsibility of the authors.

Writing this book has been a pleasure. We sincerely hope that the reader will find some pleasure in reading it.

André Vander Vorst
Isabelle Huynen

CHAPTER 1

Fundamentals

1.1 Historical background

Calculus of variations, variational equations, variational methods, variational principles, variational systems, least action, minimum energy, stability - all these concepts relate to the subject of this book. They have been used for years, and some for centuries. They have not the same meaning, however, and we shall clarify at least some of those concepts to better define our central topics. With this in view, some historical background is necessary.

Plato, describing the creation of the cosmos, said that the creator "made it spherical, equidistant from center to end, the most perfect and uniform of all forms; because he found uniformity immensely better than its opposite" [1.1]. Aristotle made another comment, about the alleged circular character of the planetary movements: "if the sky movement is the measure of all movements because it is the only one to be continuous and regular and eternal, and if, for any species, the measure is the minimum, and the minimum motion is the fastest, then, clearly, the sky movement must be the fastest of all. Among the lines which close themselves however, the circle line is the shortest, and the fastest movement is that which follows the shortest curve. Hence, if the sky moves itself into a circle and is fastest than any other, then it must be necessarily spherical" [1.1].

As we see, the concept of maximum and minimum was already present in mechanics at the time of Plato and Aristotle. In the first century A.D., Hero of Alexandria described the first minimum principle, in geometrical optics: "all what moves with no velocity change, moves along a straight line, (...) the object tries to move along the shortest distance, because it has no time for a slower movement, *i.e.* for a movement along a longest path. And the shortest line between two points is the straight line" [1.1].

Hence, the concept of extremum in science was already present 2000 years ago. Much later, in the 17th century, Pierre de Fermat expressed the law of refraction as a consequence of a minimum principle, assuming that the velocity of light decreases when the density of the medium increases. Descartes, in contrast, had proposed as "conforming to experiment" the law based on the assumption that light propagates easier and faster in media with higher density. Fermat writes: "Our postulate is based on the fact that nature operates by the easiest ways"; contrary to Descartes, "we do not consider the shortest spaces or lines, we do consider those which can be followed the most easily and in the shortest time" [1.2]. Hence Fermat used a minimum time principle.

Fermat's questions were extended to the investigation of functions maximizing or minimizing a quantity, in general an integral, like establishing the minimum length of a curve joining two points on a surface, the minimum time required for a mass to shift from one point to another, the minimum value of an area limited by a given curve, etc.

In 1744, Maupertuis wrote: "Light, when propagating through different media, does not use either the shortest path or the smallest time, (...) it follows a path which has a more actual advantage: the path on which the quantity of action is smallest. (...) When a body is shifted from one point to another, a certain action is needed: it depends on the velocity of the body and of the space it describes, it is however neither velocity nor space taken independently. The quantity of action is the larger as the speed is high and the path is long; it is proportional to the sum of the spaces each multiplied by the velocity at which the body describes them. This quantity of action is the actual expense of nature, which reduces it as far as possible in the movement of light" [1.1]. The theory, however, was approximate and suffering from an unnecessary metaphysical background.

At about the same time, the concept of maximum and minimum was extended by the Swiss mathematician Euler, under the name of calculus of variations, to a much more advanced level. The principle of least action was defined with accuracy. Theoretical and conceptual developments then continued, with the work of d'Alembert, Lagrange, Hamilton, Jacobi, Gauss, Thompson, Kelvin, Weierstrass, Hilbert, Sommerfeld, Schrödinger, Dirac, Feynman, Volterra, and others.

Extrema were of course directly linked with stability and, with no surprise, one discovers that the subject of stability has grown considerably over the years. Some bases are the variational equations of Poincaré and the methods of Liapounov. The analytical approach to the theory of stability develops from the so-called variational equations of Poincaré, with some ambiguity, however, between variationality and perturbation theory [1.3]: variational equations, variational equations of singular points, variational systems with constant coefficients, etc. Various methods were developed to reduce variational systems based on differential equations with periodic coefficients to those based on the differential equations with constants coefficients.

Sommerfeld extended the theory of Bohr's atom by observing that there had to be a certain connection between Planck's constant and the action integral of Hamilton's principle. De Broglie was influenced by the principle of least action in establishing the basis of wave mechanics, and he brought together optics and mechanics through variational principles. Schrödinger wrote his equation of quantum mechanics as an extremum of an integral obtained from Hamilton-Jacobi's theory. Dirac wondered about the link between quantum mechanics and the Lagrangian formalism, and the doctoral thesis of Feynman in 1942 was entitled "The principle of least action in quantum mechanics", paving the road towards his contributions in quantum electrodynamics for which he obtained the Nobel prize.

1.2 Extremum principle

To the best of our knowledge, the first time a variational approach was used in engineering was by the Bernoulli brothers, in the middle of the 18th century, in hydraulics. There had been quite a correspondence between them and Euler on the subject. A number of problems however, in the area common to mathematics and mechanics, were solved using the calculus of variations. The core of the methods using such calculus is based on *variations*, which means investigating an extremum by attributing variations to the arguments. The *local* aspect is dominant. We shall point out later in this chapter that this subject is not to be confused with variational principles in general. The ambiguity, however, is present in many books, such as [1.4], which states: "Many *variational problems* do appear because the laws of nature allow interpretations as *variational principles*". It must be observed that the examples cited in those references are, strictly speaking, related to the stationarity of *one* integral around a given point, yielding an extremum, either a maximum or a minimum, for that integral. We shall come back to this later.

Many problems receive their initial formulation directly in the domain of the calculus of variations. A number of examples of the stationarity of an integral can be found, such as the evaluation of:

- the shortest path from one point to another (geometrical optics), or from a point to a curve, or between two curves

- the fastest descent of a particle in movement, submitted to frictionless gravitation (brachystochrone)

- the minimum area described by a curve rotating around an axis (isoperimetric problem and geodesy).

The problem we are confronted with bears some resemblance to the one of finding the minimum or the maximum of a simple function. The problem encountered in the calculus of variations, however, is much more difficult, because instead of finding a point where a simple function $f(x)$ has a minimum, we are required to determine an argument function $y(x)$, out of an infinity of possible functions defined over an interval of x, which makes a definite integral J of the functional $F(y)$ a minimum. A further very fundamental difference between the maximum and minimum problem and the calculus of variations problem lies in the fact that a theorem (attributed to Weierstrass) always ensures the solution of the straight maximum-minimum problem, while no such general fundamental existence theorem exists for the variational calculus problem. Consequently, in the calculus of variations, the existence of a solution to any given extremum problem *requires a special proof.* In those cases where a solution exists, it turns out that a necessary condition for an extremum is that the first variation δJ of the functional must be zero. This requirement is formally analogous to the requirement for the straight minimum problem that the differential df vanish. Hence the meaning of the first variation of an integral has to be defined carefully. An Euler differential equation is obtained when the total derivative d/dx of the partial derivative $\partial F/\partial y$ is taken [1.5]. That this equation must be satisfied by $y(x)$ constitutes a necessary condition for the existence of an extremum. The differential equation arising out of the first variation of the functional J plays the same role in the calculus of variations minimum problem, as the ordinary differential quotient plays in the simple minimum problem in ordinary calculus. There are several circumstances in which it is simple to obtain the solution of the Euler equation.

So called variational methods have been used abundantly in physics [1.6], because many laws of physics can be expressed in terms of variationality. For example, in classical mechanics, it has been shown that the motion of a mechanized system is always such that the time integral of the Lagrangian function taken between two configurations of the system has an extreme value, actually a minimum. This principle is today widely known as that of least action. In its present form, another such principle is that of Fermat, which states that the propagation of light always takes place in such a way that the actual optical path, *i.e.* the length of geometric path multiplied by the index of refraction of the medium, is an extreme value, usually a minimum. If the index of refraction varies continuously from point to point in the medium, the appropriate integral must be applied.

In the case of automatic control [1.7], the names of Pontrjagin and his colleagues are well known for what is called the principle of Pontrjagin, which is a principle for determining a maximum [1.8]. They investigated the control problem, which involves finding an extremum for a functional where the function is submitted to a constraint, by determining which aspects, characterizing the simpler problem of finding an extremum for the functional, remain valid for the control problem and can be used as a basis for solving it.

For many boundary-value problems of even order, it is possible to specify an integral expression which can be formed for all functions of a certain class and which has a minimum value for just that function which solves the boundary-value problem [1.9]. Consequently, solution of the boundary-value problem is equivalent to minimizing the integral. In the Ritz method the solution of this variational problem is approximated by a linear combination of suitably chosen functions. This method shares, along with the finite-difference and the finite-element methods, a very favored position among the methods for the approximate solution of boundary-value problems. An appropriate expression for the integral can be obtained by trying to write the differential equation of the given boundary-value problem as the Euler equation of some variational problem.

In the above discussion, the problem involving only one unknown function of a single independent variable was considered. The required generalization to cover problems wherein more than one function of a single independent variable is involved can readily be made [1.4]. The same is true for the many physical problems requiring the determination of a function of several independent variables, giving rise to an extremum in the case of a multiple integral. In this case, the Euler partial differential equation is found for the problem involving several independent variables. A necessary condition for an extremum of J is that the function of the variables satisfies this equation. If the function is required by conditions of the original problem to have specified values on a boundary, a solution of Euler's equation must be obtained which satisfies the given boundary conditions. Instances of physical examples which can be replaced by a variational problem are:

- the vibrating string, by using the principle of least action

- potential problems involving Laplace equation, for which the potential function makes an integral an extremum.

For variational problems in which it is required to determine the function $u(x)$ of a variable which makes a specific integral F an extremum subject to an auxiliary condition, it may be shown that the problem is generally, but not always, equivalent to finding another function $y(x)$ of the same variable. The result will yield an extremum for the same integral, of a modified function F^*, without regard to the auxiliary condition. The modified function F^* is a specific linear combination of the function F and of the auxiliary condition [1.5].

As the reader can see, the above presentation is limited to finding functions which make *one* integral extremum. By doing so, a stationary value is obtained which is insensitive to a first-order variation of the variables, because such a first-order variation induces a second-order variation, which is negligible, in the extremum value of the integral. Variational problems, variational methods, variational principles, stationarity, least action, and perturbation methods are expressions which have been used extensively in the literature. From now on, we shall address those problems as depending upon an *extremum principle* [1.10].

1.3 Variational principle

Throughout this book, we intend to reserve the appellation *variational principle* to problems where the unknown quantity is expressed as a *ratio of integrals*. The value of the ratio will be said to be variational if there exists an extremum value for this ratio, as for the cases illustrated in the first part of this chapter. There is a fundamental difference, however, between them. If indeed the numerator and the denominator of the ratio are both variational for the same value of the function or if they both vary in the same way around the value yielding stationarity, then one feels that the stationarity might be better than in the other cases.

To illustrate our point, let us consider variational principles for eigenvalue problems. The eigenvalue equation relates differential or integral operators acting on the function:

$$\mathcal{L}(\Psi) = \lambda \mathcal{M}(\Psi) \tag{1.1}$$

and it can be proved [1.6] that the following expression is a variational principle for λ:

$$\delta[\frac{\int \Phi \mathcal{L}(\Psi)dS}{\int \Phi \mathcal{M}(\Psi)dS}] = \delta[\lambda] = 0 \tag{1.2}$$

where the function Φ is also determined by the variational principle. It satisfies indeed the equation and boundary conditions which are the adjoints of those satisfied by the function Ψ: it is the adjoint solution. When the operators $\mathcal{L}$ and $\mathcal{M}$ are self-adjoint, one has $\Phi = \Psi$ and the variational principle assumes the simpler form

$$\delta[\lambda] = \delta[\frac{\int \Psi \mathcal{L}(\Psi)dS}{\int \Psi \mathcal{M}(\Psi)dS}] = 0; \qquad \mathcal{L} \text{ and } \mathcal{M} \text{ self-adjoint} \tag{1.3}$$

The important point here is to observe that indeed variational principle (1.2) is expressed as the *ratio of two integrals*.

This is the case for a number of eigenvalue problems, like Schrödinger's equation and the vibration of a membrane, subject to the Helmholtz equation. These are problems for which the energy eigenvalues form a *discrete spectrum*. They deal with either a finite domain upon whose surface the wave equation satisfies boundary conditions, or an infinite domain in which case the wave function must go to zero at infinity.

There are also situations in which the eigenvalues form a *continuous spectrum*. Solutions satisfying the appropriate boundary conditions exist for all values of the eigenvalue within some range of values. The variational principles described above are not directly applicable in this case, because the wave functions are no longer quadratically integrable, and suitable modifications must be introduced. In such cases variational principles can be found, for instance, for phase shift, transmission and reflection of waves by a potential barrier, scattering of waves by variations of index of refraction, scattering from surfaces, radiation problems, and variation-iteration schemes [1.6],[1.11].

1.4   Microwave network theory

One feature brought to the fore by World War II was the necessity to solve rapidly new and difficult engineering problems. The MIT Radiation Laboratory grouped a number of scientists and engineers, from various disciplines of science and technology. The group was very prestigious indeed and nine members of the Laboratory later obtained a Nobel prize. In the field of variational principles, members of the group started using such principles to solve engineering microwave problems. This really was a breakthrough. One should remember that computers were not available at that time, and analytical methods, even approximate, were essential.

After World War II, Prof J. Schwinger, future Nobel prize winner, used to lecture to a small group of colleagues at the MIT Radiation Laboratory. Notes were duplicated after each lecture and handed out to the participants. The war ended before the series had finished, leaving lectures undelivered and some lecture notes unwritten. By the end of 1945, the notes were duplicated again and sent out to an accumulated mailing list. Interest in these lectures has never waned. Some of the notes were finally published more than twenty years later. They concern waveguide discontinuities [1.12]. It is quite appropriate to quote the authors' preface.

"These notes are an interesting document of the fruitful interaction of different scientific disciplines. Attitudes and methods characteristic of quantum mechanics and nuclear physics were focused on the application of electromagnetic theory to practical microwave radar problems. And out of these rather special circumstances emerged a strategic lesson of wide impact. In seeking to apply a fundamental theory at the observational level, it is very advantageous to construct an intermediate theoretical structure, a phenomenological theory, which is capable on the one hand of organizing the body of experimental data into a relatively few numerical parameters, and on the other hand employs concepts that facilitate contact with the fundamental theory. The task of the latter becomes the explanation of the parameters of the phenomenological theory, rather than the direct confrontation with raw data. The effective range formulation of low energy nuclear physics was an early postwar application of this lesson. It was a substantial return on the initial investment, for now mathematical techniques developed for waveguide problems were applied to nuclear physics. There have been other applications of these methods, to neutron transport phenomena, to sound scattering problems. And it may be that there is still fertile ground for applying the basic lesson in high energy particles physics".

Schwinger was interested in developing sound "engineering methods", which he defined as "a systematic analytical procedure for reducing complicated problems to the fundamental elements of which they are composed by the application of generalized transmission line theory, symmetry considerations, and the principle of Babinet". He used essentially two methods. One is the Integral Equation Method, in which the imposition of the particular boundary conditions which the electromagnetic field must satisfy - because of the existence of the discontinuity - leads directly to one or more integral equations for the determination of the field. He therefore obtains equivalent circuits, by means of which a discontinuity in a waveguide is replaced by a lumped parameter network in a set of transmission lines. These circuit parameters can be expressed directly in terms of the fields, so that the solution of the integral equations leads almost at once to the impedance (or admittance) elements of the equivalent circuit. He points out that a static method, with the help of conformal mapping, can solve many dynamic problems.

Schwinger called the Variational Method the means of solution in which "the impedance elements are expressed in such a form that they are stationary with respect to arbitrary small variations of the field about its true value. With the aid of such a method, one can, by judiciously choosing a trial field, obtain remarkably accurate results. Moreover, these results can be improved by a systematic process of improving the trial field, which, if carried far enough, will lead to a rigorous result. Furthermore, in solving a given problem one often uses two or more of these methods in conjunction. For example, one might solve an integral equation approximately by a static method and then use this approximate solution as a good trial field in the variational expression"[1.12].

In his well-known Waveguide Handbook [1.13], one of the twenty-eight volumes describing the research developed by the MIT Radiation Laboratory and published after World War II, Marcuvitz mentions the dominant role played by Schwinger in formulating field problems using the integral equation approach, pointing the way forward both in the setting up and the solution of a wide variety of microwave problems. These developments resulted in a rigorous and general theory of microwave structures in which conventional low-frequency electrical circuits appeared as a special case. One extremely useful aspect was the distinction made between the theoretical evaluation of microwave network parameters, which usually entails in general the solution of three-dimensional boundary-value problems and belongs in the domain of electromagnetic theory and, on the other hand, the network calculations of power distribution, frequency response, resonance properties, etc., characteristic of the "far-field" behavior in microwave structures, involving mostly algebraic problems and belonging in the domain of microwave network theory.

The main subject of the present book is clearly that of microwave network theory. It is interesting to observe that the very general variational principle demonstrated in Chapter 4 in view of solving problems of shielded and open multilayered transmission lines with gyrotropic non-Hermitian lossy media and lossless conductors [1.14] exactly follows the guidelines described by Schwinger: the proof of the variational principle is established; an equation is solved by a static method, using conformal mapping; the approximate solution is used as a good trial field in the variational expression; the trial field is improved by a systematic process, and very accurate results are obtained with a remarkably small number of iterations. Hence the necessary computing time is drastically reduced, as will be shown.

The variational approach can be applied to a wider variety of problems than can perturbation theory. It is indeed not necessary, as it is with perturbation theory, for the departure from a simple known case to be small. As an example, the variational approach can be applied to deformed waveguides and cavities, and to cavities and waveguides containing various dielectrics. It can be used successfully, unlike perturbation theory, even when the departure from a simple case is large [1.15]. Complicated waveguides and cavities can be treated, and singularities such as those occurring at the corners of re-entrant conducting surfaces, or at corners of blocks of dielectric, can be dealt with.

The variational approach does have its limitations. It can only be applied to the evaluation of one eigenvalue at a time, unlike an exact method yielding a characteristic equation giving implicitly all the eigenvalues, or the perturbation method which can give general formulae for eigenvalue shifts applicable to all the modes, or at least to all the modes pertaining to a given class. Furthermore, it should never be forgotten that, precisely because of the variationality of the eigenvalue, which is a variationality *with respect to the fields*, the values of the fields in the integrals whose ratio is variational can be rather different from the unknown actual values, at least in some parts of the domain of interest.

It is now time to show the difference between *perturbation theory* and *variational theory*. Perturbation theory is valid only when a small change is introduced in a system. It evaluates the modification in the parameter of interest due to the perturbation with respect to the value of that parameter in the absence of perturbation. Taking cavities as an example, and considering the resonant frequency as the parameter of interest, perturbation theory can be used to calculate the resonant frequency of a cavity filled with gas compared to that of an empty cavity, which implies a very small change over a large volume. It can also be used to calculate the variation in resonance frequency when a small solid body is introduced, which is a large change over a small volume, or when part of a perfect conductor is replaced by a metal with a large but finite conductivity, which means a change not over a volume but over a surface, or when the cavity boundary is slightly modified [1.16].

The reason why perturbation theory can be applied only for small changes, which is not the case for variation theory, is a fundamental question. In perturbation theory an approximation is always made: *a term is neglected*, because it is small due to the fact that the change is small. This term would not be small if the change was not just a perturbation. Usually the neglected term is the integral of a quantity which may be of importance only in a small part of the domain of interest, so that its integration over the whole domain is small and can be neglected to the first order. As an example, when establishing the value of the first-order variation $\delta\omega/\omega$ of the resonant frequency of a cavity [1.15], the neglected term is an integral over the whole volume of the cavity of a quantity which is significant only in the vicinity of, for instance, a small solid sample introduced in the cavity, or a small deformation of the surface of the cavity. As a consequence, *the equation* which is solved for calculating the parameter of interest *is not exact*: it is valid only to the first-order. Hence the small size of the change is essential for the practical use of perturbational methods. The perturbation method yields adequate values because it calculates the effect of a perturbation on the unperturbed known system.

This is not the case when using a variational approach: *the equation* solved for calculating the parameter of interest *is exact*. Only the exact values of the fields throughout the surface or the volume are unknown. The variationality ensures a very accurate value for a *circuit parameter*, like the complex propagation constant and impedance of a transmission line, or the complex resonant frequency of a cavity. This high degree of accuracy is obtained by using field values which may be inaccurate; this is due to two reasons. One is because the integrations are made over the whole surface, or volume, or contour. The other is because the result is expressed as a ratio of integrals, with numerator and denominator varying the same way under the influence of a change of the fields, for a rather wide variation of those fields. There is a very fundamental reason why the quality of the two methods differ: there is an energy relation underneath the variational approach, which is not the case for the perturbation method.

Rumsey, another member of the MIT Radiation Laboratory who later obtained a Nobel prize, defined in 1954 a physical observable, the *reaction*, to simplify the formulation of boundary value problems in electromagnetic problems [1.17]. To illustrate the value of this new concept, he used it to obtain formulas for scattering coefficients, transmission coefficients, and aperture impedances. The formulas so obtained have a stationary character and thus the results could also be obtained from a variational approach. Rumsey pointed out that this physical approach, conceptually simple, is general while the variational approach has to be worked out for each problem. His point of view is that of an experimenter, whose objective is to use the theory to correlate his measurements. He pointed out the fact that an expression is stationary for variations of an assumed distribution about the correct distribution does not justify the assumption that it will yield the best approximation when the assumed distribution is completely arbitrary. It should be noted however that there is no way of deciding which approximation is "best". Harrington [1.18] showed that many of the parameters of interest in electromagnetic engineering are proportional to reactions, for instance, the impedance parameters of a multiport network. The paper by Rumsey limits the reaction concept to isotropic media and fields contained in a finite volume. In the last part of his paper, however, and in an addendum published a little later, he shows that the reaction concept can be extended to anisotropic media by using a more general form of the reciprocity theorem. An argument developed later, however, [1.19] questions the use of the reaction method for gyrotropic media such as magnetized ferrites, and magneto-ionic media, but states it is well suited to isotropic materials and materials with crystalline anisotropy, lossy or lossless. The dielectric constant and the permeability are, at most, symmetric tensors. Today, there is still place for research on the limitations of the reaction concept.

1.5 Variational expressions for waveguides and cavities

The next milestone is the excellent paper published by Berk in 1956 [1.19]. The author presented variational expressions for propagation constants of a waveguide and for resonant frequencies of a cavity, directly in terms of the field vectors. Those situations occur when the electromagnetic problem cannot depend from a scalar formulation satisfying Helmholtz equation and are typified by the presence of inhomogeneous or anisotropic matter. In these cases the need for vector variational principles is apparent. Such variational formulas are presented and derived directly in terms of the field vectors. They are valid for media whose dielectric constant and permeability are Hermitian tensors, restricted, however, to lossless media. It should be underlined that the practical use of a formula developed for lossless gyrotropic media is quite limited, because gyrotropy implies losses, except in very narrowband circumstances. Furthermore, all the formulations are limited to closed waveguides and resonators. Several examples illustrated the advantages of approximations based on the variational expressions of the paper. They included the cut-off frequency of the fundamental mode of a rectangular waveguide in which a vertical dielectric slab is placed respectively adjacent to one of the walls and symmetrically in the middle of the guide, as well as a vertical ferrite slab placed off center.

In 1971, Gardiol and Vander Vorst [1.20], analyzing E-plane resonance isolators, used a variational principle for the propagation constant in lossy gyrotropic ferrites derived by Eidson in an internal report. The formulation involves six scalar quantities corresponding to the electric and magnetic field components in the structure.

In 1994, a general variational principle had been established by Huynen and Vander Vorst, valid for planar lines, open or shielded, containing an arbitrary number of layers and of conducting sheets [1.14]. It yields an explicit variational expression for the propagation constant as a function of the three scalar components of only one field. The materials may be gyrotropic and lossy. The formulation is general enough to characterize, for instance, magnetostatic modes in Yttrium-Iron-Garnet (YIG) films used as gyrotropic substrates. The efficiency of the method is due to both the explicit formulation and the fact that the error made on the trial field is compensated by an exact analytical integration of the fields over the whole space. It yields results which agree very well with new experimental data on slot-lines and YIG-resonators in a microstrip configuration. The conductors, however, do have to be lossless. The effect of conductor losses must be calculated by another method, such as a perturbation method. A spectral domain formulation of the principle [1.14] has also been established [1.21]. In combination with conformal mapping, it drastically reduces the complexity of the numerical computation and leads to rapidly convergent results even when higher order modes are considered. The formulation applies to open and shielded multilayered lines with coplanar conductors lying at different interfaces between the layers. It is also valid for gyrotropic non-Hermitian lossy media. Results are shown to be in excellent agreement with previously published results, calculated by other methods, as well as with new measurements made by the authors. The propagation constant is variational with respect to the fields and not to the material properties. Hence, a new method for characterizing the properties of planar materials with a very good accuracy, over a broad frequency range at frequencies up to 100 GHz, was established [1.22]. Chapter 4 is essentially devoted to the demonstration of this principle, validated by some applications, while Chapter 5 describes a number of variational applications.

Finally, in 1998, Huynen *et al.* proved the stationary character of the magnetic energy in the case of a resonator containing lossy gyrotropic media and supporting microwave magnetostatic waves [1.23]. A variational expression is obtained for the input impedance of the circuit. It is used for calculating the input reflection coefficient of a planar multilayered magnetostatic wave straight-edge resonator. The model is an efficient tool for designing low-noise wide-band YIG-tuned oscillators.

1.6   New theoretical developments

Applications of the variational approach are numerous. A number of them can be found in the two editions of the book by Collin [1.24][1.25], with new material in the second edition. They essentially relate to waveguide discontinuities, such as diaphragms, various discontinuities, junctions, dielectric steps, input impedance of waveguide, inductive posts, ferrite slabs, etc. The author uses the Hertzian potential approach, leading to just two scalar field components, applicable only to the most elementary problems. For more complicated problems, the resulting eigenvalue equations are not of the standard type, because the eigenvalue either does not appear in a linear form or is present in the boundary or interface conditions. The author also offers an excellent outline of the method for that class of problems.

For microstrip-like transmission lines, the first explicit variational principles, developed by Yamashita, were quasi-static ones, providing values for quasi-transverse-electromagnetic (TEM) capacitances [1.26].

In 1971, English and Young [1.27] obtained a variational formulation for waveguide problems in terms of three scalar field components (the E formalism), while the expressions developed by Berk [1.19] were in terms of six scalar field components (the EH formalism). Because the parameter of interest - the propagation constant β - appears in their functional equation in quadratic form with first- and second-order powers, they have to apply the variational method in a reverse way: solve for the frequency, which is normally known, in terms of the propagation factor, which is normally unknown.

Lindell [1.28] defined the eigenvalue problem in a less restrictive manner so that different parameters involved in the problem can be interpreted as eigenvalues. These general eigenvalues are called non-standard eigenvalues. Specifically, the basis of the method is that it uses any existing stationary functional for a standard eigenvalue, solves it for any parameter in the functional, and obtains a stationary functional for that parameter, which is by definition a non-standard eigenvalue of the problem. A unified theory for obtaining stationary functionals for different non-standard eigenvalue problems, based on a mathematical principle, is presented. The problems are classified in terms of the complexity of their functional equation. Because there may exist many parameters each recognizable as a nonstandard eigenvalue of the problem, there may exist different functionals giving a choice of methods of different complexity in solving the same problem. Several examples are presented, like the cutoff frequency problem of a waveguide with reactance boundaries, comparing different formulations of the problem, the azimuthally magnetized ferrite-filled waveguide propagation problem, the corrugated waveguide, and the cavity with a homogeneous insert. (A correspondence on the possible fallacy of the proposed principle lasted for two years in the IEEE Transactions of the Society Microwave Theory and Techniques (MTT), in September 1983 and April 1984). Later, the method was applied to the calculation of attenuation in optical fibers [1.29].

In 1990, Baldomir and Hammond [1.30] showed that a geometrical approach to electromagnetics unifies the subject of using methods which are coordinate-free and based on geometry rather than algebraic equations, offering guidance in the choice of simple numerical methods of calculation. They show that the principle of least action is a particular case of the geometrical symmetry of the configuration space. Applications are illustrated in electrostatics, magnetostatics, and electrodynamics.

Richmond discussed the interesting subject of the variational aspects of the moment method [1.31]. Variational techniques were in use prior to the introduction of the moment method. Since then, the moment method has been widely adopted for solving electromagnetic problems via the integral-equation approach. At the same time, interest in variational methods almost vanished. Richmond raises several very good questions. *Under what conditions* does the moment method possess the variational property? When the moment method is variational, precisely *which quantities are stationary*? Do the variational moment methods offer *any significant advantage* over the non-variational moment methods? A limitation of the paper is that it is restricted to time-harmonic problems involving perfectly conducting antennas and scatterers, formulated with the electric field integral equation. The author also points out that his paper does not consider other important techniques such as finite-differences, finite-elements, T-matrix, conjugate gradient, or least squares. He uses the terms "moment method" and "Galerkin's method" as they are defined by Harrington [1.32]. Richmond shows that admittance, impedance, gain and radar cross section can be expressed in terms of the self-reaction or the mutual reaction, in Rumsey's sense [1.17], of the electric current distribution(s) induced on the body. Thus the question of the stationarity of these important quantities reduces to the question of the stationarity of the reaction. Subject to the above restrictions and assuming the usual reciprocity theorems are applicable, it is shown that the reaction is stationary with Galerkin's method, while it is not when the non-Galerkin moment method is applied in the customary manner. A few numerical results are included to compare the performance of the variational and non-variational moment methods. A simple example is that of a two-dimensional problem involving a perfectly conducting circular cylinder with transverse-magnetic polarization. The self-reaction per unit length is calculated as a function of a parameter p which vanishes for the exact solution, via the Galerkin and the non-Galerkin moment method, respectively. Both methods yield precisely the correct reaction when parameter p is zero. The Galerkin result, however, displays zero slope at $p = 0$ and thus maintains good accuracy even when the basis function takes on a rather large error. On the other hand, the non-Galerkin result displays a large nonzero slope when $p = 0$, and therefore quickly loses accuracy as the basis function departs from the correct function. Furthermore Wandzura [1.33] proved that the Galerkin method, when applied to scattering problems, gives more rapid solution convergence than more general moment methods.

As said before, most of the results obtained when using a variational approach are based on a microwave network point of view, for quantities such as propagation constant, resonant frequency, and input impedance. Few consider field computations, and this is particularly true for problems involving anisotropic media. Jin and Chew [1.34] have presented a formulation with specific functionals for general electromagnetic problems involving anisotropic media. Such a formulation provides a foundation for the development of the finite-element method for the analysis of such problems. Specific functionals are stated and their validity is proven, rather than constructing them from the original boundary-value problems. The property of the associated numerical system is discussed for the following three cases: (i) lossy problems involving anisotropic media having symmetric permittivity and permeability; the resultant numerical system obtained from the finite-element discretization is symmetric (ii) lossless problems involving anisotropic media having Hermitian permittivity and permeability; the resulting numerical system is Hermitian (iii) general problems involving general anisotropic problems; the resulting numerical system is neither symmetric nor Hermitian.

The variational aspects of the reaction in the method of moments have been discussed again by Mautz in 1994 [1.35]. He refers to Richmond [1.31], who pointed out that the reaction between two electric surface current densities J_1 and J_2 (the integral of the dot product of J_1 with the electric field produced by J_2) is variational with Galerkin's method but is not necessarily variational when the non-Galerkin moment method is applied in the manner where one set of functions is used to expand both J_1 and J_2 and another set of functions is used to test both the integral equation for J_1 and that for J_2. Mautz shows that the same reaction is variational when the non-Galerkin method is applied in the manner of operation where the set of functions used to expand J_2 is the set of functions used to test the integral equation for J_1, and the set of functions used to test the integral equation for J_2 is the set of functions used to expand J_1. Hence a non-Galerkin formulation may be variational, which is illustrated by calculating the reaction with four non-Galerkin methods, including two variational and two non-variational results. This operating way may be more appropriate than Richmond's if a variational result is desired. The paper establishes the conditions under which a symmetric product is variational.

Liu and Webb derived a variational formulation for the propagation constant satisfying the divergence-free condition in lossy inhomogeneous anisotropic reciprocal or nonreciprocal waveguides whose media tensors have all nine components [1.36]. Their formulation is implemented using the finite-element method. The variational expressions are in the form of standard generalized eigenvalue equations, where the propagation constant appears explicitly as the eigenvalue. It is also shown that for a general lossy nonreciprocal problem the variational functional exists only if the original and adjoint waveguide are mutually bi-directional, *i.e.*, for each mode with the propagation constant γ in the original waveguide there exists a mode with propagation constant $-\gamma$ for the adjoint waveguide. On the other hand, for a general lossy reciprocal problem the variational functional exists only if the waveguide is bi-directional, *i.e.*, if modes with propagation constant γ and $-\gamma$ exist simultaneously for the same waveguide.

Finally, Huynen and Raida recently compared finite-element methods with variational analytical methods for planar guiding structures [1.37]. They compared an explicit variational principle for the propagation constant of guiding structures with the finite-element method using a functional involved in the explicit variational principle formalism. They show that combining the finite-element method with this functional provides stationary values of the propagation constant with respect to trial-discretized fields, provided, however, that the media are either isotropic or lossless gyrotropic. Comparative results show how the convergence is influenced by the differing nature of the explicit unknown (field for finite elements, propagation constant for the explicit variational principle). The performances of both methods are compared in terms of CPU time, and accuracy. It is concluded that the variational principle has to be preferred for conventional conductor shapes, when analytical expressions in the spatial or the spectral can be derived *a priori* for trial quantities. Exact fields can be derived by using the stationarity of the propagation constant to obtain successive improvements of field expressions.

1.7   Conclusions

From all those considerations, it appears wise to consider that a resolution method should be formulated using one of the two approaches: the variational approach and Galerkin's approach. Although Galerkin's approach is conceptually simpler, the authors recommend using the variational approach because, as well as yielding a more elegant formulation, it has a solid foundation in physics and mathematics and can provide physical insights to some difficult concepts such as essential and natural conditions or the choice of expansion (basis) functions. However, unlike Galerkin's approach, which starts directly with differential equations, the variational approach starts from a variational formulation. The applicability of the approach depends, of course, directly on the availability of such a variational formulation. On the other hand, the link between the use of a variational principle based on an analytical function and the use of a direct numerical calculation still offers some unexplored fields of investigation.

1.8   References

[1.1]   J. Mawhin, "Le principe de moindre action et la finalité", *Revue d'éthique et de théologie morale, Le supplément*, Editions du Cerf, Cahier no. 205, pp. 49-82, Jun. 1998.

[1.2]   P. de Fermat, "Méthode pour la recherche du maximum et du minimum", *Oeuvres*. Paris: Gauthier-Villars, 1891-1922, Tome 3, pp.121-156.

[1.3]   N. Minorski, *Nonlinear Oscillators*. Princeton, N.J.: Van Nostrand, 1962.

[1.4]   E. Roubine, *Mathematics Applied to Physics*. Berlin: Springer-Verlag, 1970.

[1.5]   L.P. Smith, *Mathematical Methods for Scientists and Engineers*. New York: Dover, 1961, Ch. 13, pp. 404-409.

[1.6]   P.M. Morse, H. Feshbach, *Methods of Theoretical Physics*. New York: McGraw-Hill, 1953.

[1.7]   L.D. Berkovitz, "Variational methods in problems of control and programming", *J. Math. Analysis and Applications*, Vol. 3, no. 1, pp. 145-169, Aug. 1961.

[1.8]   L.S. Pontrjagin, V.G. Boltyanskii, R.V. Gamkrelidze, E.F. Mischenko, *The Mathematical Theory of Optimal Processes* (translated from Russian). New York: Interscience Publishers, 1962.

[1.9]   L. Collatz, *The Numerical Treatment of Differential Equations*. Berlin: Springer-Verlag, 1966.

[1.10] R.P. Feynman, R.B. Leighton, M. Sands, *The Feynman Lectures in Physics*, Vol. II. Reading, Mass.: Addison-Wesley, 1964.

[1.11] A. Vander Vorst, R. Govaerts, "Application of a variation-iteration method to inhomogeneously loaded waveguides", *IEEE Trans. Microwave Theory Tech.*, vol. MTT-18, pp. 468-475, Aug. 1970.

[1.12] J. Schwinger, D.S. Saxon, *Discontinuities in Waveguides. Notes on Lectures by Julian Schwinger*. New York: Gordon and Breach, 1968.

[1.13] N. Marcuvitz, *Waveguide Handbook*. Lexington, Mass.: Boston Technical Publishers, 1964.

[1.14] I. Huynen, A. Vander Vorst, "A new variational formulation, applicable to shielded and open multilayered transmission lines with gyrotropic non-hermitian lossy media and lossless conductors", *IEEE Trans. Microwave Theory Tech.*, vol. MTT-42, pp. 2107-2111, Nov. 1994.

[1.15] R.A. Waldron, *Theory of Guided Electromagnetic Waves*. London: Van Nostrand, 1969.

[1.16] J.C. Slater, *Microwave Electronics*. Princeton, N.J.: Van Nostrand, 1950.

[1.17] V.H. Rumsey, "Reaction concept in electromagnetic theory", *Phys. Rev.*, vol. 94, no 6, pp. 1483-1491, June 15, 1954.

[1.18] R.F. Harrington, *Time-Harmonic Electromagnetic Fields*. New York: McGraw-Hill, 1961.

[1.19] A.D. Berk, "Variational principles for electromagnetic resonators and waveguides", *IRE Trans. Antennas Propagat.*, vol. AP-4, no 2, pp. 104-111, Apr. 1956.

[1.20] F.E. Gardiol, A. Vander Vorst, "Computer analysis of E-plane resonance isolators", *IEEE Trans. Microwave Theory Tech.*, vol. MTT-19, pp. 315-322, Mar. 1971.

[1.21] I. Huynen, D. Vanhoenacker-Janvier, A. Vander Vorst, "Spectral domain form of new variational expression for very fast calculation of multilayered lossy planar line parameters", *IEEE Trans. Microwave Theory Tech.*, vol. MTT-42, pp. 2009-2106, Nov. 1994.

[1.22] M. Fossion, I. Huynen, D. Vanhoenacker, A. Vander Vorst, "A new and simple calibration method for measuring planar lines parameters up to 40 GHz", *Proc. 22nd Eur. Microw. Conf.*, Helsinki, pp. 180-185, Sept. 1992.

[1.23] I. Huynen, B. Stockbroeckx, G.Verstraeten, "An efficient energetic variational principle for modelling one-port lossy gyrotropic YIG straight-edge resonators", *IEEE Trans. Microwave Theory Tech.*, vol. MTT-46, pp. 932-939, July 1998.

[1.24] R.E. Collin, *Field Theory of Guided Waves*. New York: McGraw-Hill, 1960.

[1.25] R.E. Collin, *Field Theory of Guided Waves*, 2nd edition. NewYork: IEEE Press, 1991.

[1.26] E. Yamashita, "Variational method for the analysis of microstrip-like transmission lines", *IEEE Trans. Microwave Theory Tech*, vol. MTT-16, pp. 529-535, Aug. 1968.

[1.27] W. English, F. Young, "An E vector variational formulation of the Maxwell equations for cylindrical waveguide problems", *IEEE Trans. Microwave Theory Tech.*, vol. MTT-19, pp. 40-46, Jan. 1971.

[1.28] I.V. Lindell, "Variational methods for nonstandard eigenvalue problems in waveguide and resonator analysis", *IEEE Trans. Microwave Theory Tech.*, vol. MTT-30, pp. 1194-1204, Aug. 1982.

[1.29] M. Oksanen, H. Mäki, I.V. Lindell, "Nonstandard variational method for calculating attenuation in optical fibers", *Microwave Optical Tech. Lett.*, vol. 3, pp. 160-164, May 1990.

[1.30] D. Baldomir, P. Hammond, "Geometrical formulation for variational electromagnetics", *IEE Proc.*, vol. 137, PtA, pp. 321-330, Nov. 1990.

[1.31] J.H. Richmond, "On the variational aspects of the moment method", *IEEE Trans. Microwave Theory Tech.*, vol. MTT-39, pp. 473-479, April 1991.

[1.32] R.F. Harrington, *Field Computation by Moment Methods*. New York: MacMillan, 1968.

[1.33] S. Wandzura, "Optimality of Galerkin method for scattering computations", *Microwave Optical Tech. Lett.*, vol. 4, pp. 199-200, Apr. 1991.

[1.34] J. Jin, W.C. Chew, "Variational formulation of electromagnetic boundary-value problems involving anisotropic media", *Microwave Optical Tech. Lett.*, vol. 7, pp. 348-351, June 1994.

[1.35] J.R. Mautz, "Variational aspects of the reaction in the method of moments", *IEEE Trans. Antennas Propagat.*, vol. 42, pp. 1631-1638, Dec. 1994.

[1.36] Y. Liu, K.J. Webb, "Variational propagation constant expressions for lossy inhomogeneous anisotropic waveguides", *IEEE Trans. Microwave Theory Tech.*, vol. 43, pp. 1765-1772, Aug. 1995.

[1.37] I. Huynen, Z. Raida, "Comparison of finite-element method with variational analytical methods for planar guiding structures", *Microwave and Optical Tech. Lett.*, vol. 18, no. 4, pp. 252-258, July 1998.

CHAPTER 2

Variational principles in electromagnetics

2.1 Basic variational quantities

2.1.1 Introduction

It is well known that exact solutions of the equations of physics may be obtained only for a limited class of problems. This is true whether the formulation of the equation is differential or integral. So, we are faced with the task of developing approximate techniques of sufficient power to handle most of the problems. In this book, we are interested with distributed circuits, operating mostly at centimeter and millimeter wavelength. These circuits are found in high frequency or high speed electronic circuits used in communications and in computer systems, where they need to be synthesized. Hence, the approximate techniques have to be powerful indeed, with respect to both accuracy and speed of calculation.

Perturbation methods are commonly used. The general theory is well described in the literature, as well as a variety of its applications. Perturbations are deviations from exactly soluble situations. They may be surface perturbations, referring to deviations in the boundary surface or boundary conditions, or both, from the exactly soluble case. Such techniques may obviously be used to solve a number of transmission line or cavity problems. They may also be volume perturbations, which destroy the separability of the problem. In the perturbation method, the volume or surface perturbations are assumed to be small and expansions in powers of a parameter measuring the size of the perturbation may be made, the leading term being the solution in the absence of any perturbation.

Perturbation methods are especially appropriate when the problem closely resembles one which is exactly solvable. It presumes that one may change from the exactly solvable situation to the problem under consideration in a gradual fashion - the difference is not singular in character. This requires the perturbation to be a continuous function of a parameter, measuring the importance of the perturbation. When this is the case, it is possible to develop formulas which describe the change in the physical situation as the parameter varies from zero. Perturbation formulas may be shown to be the consequence of the application an iterative procedure to the integral formulation of the problem. They have been developed for the determination of eigenvalues and eigenfunctions, and for problems in which the eigenvalues form a continuous spectrum, which is typical of scattering and diffraction, in a variety of situations [2.1].

When the perturbation is large, perturbation methods become tedious and the expressions which are developed so complex that the results lose their physical meaning. It this case the variational method may be more appropriate [2.2]. For this method, the equations have to be put in a variational form, usually a variational integral, which implies finding a quantity involving the unknown function which is to be stationary upon variation of the function. This quantity is a scalar number and is given by the ratio of two integrals. In practice, a function will be used - called the trial function - involving one or more parameters. It is inserted for the unknown function into the variational principle. The function may be varied by changing the value of the parameters and the procedure may be improved by introducing additional parameters. Also, the original trial function may be improved by making the trial function the first term in an expansion in a complete set of functions which are not necessarily mutually orthogonal.

When the trial function differs from the correct function by a given quantity, the variational form of the equations differs from the true value by an amount proportional to the second-order, and, in the neighborhood of the stationary values, the variational form is less sensitive to the details of the trial function than it is elsewhere. Because of the stationary character of the variational expression and its first-order insensitivity to the errors in the trial functions, it is often possible to obtain excellent estimates of the unknown with a relatively crude trial function. This property is of great practical importance. The quantity of interest will be, in our case, the propagation constant of a transmission line, the reflection coefficient of a circuit or the resonant frequency of a cavity. In electromagnetics, there are essentially two physical quantities which can be extremely useful as variational entities: energy and eigenvalues. In this section we shall introduce their variational properties in general terms. Details will be available later in this chapter, and especially in Chapter 3 and 4.

2.1.2   Energy

In this book, we are essentially interested with transmission lines and resonators. They are indeed the basic elements of distributed circuits. Hence, we are going to develop variational methods to calculate complex impedances and propagation constants, in a variety of configurations. It is often possible to arrange the form of a variational principle so that the value of the variational quantity for the exact unknown has a physical significance, for instance energy. This will be illustrated by the very simple example of the capacitance of a transmission line per unit length [2.3].

The electrostatic energy per unit length stored in the field surrounding the two conducting surfaces of a two-conductor transmission line is given by

$$W_e = \frac{\varepsilon}{2} \int \int \left[\left(\frac{\partial \Phi}{\partial x} \right)^2 + \left(\frac{\partial \Phi}{\partial y} \right)^2 \right] dx \, dy = \frac{1}{2} C_0 V^2 \tag{2.1}$$

where V is the potential difference between the two conductors and ε is assumed to be constant. It can easily be proved that, when calculating the first-order variation in W_e to a first-order variation in the functional form of Φ, subject to the condition that the potential difference between the two conductors is V, the first-order variation vanishes, provided Φ satisfies the equation

$$\frac{\partial^2 \Phi}{\partial x^2} + \frac{\partial^2 \Phi}{\partial y^2} = \nabla_t^2 \Phi = 0 \tag{2.2}$$

Since we know that Φ is a solution of Laplace's equation, we obtain the following variational expression for the capacitance:

$$C_0 = \frac{1}{\varepsilon V^2} \int \int \nabla_t \Phi \cdot \nabla_t \Phi \, dx \, dy \tag{2.3}$$

The integral is always positive and, hence, the stationary value is an absolute minimum. So, if we substitute an approximate value for $\nabla_t \Phi$ into the integral, the calculated value C for the capacitance will always be too large, and the value for the characteristic impedance, equal to $1/C_0 v$ where v is the speed of light in ε, will always be too small. In practice, if the capacitance is a function of a number of variational parameters, the best possible solution for C is obtained by choosing the minimum value of C that can be produced by varying the parameters. This is obtained by treating the parameters as independent variables and equating all the derivatives of C with respect to the parameters to zero. This yields a set of homogeneous equations which, together with the boundary conditions, gives a solution for the parameters and, as a consequence, for the minimum value of C. In this way, what has been demonstrated is a *variational principle for an upper bound on the capacitance and, hence, a lower bound for the characteristic impedance.*

For the same configuration, energetic considerations can also yield a *variational principle for the lower bound of the capacitance and, hence, an upper bound for the characteristic impedance.* As a matter of fact, a variational principle is easily developed for an upper bound of $1/C$, which yields a lower bound for the capacitance. To do so, instead of calculating *directly* the energy, one solves the problem in terms of the unknown charge distribution on the conductors. The problem is then to find the potential function which satisfies Poisson's equation. The method utilizes the Green's function technique for solving boundary-value problems, by solving the particular expression of Poisson's equation for a unit charge located on a conductor, in such a way that the boundary conditions for the potential are satisfied. It can be shown

[2.4], as detailed in Chapter 3, that a variational principle is obtained for $1/C$ by multiplying the solution of Poisson's equation, expressed as a function of the Green's function integrated over one conductor, by the unknown charge density and integrating over the conductor. This is easily done. It yields a variational principle for $1/C$, in the form of a ratio of integrals in which both numerator and denominator are always positive, and the stationary value is an absolute minimum. Hence it is an upper bound for the capacitance and a lower bound for the characteristic impedance.

Obtaining a variational principle for an upper bound of a physical quantity and one for a lower bound for the same quantity is extremely interesting: *comparing the upper and lower bounds yields the maximum possible error in the approximate values.*

2.1.3   Eigenvalues

The general operator notation will be used here, because it reveals most clearly the technique employed in forming the variational principle. As said in Chapter 1, the eigenvalue equation relates differential or integral operators acting on the function:

$$\mathcal{L}(\Psi) = \lambda \mathcal{M}(\Psi) \qquad (1.1)$$

We shall now prove that the following expression is a variational principle for λ [2.5]:

$$\delta\left[\frac{\int \Phi \mathcal{L}(\Psi)dS}{\int \Phi \mathcal{M}(\Psi)dS}\right] = \delta[\lambda] = 0 \qquad (1.2)$$

where function Φ is also determined by the variational principle. It will be defined in Subsection 2.1.4. A square bracket is placed around λ to indicate that the quantity to be varied is not λ, which is the unknown exact eigenvalue. The integration is over all the volume determined by the independent variable upon which Ψ and also Φ depend. Equation (1.2) is easily obtained by multiplying (1.1) by Φ, as yet arbitrary, integrating and solving for λ. It is obvious that, if the exact Φ is inserted into (1.2), the exact λ is obtained.

To show that (1.1) follows from (1.2), we consider the equation

$$[\lambda] \int \Phi \mathcal{M}(\Psi)\, dV = \int \int \Phi \mathcal{L}(\Psi)\, dV \qquad (2.4)$$

Varying Φ and λ, *i.e.* performing the variation, yields

$$\delta[\lambda] \int \Phi \mathcal{M}(\Psi)\, dV + [\lambda] \int \delta\Phi \mathcal{M}(\Psi)\, dV = \int \delta\Phi \mathcal{L}(\Psi)\, dV \qquad (2.5)$$

Inserting the condition $\delta[\lambda] = 0$ and replacing $[\lambda]$ by λ elsewhere, since the effect of the variation is only calculated to first-order, we obtain

$$\int \delta\Phi \left[\mathcal{L}(\Psi) - \lambda\mathcal{M}(\Psi)\right]\, dV = 0 \qquad (2.6)$$

Since $\delta\Phi$ is arbitrary, equation (1.1) follows.

One may then wonder about the equation satisfied by Φ, which is also determined by the variational principle. It may be shown easily [2.2] that Φ satisfies the equation and boundary conditions which are the adjoints of those satisfied by Ψ, hence Φ is the solution adjoint to Ψ. When the operators $\mathcal{L}$ and $\mathcal{M}$ are self-adjoint, one has $\Phi = \Psi$ and the variational principle assumes the simpler form

$$\delta[\lambda] = \delta[\frac{\int \Psi\mathcal{L}(\Psi)dS}{\int \Psi\mathcal{M}(\Psi)dS}] = 0 \qquad (1.3)$$

One important point here is to observe that the variational principle (1.2) is a *ratio of two integrals.*

Other variational principles for λ may be obtained: there may indeed be many ways to formulate a problem. A number of examples are given in [2.2], to be used in a variety of physical problems. The formulation used here, as well as in the reference just mentioned, is very abstract. The advantage is that it reveals the technique from which variational principles are formed. Specific examples will follow later in this chapter, and in Chapter 3.

2.1.4 Iterating for improving the trial function

Generally, for estimating the accuracy of the results obtained by a variational method, the user simply inserts additional variational parameters and observes the convergence of the quantity of interest with the number of such parameters. In principle, such a method must involve an infinite number of parameters, for one is certain of the answer only when all of a complete set of functions has been employed as trial functions in the variational integrals. The variation-iteration method, developed by Morse and Feshbach [2.5], is a superb technique which not only provides an estimate of the error, by giving both an upper and a lower bound to quantities being varied, but also results in a method for systematically improving upon the trial function. We have used the method with success in the past, for calculating modes, including higher-order modes, of propagation in waveguides loaded with two-dimensional inserts, as well as for resonant frequencies in cavities loaded with three-dimensional inserts, lossy or not [2.6]-[2.8]. Our first presentation already illustrated the fact that the method was quite powerful. Indeed the chairman of the session at which we presented our method [2.6], A. Wexler, pointed out that this was the first numerical solution for the vector second-order eigenvalue equation with partial derivatives, in a two-dimensional inhomogeneous medium. It should be noticed that those results were obtained more than 30 years ago, at a time when computers were not at all as powerful and friendly as they are today. A reviewer of one of our papers at that time mentioned that he believed that Schwinger had used the method at the MIT Radiation Laboratory, during World War II.

The method can be rapidly outlined. We limit ourselves to positive-definite self-adjoint operators. Using a formal operator language, the eigen-value problem is characterized by

$$\mathcal{L}\chi_p = \lambda_p \mathcal{M}\chi_p \qquad (p = 0, 1, 2, \ldots) \tag{2.7}$$

where $\mathcal{L}$ and $\mathcal{M}$ are positive-definite self-adjoint operators
χ_p is an unknown eigenfunction
λ_p is the corresponding unknown eigenvalue ($\lambda_0 \leq \lambda_1 \leq \lambda_2 \leq \ldots$)

When the operators are not positive-definite, the method may still be used after some adjustments are made, however with a slower convergence.

To solve the eigenvalue equation, a method like the Rayleigh-Ritz proce-dure (see Section 2.4) uses a limited series expansion in terms of the known eigenfunctions of another eigenvalue equation. When the difference between the two equations is not too significant, the Rayleigh-Ritz method shows a reasonable convergence. However it is often used in other cases, where the difference is quite significant. The convergence of this method is then very slow, which is a major drawback.

On the other hand, starting with an initial trial function, the variation-iteration method provides a process of iteration, which improves this function until the required accuracy is obtained. The accuracy is checked by comparing the upper and lower bounds of the eigenvalue until they are close enough. We assume here non-degenerate eigenvalues. Let Φ_0 be the initial trial wave function which, of course, does not satisfy (2.7). The unknown eigenfunctions form a complete, orthogonal set in terms of which the trial function can be expanded:

$$\Phi_0 = \sum_{p=0}^{\infty} a_p \chi_p \tag{2.8}$$

We then define the nth iterate (n is an integer) Φ_n by

$$\Phi_n = \mathcal{L}^{-1}\mathcal{M}\Phi_{n-1} \tag{2.9}$$

Hence, from (2.7) to (2.9)

$$\Phi_n = \sum_{p=0}^{\infty} \frac{a_p}{\lambda_p^n} \chi_p \tag{2.10}$$

which shows that the set Φ_n converges to χ_0 by elimination of the unwanted components contained in the trial function, if λ_0 is smaller than λ_{p+1}. When $\mathcal{L}$ and $\mathcal{M}$ are self-adjoint, the eigenfunctions are given by

$$\lambda_p = \frac{\int \chi_p \mathcal{L}\chi_p \, dV}{\int \chi_p \mathcal{M}\chi_p \, dV} \tag{2.11}$$

which can also be written

$$\lambda_p = \frac{\int \chi_p \mathcal{M} \chi_p \, dV}{\int \chi_p \mathcal{M} \mathcal{L}^{-1} \mathcal{M} \chi_p \, dV} \tag{2.12}$$

by noting that

$$\chi_p = \lambda_p \mathcal{L}^{-1} \mathcal{M} \chi_p \tag{2.13}$$

It is shown in [2.5] that (2.12) and (2.13) express a variational principle for λ. Inserting the iterates into (2.12) and (2.13) leads to the following approximations for λ_0, the lowest eigenvalue:

$$\begin{aligned} \lambda_0^{n-1/2} &= \frac{\int \phi_{n-1} \mathcal{M} \phi_{n-1} \, dV}{\int \phi_{n-1} \mathcal{M} \mathcal{L}^{-1} \mathcal{M} \phi_{n-1} \, dV} \\ &= \frac{\int \phi_{n-1} \mathcal{M} \phi_{n-1} \, dV}{\int \phi_{n-1} \mathcal{M} \phi_n \, dV} \end{aligned} \tag{2.14}$$

$$\lambda_0^n = \frac{\int \phi_n \mathcal{L} \phi_n \, dV}{\int \phi_n \mathcal{M} \phi_n \, dV} = \frac{\int \phi_n \mathcal{M} \phi_{n-1} \, dV}{\int \phi_n \mathcal{M} \phi_n \, dV} \tag{2.15}$$

The half-integral value of the superscript is made clear when noting that the set of approximate eigenvalues, including both integral and half-integral superscripts, forms a monotonic decreasing sequence, approaching the exact value λ_0 *from above*, if some amount of λ_0 was present in the trial functions:

$$\lambda_0^{n-1/2} \geq \lambda_0^n \geq \lambda_0^{n+1/2} \geq \ldots \geq \lambda_0 \tag{2.16}$$

and

$$\lambda_0^n, \lambda_0^{n+1/2} \xrightarrow{n \to \infty} \lambda_0, \quad \text{if} \quad \int \chi_0 \mathcal{M} \chi_0 \, dV \neq 0 \tag{2.17}$$

The formal proof of this statement has been given earlier [2.7]. It should be noticed here that *one* iterate leads to the evaluation of *two* successive approximate eigenvalues: one (half-integral superscript) by introducing once the new iterate into the second expression (2.15), and one (integral superscript) by introducing it three times into the same second expression (2.15). An extrapolation method can be used even after only one iterate from a given trial function.

Furthermore a *lower bound* can also be found for λ_0. If the iterations have proceeded far enough so that $\lambda_0^{n+1} \leq \lambda_1$ then the following inequality holds:

$$\lambda_0^{n+1} \geq \lambda_0 \geq \lambda_0^{n+1} \left[1 - \frac{\lambda_0^{n+1/2} - \lambda_0^{n+1}}{\lambda_1 - \lambda_0^{n+1}} \right] \tag{2.18}$$

Only two successive iterates are required for this lower bound. If three successive iterates are calculated and if $\lambda_0^{n+1}\lambda_0^{n+3/2} \leq \lambda_1^2$, then

$$\lambda_0 \geq \lambda_0^{n+3/2}\left[1 - \lambda_0^{n+1}\frac{\lambda_0^{n+1/2} - \lambda_0^{n+3/2}}{\lambda_1^2 - \lambda_0^{n+3/2}\lambda_0^{n+1}}\right] \tag{2.19}$$

For this method to be used, an estimation of λ_1 is necessary. In many transmission lines and cavities problems, this estimation is available. Otherwise, a procedure for finding an approximate value of λ_1 is outlined in [2.5].

As mentioned, an *extrapolation method* is available to obtain a more accurate estimation of the exact answer. Assuming that, after some iterations, the only contamination of the eigenfunctions χ_0 is the next eigenfunction χ_1, one lets

$$\phi_n = \chi_0 + b\chi_1 \tag{2.20}$$

where b^2 is small. Hence, one has

$$\phi_{n+1} = \frac{\chi_0}{\lambda_0} + b\frac{\chi_1}{\lambda_1} \qquad \text{and} \qquad \phi_{n+2} = \frac{\chi_0}{\lambda_0^2} + b\frac{\chi_1}{\lambda_1^2} \tag{2.21}$$

These three iterates lead to expressions of $\lambda_0^{n+1/2}$, λ_0^{n+1}, and $\lambda_0^{n+3/2}$ in terms of the unknown quantities λ_0, b^2, and λ_0/λ_1. Hence, calculating three successive approximate eigenvalues by (2.14) and (2.15), the extrapolated value λ_0 (together with b^2, and λ_0/λ_1) can be calculated. It has been shown [2.5] that the condition $b^2 \ll 1$ is not difficult to satisfy if the trial function is properly chosen (in view of the geometry of the problem). It was our experience that the computation time, including iteration, calculation of the approximate eigenfunctions, extrapolation, and calculation of the lower bound, was much lower than when using the Rayleigh-Ritz method.

2.2 Methods for establishing variational principles

2.2.1 Adequate rearrangement of Maxwell's equations

In the preceding section, we have used a formal operator language. It is advantageous because it is compact and clearly illustrates the principles involved in the method. In practice, variational expressions for resonant cavities and propagation constants can be obtained *directly in terms of the field vectors*, by rearranging Maxwell's equations. In Chapter 4, we shall develop in detail a very general variational principle for multilayered planar transmission lines on lossy gyrotropic or dielectric substrates. In this section, however, we intend to show how this can be done, in general terms, for the resonant frequency of a cavity as well as for the propagation constant of a transmission line. We follow the excellent presentation by Berk [2.9] for a resonator and a waveguide, for simplicity, bearing in mind that the variational principles which will be demonstrated here are valid for *lossless cases* only. The general variational principle, to be developed in Chapter 4, is valid for multilayered planar transmission lines on *lossy* gyrotropic or dielectric substrates.

We consider a resonator with perfectly conducting walls enclosing a medium of permittivity ε and permeability μ. Both ε and μ may be tensors, noted $\overline{\overline{\varepsilon}}$ and $\overline{\overline{\mu}}$ respectively, and functions of position. We call ω the resonant angular frequency. The following is asserted to be a variational expression for ω, provided $\overline{\overline{\varepsilon}}$ and $\overline{\overline{\mu}}$ are Hermitian, *i.e.* provided that no losses are present, so that ω^2 is a real scalar:

$$\omega^2 = \frac{\displaystyle\int_V (\nabla \times \overline{E}^*) \cdot (\overline{\overline{\mu}}^{-1} \cdot \nabla \times \overline{E})\, dV}{\displaystyle\int_V \overline{E}^* \cdot (\overline{\overline{\varepsilon}} \cdot \overline{E})\, dV} \tag{2.22}$$

The integrals are over the volume V of the resonator, $\overline{\overline{\mu}}^{-1}$ is the inverse of $\overline{\overline{\mu}}$, and $\overline{E}^*$ is the conjugate of $\overline{E}$. To prove this assertion, we must show that those field configurations $\overline{E}$ and $\overline{E}^*$ which render ω^2 stationary are solutions of

$$\nabla \times \left[\overline{\overline{\mu}}^{-1} \cdot (\nabla \times \overline{E})\right] - \omega^2 \overline{\overline{\varepsilon}} \cdot \overline{E} = 0 \tag{2.23}$$

and of its complex conjugate, and have vanishing tangential components at the boundary. This is indeed the case, because, when varying $\overline{E}$ and $\overline{E}^*$ in (2.22) we obtain, utilizing the Hermitian character of $\overline{\overline{\varepsilon}}$ and $\overline{\overline{\mu}}$, the following expression for the variation of ω^2

$$\left[\int_V \overline{E}^* \cdot (\overline{\overline{\varepsilon}} \cdot \overline{E})\, dV\right] \delta\omega^2$$
$$= \int_V \delta\overline{E}^* \cdot \left(\nabla \times \left[\overline{\overline{\mu}}^{-1} \cdot (\nabla \times \overline{E})\right] - \omega^2 \overline{\overline{\varepsilon}} \cdot \overline{E}\right) dV$$
$$- \oint_S \delta\overline{E}^* \cdot \left(\overline{n} \times \left[\overline{\overline{\mu}}^{-1} \cdot (\nabla \times \overline{E})\right]\right) \cdot dS \tag{2.24}$$
$$+ \int_V \delta\overline{E} \cdot \left\{\nabla \times \left[\overline{\overline{\mu}}^{-1} \cdot (\nabla \times \overline{E})\right] - \omega^2 \overline{\overline{\varepsilon}} \cdot \overline{E}\right\}^* dV$$
$$- \oint_S \delta\overline{E} \cdot \left(\overline{n} \times \left[\overline{\overline{\mu}}^{-1} \cdot (\nabla \times \overline{E})\right]^*\right) \cdot dS$$

The second and fourth integrals are over the boundary of the cavity. They were obtained after using a vector identity. The variation of ω^2 vanishes, provided $\overline{E}$ satisfies (2.23), $\overline{E}^*$ satisfies the complex conjugate of (2.23), and the surface integrals in (2.24) vanish. Equation (2.22) is thus a variational formulation of the problem defined by (2.23) and the boundary condition $\overline{n} \times \overline{E} = 0$. Admissible trial fields must have vanishing tangential components at the boundary, be continuous together with their first derivatives and possess finite second derivatives everywhere in the cavity except at surfaces where $\overline{\overline{\varepsilon}}$ and $\overline{\overline{\mu}}$ are discontinuous. Equation (2.22) can be modified so that trial vectors $\overline{E}$ are *not* required to satisfy the boundary condition $\overline{n} \times \overline{E} = 0$ at the wall of the resonator. This can be achieved by adding appropriate terms to the numerator of (2.22) [2.9]. In this section, however, we do not enter into details, we only illustrate the main guidelines.

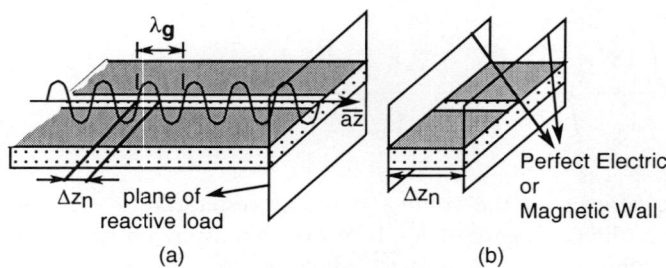

Fig. 2.1 *Geometry of transmission line modeled as equivalent resonators*

As an example of the efficiency of a formula like (2.22), we consider a planar transmission line, bound by a purely reactive load (Fig. 2.1a). Under these conditions, we know that the longitudinal dependence of fields along the line is periodical (Fig. 2.1), with maxima and zeros located at specific abscissas z_n given by

$$z_n = n\frac{\lambda_g}{2} = n\frac{\pi}{\beta} = n\pi\frac{c_0}{\sqrt{\varepsilon_{eff}}\,\omega} \tag{2.25}$$

where β is the propagation constant of the transmission line
       c_0 the light velocity in vacuum
       $c_0/\sqrt{\varepsilon_{eff}}$ is the effective light velocity in the line

Hence, the structure can be divided into stubs of finite length, bound by either perfect electric walls (PEW) or perfect magnetic walls (PMW) (Fig. 2.1b). Fixing the length of each stub equal to L, there is a relationship, derived from (2.25), between L and the operating frequency:

$$L = \Delta z_n = \frac{\lambda_g}{2} = \pi\frac{c_0}{\sqrt{\varepsilon_{eff}}\,\omega} \tag{2.26}$$

Applying formula (2.22) in the volume between two successive PEWs or PMWs, we expect to get a value for the resonant frequency of this equivalent resonator, which approaches the exact value satisfying (2.26).

Figure 2.2 shows, for a slot-line, and for different lengths of resonator considered, the resonant frequency obtained by the variational formula (2.22) (the effective dielectric constant of the slot-line has been computed by another method) and by the formula (2.26). An excellent agreement is observed, which illustrates the efficiency of such a variational formula. A similar result, not shown here, is also observed for other line topologies.

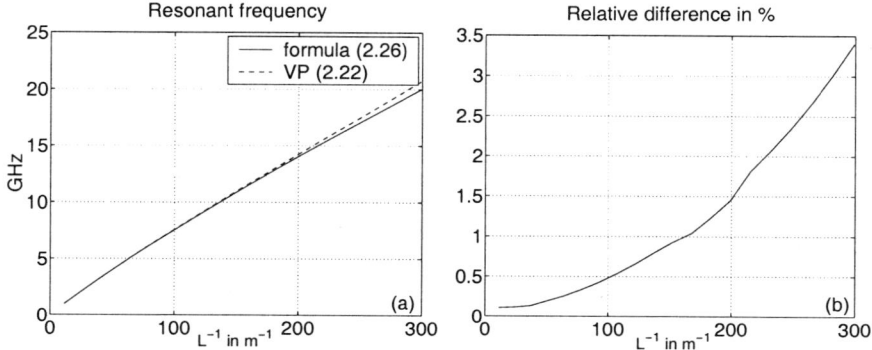

Fig. 2.2 *Resonant frequency of equivalent resonators, computed using respectively variational principle (2.22) and formula (2.26) (a) comparison; (b) relative difference*

When the distribution of matter within the cavity is discontinuous, (2.22) can be further modified so that trial fields will *not* be required to have continuous tangential components of $\overline{E}$ and of $\overline{\overline{\mu}}^{-1} \cdot (\nabla \times \overline{E})$. The modification consists of adding to the numerator of (2.22) the term [2.9]

$$- \oint_S \overline{n} \cdot \left(\overline{E}_+^* \times [\overline{\overline{\mu}}^{-1} \cdot (\nabla \times \overline{E}_+)] - \overline{E}_-^* \times [\overline{\overline{\mu}}^{-1} \cdot (\nabla \times \overline{E}_-)] \right) dS$$

$$- \oint_S \overline{n} \cdot \left(\overline{E}_+ \times [\overline{\overline{\mu}}^{-1} \cdot (\nabla \times \overline{E}_+)]^* - \overline{E}_- \times [\overline{\overline{\mu}}^{-1} \cdot (\nabla \times \overline{E}_-)]^* \right) dS$$

$$(2.27)$$

where the subscripts +, - refer to values on opposite sides of the surface of discontinuity, respectively, and the integrals are over such a surface. The addition of such terms into (2.22) enables one to expand the class of admissible trial functions. Similar variational expressions can be obtained in terms of the magnetic field and of both field vectors. They can, of course, be reduced in the special case when the electromagnetic problem can be expressed in terms of a single scalar field. This will be the case for rectangular waveguides, as illustrated in Section 2.5.

In contrast to the preceding formulas which are in terms of the electric field, we shall now develop an expression in terms of mixed fields, *i.e.*, both field vectors, valid for the propagation of a waveguide [2.9]. Consider a waveguide with perfectly conducting walls, possibly enclosing anisotropic matter, whose distribution may be a function of the transverse coordinates but not of the coordinate along the direction of propagation. If z is this coordinate, the field vectors may be expressed as $\bar{e}(x,y)e^{-\gamma z}$ and $\bar{h}(x,y)e^{-\gamma z}$ where γ is the propagation constant. Vectors $\bar{e}$ and $\bar{h}$ are three-dimensional, depending only on x and y. They satisfy the following relations obtained by substituting the field vectors in Maxwell's equations:

$$\nabla \times \bar{e} + j\omega\bar{\bar{\mu}} \cdot \bar{h} = \gamma\bar{a}_z \times \bar{e} \tag{2.28}$$

$$\nabla \times \bar{h} - j\omega\bar{\bar{\varepsilon}} \cdot \bar{e} = \gamma\bar{a}_z \times \bar{h} \tag{2.29}$$

where $\bar{a}_z$ is the unit vector in the z-direction. Pre-multiplying (2.28) by $\bar{h}^*$, (2.29) by $\bar{e}^*$, integrating over the cross section of the waveguide, and subtracting, we obtain the variational expression

$$\gamma = j\beta = \frac{j\omega \int_S \bar{e}^* \cdot (\bar{\bar{\varepsilon}} \cdot \bar{e})\, dS + j\omega \int_S \bar{h}^* \cdot (\bar{\bar{\mu}} \cdot \bar{h})\, dS}{\int_S \bar{h}^* \cdot (\bar{a}_z \times \bar{e})\, dS - \int_S \bar{e}^* \cdot (\bar{a}_z \times \bar{h})\, dS}$$
$$-\frac{\int_S \bar{e}^* \cdot (\nabla \times \bar{h})\, dS - \int_S \bar{h}^* \cdot (\nabla \times \bar{e})\, dS}{\int_S \bar{h}^* \cdot (\bar{a}_z \times \bar{e})\, dS - \int_S \bar{e}^* \cdot (\bar{a}_z \times \bar{h})\, dS} \tag{2.30}$$

That it is indeed a variational expression can be shown by evaluating the variation of γ and observing that it vanishes, provided that $\bar{e}$ and $\bar{h}$ satisfy (2.28) and (2.29) and the trial tangential component of $\bar{e}$ vanishes at the walls of the waveguide. Thus, trial fields $\bar{e}$ and $\bar{h}$ must be continuous and possess first derivatives throughout the waveguide. At the boundary, the tangential component of $\bar{e}$ must vanish, but $\bar{h}$ is arbitrary. When discontinuities are present in the distribution of matter within the waveguide, the tangential components of both $\bar{e}$ and $\bar{h}$ must be continuous at the surfaces of these discontinuities. The variational expression (2.30) is in terms of mixed fields. Variational expressions in terms of only $\bar{e}$, or of only $\bar{h}$, can also be obtained, by eliminating either $\bar{h}$ or $\bar{e}$ using (2.28) or (2.29).

Part of research, when developing a variational principle, is in finding attractive functionals [2.10]. In particular, the number of scalar field components is of interest, and it may happen that an explicit expression requires more scalar components than an implicit expression. We do not enter into details here and shall be happy by having shown that, in the two specific cases of a resonant cavity and a waveguide, respectively, rearranging Maxwell's equations may yield variational principles. These cases have been limited to *lossless* situations.

2.2.2   Explicit versus implicit expressions

Expressions (2.22) and (2.30) provide a straightforward evaluation of the resonant frequency of a cavity and of the propagation constant of a waveguide, respectively. The expressions are *explicit*, and all that is needed is a calculation. This is, of course, simpler and faster than the classical extraction of the root of the determinantal equation in the Spectral Domain Galerkin's method, as will be shown in Chapter 4. Furthermore expressions (2.22) and (2.30) are a dynamic formulation, while usual explicit variational formulations are quasi-static, providing values for quasi-TEM parameters such as capacitances and inductances.

Variational functionals have been developed for configurations other than closed waveguides and cavities, for instance for calculating attenuation in an optical fiber [2.11]. The analysis starts by deriving a stationary functional for a lossy open waveguide. The functional is written in terms of longitudinal electric and magnetic components. The theory is then applied to round isotropic step-index and weakly-guiding fibers with various loss profile distributions, which yields a variational expression for ω^2 as a function the propagation constant γ. Hence, the obtained expression is *implicit*. In this case, the variational process starts from a given propagation constant, for which an approximation for the parameter ω^2 can be calculated through optimization of trial complex electric and magnetic fields. This may lead to a very complicated treatment of the problem unless the functional can be simplified, which is the case for weakly-guiding optical fibers. Several other implicit expressions, with the same defect, have been published in the literature [2.12]-[2.14]. The optical fiber has been chosen here as an example only. It is worth noting that implicit variational expressions are generally obtained from the reaction concept, as will be shown in Section 2.3.

2.2.3 Fields

We have seen that excellent accuracy can be obtained when using a variational expression to calculate a scalar quantity, like the resonant frequency of a resonator or the propagation constant of a transmission line. One may then wonder about the accuracy of the fields (improved trial functions) used in the functional. The answer is that these fields can differ quite substantially from the actual (unknown) fields, at least in some regions of the space of interest. The reason is obvious. The variational principle is a ratio of integrals, and the integrals average their integrands over the specific domain. Because the ratio is variational, the integrals are not very sensitive to particular values of the fields, even when they are very distinct from the actual solution. Hence, using the trial fields in the functional, even improved, as an approximation of the actual unknown fields may be quite misleading, and one should be very careful in doing so.

2.3 Comparison with other methods

2.3.1 Reaction

The reaction concept introduced by Rumsey [2.15] is efficient in obtaining variational principles for a wide variety of electromagnetic parameters. Using Rumsey's notations, two sources located in different areas of space and denoted by a and b generate their associated fields, denoted by A and B respectively. Hence, the reaction of field A on source b has the form

$$\langle A, b \rangle = \int \{ \overline{E}^a \cdot \overline{dJ}^b - \overline{H}^a \cdot \overline{dM}^b \} \tag{2.31a}$$

where fields $\overline{E}^{a,b}$ and $\overline{H}^{a,b}$ generated by sources a, b respectively, satisfy

$$\nabla \times \overline{E}^{a,b} = -j\omega \overline{\overline{\mu}} \cdot \overline{H}^{a,b} + \overline{M}^{a,b} \tag{2.31b}$$

$$\nabla \times \overline{H}^{a,b} = j\omega \overline{\overline{\varepsilon}} \cdot \overline{E}^{a,b} + \overline{J}^{a,b} \tag{2.31c}$$

where $\overline{M}$ is a magnetic current (V/m^2) and $\overline{J}$ is an electric current (A/m^2). Using the reaction concept, the Lorentz reciprocity principle is then rewritten for isotropic media as

$$\langle A, b \rangle = \langle B, a \rangle \tag{2.32}$$

Equation (2.32) states that the reaction of field A on source b is equivalent to the reaction of field B on source a. When the medium is not isotropic, Rumsey proposes a correction to (2.32):

$$\langle A, b \rangle = \langle \hat{B}, a \rangle \tag{2.33}$$

where $\langle \hat{B}, a \rangle$ is written for the case corresponding to the same sources as $\langle A, b \rangle$ but where the fields generated by the b source are solutions of Maxwell's equations for a medium having transposed permittivity and permeability tensors:

$$\langle \hat{B}, a \rangle = \int \{ \overline{E}^{b'} \cdot \overline{dJ}^a - \overline{H}^{b'} \cdot \overline{dM}^a \} \tag{2.34a}$$

with $\overline{E}^{b'}$ and $\overline{H}^{b'}$ satisfying

$$\nabla \times \overline{E}^{b'} = -j\omega \overline{\overline{\mu}}^T \cdot \overline{H}^{b'} + \overline{M}^b \tag{2.34b}$$

$$\nabla \times \overline{H}^{b'} = j\omega \overline{\overline{\varepsilon}}^T \cdot \overline{E}^{b'} + \overline{J}^b \tag{2.34c}$$

while the fields $\overline{E}^a$ and $\overline{H}^a$ generated by source a still satisfy (2.31b,c). This formalism has been extensively used by Harrington and Villeneuve [2.16]. They derive equivalent circuit formulations based on reaction, which apply in case of gyrotropic media. When the tensors are simply symmetric, they satisfy

$$\overline{\overline{\varepsilon}} = \overline{\overline{\varepsilon}}^T \tag{2.35a}$$

$$\overline{\overline{\mu}} = \overline{\overline{\mu}}^T \tag{2.35b}$$

and the fields $\overline{E}^{b'}$ and $\overline{H}^{b'}$ are equal indeed to the fields $\overline{E}^{b}$ and $\overline{H}^{b}$ satisfying (2.31b,c). Hence (2.34a) reduces in this case to (2.31a) rewritten for the B field.

The major interest in the reaction is that it is helpful to find approximate field or source distribution (trials), which have almost the same effect as the actual (correct) ones, on a portion of a surface or volume. As explained by Rumsey, the effect of a test source x is compared to the effect of an exact source c and of a trial source a, by forming the two reactions

> $\langle C, x \rangle$ measuring the effect of the field generated by the test source x on the true (unknown) current distribution c
>
> $\langle A, x \rangle$ measuring the effect of the field generated by the test source x on the trial (approximated) current distribution a.

It is obvious that the best trial source is the one which has the same effect on the test source as the exact one, that is which satisfies

$$\langle A, x \rangle = \langle C, x \rangle \tag{2.36}$$

Rumsey says that "a and c should look the same to an arbitrary test source x".

Unfortunately, neither the test source x nor the true current distribution c are known. The best choice is then to equate x to the trial source distribution a and impose that the reaction of the field created by a on itself is the same as the reaction of the field created by a on the actual correct source distribution c:

$$\langle A, a \rangle = \langle A, c \rangle \tag{2.37}$$

Such a condition will of course be of interest only if a particular behavior of either the trial source or the exact one is known *a priori* in the portion of space where the reaction is calculated. This is indeed the case for the Spectral Domain Galerkin's procedure applied to planar lines, as will be shown in Chapter 3. Assuming that the trial source is chosen such that it satisfies (2.37):

$$\langle A, b \rangle = \langle A_0, b \rangle = \langle A, b_0 \rangle \tag{2.38}$$

the reaction written for the approximate field is

$$\langle A, b \rangle = \langle A_0, b_0 \rangle + \langle \delta A, b_0 \rangle + \langle A_0, \delta b \rangle + \langle \delta A, \delta b \rangle \tag{2.39}$$

where the subscript 0 denotes exact quantities.

2.3.1.1   Variational behavior of reaction

Harrington demonstrates in [2.17] that (2.39) is variational about A and b. Indeed using (2.38), the reaction $\langle A, b \rangle$ is rewritten as

$$\langle A, b \rangle = \langle A_0, b \rangle = \langle A_0, b_0 \rangle + \langle A_0, \delta b \rangle \tag{2.40a}$$

$$= \langle A, b_0 \rangle = \langle A_0, b_0 \rangle + \langle \delta A, b_0 \rangle \tag{2.40b}$$

Introducing the right-hand sides of equation (2.40a,b) into (2.39) and recombining we obtain

$$\langle A, b \rangle = \langle A_0, b_0 \rangle - \langle \delta A, \delta b \rangle \tag{2.41}$$

which demonstrates the stationary character of reaction $\langle A, b \rangle$ about the exact sources and field distributions:

$$\langle \delta A, b_0 \rangle = \langle A_0, \delta b \rangle = 0 \tag{2.42}$$

2.3.1.2   Self-reaction and implicit variational principles

When b is replaced by a in (2.31a) and (2.38), it covers the case where source b is identical to source a and located at the same place (only one source is present). The product $\langle A, a \rangle$ is then called "self-reaction". It has some additional interesting properties. Assuming that the trial source and associated field generated by the source are functions of an unknown parameter p, and knowing *a priori* that the value of the actual reaction $\langle A_0, a_0 \rangle$ is zero, it is easy to demonstrate that making the trial reaction $\langle A, a \rangle$ vanish, renders the parameter p stationary about the trial source and associated field. The trial reaction $\langle A, a \rangle$ is imposed to be equal to the actual one:

$$\langle A, a \rangle = \langle A_0, a_0 \rangle = 0 \tag{2.43}$$

and is developed around the actual values of the parameter p as well as of the field and sources (second-order variations are neglected):

$$\langle A, a \rangle = \langle A_0, a_0 \rangle + \delta p \left. \frac{\partial \langle A_0, a_0 \rangle}{\partial p} \right|_{p=p_0}$$
$$+ \langle \delta A, a_0 \rangle |_{p=p_0} + \langle A_0, \delta a \rangle |_{p=p_0} \tag{2.44}$$

Since the reaction has been demonstrated to be variational about the source and field, whatever the sources a and b are, (2.42) is satisfied for the self-reaction case also, and (2.44) is finally equivalent to

$$0 = \delta p \left. \frac{\partial \langle A_0, a_0 \rangle}{\partial p} \right|_{p=p_0} \tag{2.45}$$

by virtue of (2.43).

It has to be noted that the stationary character of p is derived by Harrington, assuming reaction (2.43) vanishes. This assumption, however, is not necessary, as shown by (2.44). It is sufficient to ensure that $\langle A, a \rangle = \langle A_0, a_0 \rangle$ and that the reaction is stationary.

2.3.2   Method of moments

The method of moments (MoM) can be related to variational theory and the reaction concept. It is widely used in electromagnetics and other engineering areas for solving a variety of problems, such as: determining current distributions on antennas and obstacles, and field and current distributions on planar transmission lines, by using the Galerkin variant of the MoM. The MoM usually searches for a field distribution solution of a given problem inside a given domain, which satisfies specific conditions imposed on the boundaries of the domain. The MoM and its application to basic electromagnetic problems was first presented by Harrington [2.18]. A collection of papers covering various applications of the method in electromagnetics is found in [2.19]. In this section, we report on the main features of the MoM applied to scalar quantities, as in [2.18], and point out the variational functionals associated with the method.

The MoM usually solves integral equations over a given domain, subject to known boundary conditions. This is in fact an alternate way to solve an inhomogeneous problem in space V_x subject to inhomogeneous boundary conditions on surface Ω_x. The linear integral equation to be considered is

$$\mathcal{L}(\overline{f}) = \overline{g} \qquad \text{on } \Omega_x \tag{2.46}$$

where $\mathcal{L}$ is the linear operator associated with the linear integral equation
$\overline{f}$ is the spatial distribution of scalar or vector quantity to be determined
$\overline{g}$ is the scalar or vector condition imposed on the boundary

The unknown function f is expanded into a series of functions $\{f_n\}$ which are supposed to form *a complete set for the problem*. Hence the unknown solution f can be described *without error* by an infinite series expansion of those basis functions:

$$f = \sum_{n=1}^{\infty} \alpha_n f_n \tag{2.47}$$

Entering (2.47) into (2.46) yields

$$\sum_{n=1}^{\infty} \mathcal{L}(\alpha_n f_n) = g \tag{2.48}$$

Assuming that a suitable inner product has been defined, the two sides of (2.48) are multiplied by a set of known weighting functions $\{w_k\}$ and the inner product is taken, which yields a system of linear equations for the unknowns α_n:

$$\sum_{n=1}^{\infty} \alpha_n \langle w_k, \mathcal{L}(f_n) \rangle = \langle w_k, g \rangle \qquad \text{with } k = 1, \ldots, \infty \tag{2.49}$$

The result obtained may obviously depend on the choice made for the expansion functions $\{f_n\}$ and the weighting functions $\{w_k\}$. A particular choice is $\{w_k\} = \{f_n\}$. The MoM is then referred to as Galerkin's procedure with the characteristic equations

$$\sum_{n=1}^{\infty} \alpha_n \langle f_k, \mathcal{L}(f_n) \rangle = \langle f_k, g \rangle \qquad \text{with } k = 1, \ldots, \infty \tag{2.50}$$

Up to this point, the solution of (2.49) is exact. It is indeed equivalent to solving the equation

$$\langle w, \mathcal{L}(f) \rangle = \langle w, g \rangle \tag{2.51a}$$

which, using (2.47), can be rewritten as

$$\sum_{k=1}^{\infty} \sum_{n=1}^{\infty} \omega_k \alpha_n \langle w_k, \mathcal{L}(f_n) \rangle = \sum_{k=1}^{\infty} \omega_k \langle w_k, g \rangle \tag{2.51b}$$

where we define w similarly to (2.47) as

$$w = \sum_{n=1}^{\infty} \omega_n w_n \tag{2.51c}$$

It is obvious that a solution satisfying (2.48) will automatically satisfy (2.51b) and, hence, can be viewed as the exact solution. This is true only because the infinite series is assumed to be formed by a *complete* set of basis functions. In practice, however, the series expansion is limited to N terms, so that an error is made on the description of the function f. After truncating (2.47)

$$f_t = \sum_{n=1}^{N} \alpha_n f_n \tag{2.52a}$$

the linear system of N equations resulting from (2.49) is solved for the N unknowns α_n:

$$\sum_{n=1}^{N} \alpha_n \langle w_k, \mathcal{L}(f_n) \rangle = \langle w_k, g \rangle \qquad \text{with } k = 1, \ldots, N \tag{2.52b}$$

2.3.2.1   Variational character of error made

First, it can be demonstrated that the solution of (2.52b) minimizes the error made on the left-hand of equation (2.46) [2.20]. Defining this error as

$$\Delta = \mathcal{L}(f_t) - g \tag{2.53a}$$

and multiplying with w results in

$$
\begin{aligned}
\langle w, \Delta \rangle &= \langle w, \mathcal{L}(f_t) - g \rangle \\
&= \sum_{k=1}^{N} \sum_{n=1}^{N} \omega_k \alpha_n \langle w_k, \mathcal{L}(f_n) \rangle - \sum_{k=1}^{N} \omega_k \langle w_k, g \rangle
\end{aligned}
\tag{2.53b}
$$

Taking the first derivative of the right-hand side of (2.53b) with respect to ω_k and cancelling the result, yields equation (2.52b). Hence solving (2.53b) is equivalent to minimizing the error product $\langle w, \Delta \rangle$. It can then be demonstrated [2.21] that the solution of (2.52b) minimizes the least-squares error made on the left-hand of equation (2.46). Defining this error as

$$
\Delta = \langle \mathcal{L}(f_t) - g, \mathcal{L}(f_t) - g \rangle
\tag{2.54a}
$$

results in

$$
\begin{aligned}
\Delta &= \sum_{k=1}^{N} \sum_{n=1}^{N} \omega_k \alpha_n \langle \mathcal{L}(f_k), \mathcal{L}(f_n) \rangle - \sum_{k=1}^{N} \omega_k \langle \mathcal{L}(f_k), g \rangle \\
&- \sum_{n=1}^{N} \alpha_n \langle g, \mathcal{L}(f_n) \rangle + \langle g, g \rangle
\end{aligned}
\tag{2.54b}
$$

Taking the first derivative of the right-hand side of (2.54b) with respect to ω_k, α_n and cancelling the result, yields both equations

$$
\sum_{n=1}^{N} \alpha_n \langle \mathcal{L}(f_k), \mathcal{L}(f_n) \rangle - \langle \mathcal{L}(f_k), g \rangle = 0
\tag{2.55a}
$$

$$
\sum_{k=1}^{N} \omega_k \langle \mathcal{L}(f_k), \mathcal{L}(f_n) \rangle - \langle g, \mathcal{L}(f_n) \rangle = 0
\tag{2.55b}
$$

This is equivalent to using the moment method with

$$
w_k = \mathcal{L}(f_k)
\tag{2.56}
$$

Hence applying the moment method with the set of weighting functions (2.56) is equivalent to minimizing the error functional defined by (2.54a).

2.3.2.2 Variational functionals associated with the MoM

A. Inner product

Assuming that f is the solution of the linear integral problem (2.46) with associated $\mathcal{L}$ operator, the adjoint operator of $\mathcal{L}$, noted $\mathcal{L}^a$, is defined by the property

$$
\langle \mathcal{L}(a), b \rangle = \langle a, \mathcal{L}^a(b) \rangle
\tag{2.57}
$$

When the MoM is used to find f, a variational functional exists for the expression

$$\rho = \langle f, h \rangle \tag{2.58}$$

This is also a measure of the error made, since the distribution h is *a priori* known. Harrington [2.18] states that the functional

$$\rho = \frac{\langle f, h \rangle \langle f^a, g \rangle}{\langle \mathcal{L}(f), f^a \rangle} \tag{2.59}$$

is a stationary formula for (2.58) provided that an adjoint function f^a is defined, which is a solution of the adjoint problem

$$\mathcal{L}^a(f^a) = h \tag{2.60a}$$

and has the form

$$f^a = \sum_{n=1}^{\infty} \beta_n w_n \tag{2.60b}$$

The MoM formulation of this adjoint problem is

$$\sum_{n=1}^{N} \beta_n \langle f_k, \mathcal{L}^a(w_n) \rangle = \langle f_k, h \rangle \qquad \text{with } k = 1, \ldots, N \tag{2.60c}$$

Taking the first-order derivative with respect to α_n, β_n in expression (2.59) yields equation (2.51b). This means that the MoM cancels the first-order error made on ρ, provided that ρ is calculated by (2.59). Since g is exact, this also means that the error made on ρ is only due to f. This, however, *does not mean that the error made on the solution f is of the second-order*. When the operator is self-adjoint, the following identities are valid:

$$f^a = f \tag{2.61a}$$

$$\mathcal{L}^a = \mathcal{L} \tag{2.61b}$$

$$h = g \tag{2.61c}$$

and the variational functional simplifies into

$$\rho = \frac{\langle f, g \rangle \langle f, g \rangle}{\langle \mathcal{L}(f), f \rangle} \tag{2.62}$$

which finally reduces to

$$\rho = \langle f, \mathcal{L}(f) \rangle \tag{2.63}$$

because the solution of MoM using Galerkin procedure satisfies $\langle f, \mathcal{L}(f) \rangle = \langle f, g \rangle$, even when it is not the exact solution of problem (2.46).

B. Functionals

Harrington [2.21] proposes other variational functionals. One is

$$J = \langle h, f \rangle + \langle f^a, g \rangle - \langle f^a, \mathcal{L}(f) \rangle \tag{2.64}$$

Inserting the expressions for f^a and f given by (2.47) and (2.51c), respectively, and taking the first-order derivative with respect to α_k and β_n, yields

$$\frac{\partial J}{\partial \alpha_k} = \langle h, f_k \rangle - \langle \sum_{n=1}^{\infty} \beta_n w_n, \mathcal{L}(f_k) \rangle = \langle h, f_k \rangle - \sum_{n=1}^{\infty} \beta_n \langle \mathcal{L}^a(w_n), f_k \rangle \tag{2.65a}$$

$$\frac{\partial J}{\partial \beta_n} = \langle w_n, g \rangle - \langle w_n, \mathcal{L}(\sum_{k=1}^{\infty} \alpha_k f_k) \rangle = \langle w_n, g \rangle - \sum_{k=1}^{\infty} \alpha_k \langle w_n, \mathcal{L}(f_k) \rangle \tag{2.65b}$$

Limiting the summations to N and requiring the right-hand side of (2.65a,b) to be zero yields the system of equations (2.52b) and (2.60c) generated by the MoM to solve respectively problem (2.46) and its adjoint (2.60a). Again, solving problem (2.46) and its adjoint by using MoM is equivalent to minimizing the functional (2.54). At the MoM solution, J is stationary about the set of coefficients associated with the set of trial functions $\{f_k\}$ and $\{w_n\}$, and hence about the distributions f and w.

C. Non self-adjoint operators and non-Galerkin method

Another generalization of the problem is given by Peterson *et al.* [2.22]. They clarify the problem of the variationality of the MoM in the case of non self-adjoint or adjoint operators, without using the Galerkin simplification. They prove that the inner product between the test function and the linear operator applied to the trial solution is variational, in the sense that the error made is of the second-order with respect to any variation of test and trial functions. They assume that the solution f for the following scalar problem equivalent to (2.46) is:

$$\mathcal{L}(f) = g \tag{2.66}$$

is approximated by the trial expansion f_t

$$f \approx f_t = \sum_{n=1}^{N} \alpha_n f_n \tag{2.67a}$$

with as associated error

$$\varepsilon_f = f - f_t \tag{2.67b}$$

They then investigate the functional

$$I = \langle f, h \rangle \tag{2.68}$$

which is identical to functional ρ of expression (2.58) defined by Harrington [2.18], where h is a given function. Defining w as the solution of the adjoint problem:

$$\mathcal{L}^a(w) = h \qquad (2.69a)$$

with as approximate solution

$$w \approx w_t = \sum_{n=1}^{N} \beta_n w_n \qquad (2.69b)$$

and associated error

$$\varepsilon_w = w - w_t \qquad (2.69c)$$

the exact functional I can be expressed as

$$I = \langle f, h \rangle = \langle f, \mathcal{L}^a(w) \rangle = \langle \mathcal{L}(f), w \rangle = \langle g, w \rangle \qquad (2.70)$$

where (2.69a,b) and (2.57) and the scalar form of (2.46) have been used. The solution of problems (2.66) and (2.69a) by the MoM can be formulated adequately by choosing the weighting functions w_t of the problem (2.66) as trial functions f^a for the adjoint problem (2.60a), and *vice − versa*.

This forms the two previous sets of equations (2.52b) and (2.60c). When approximate solutions (2.67a) and (2.69b) are introduced into (2.70), an approximate value of the functional I, denoted by I_t , is obtained:

$$I_t = \langle f_t, h \rangle = \langle f_t, \mathcal{L}^a(w_t) \rangle = \langle \mathcal{L}(f_t), w_t \rangle \qquad (2.71a)$$
$$= \langle f_t, \mathcal{L}^a(w) \rangle = \langle \mathcal{L}(f_t), w \rangle \qquad (2.71b)$$

The identity $\langle f_t, h \rangle = \langle f_t, \mathcal{L}^a(w_t) \rangle$ results from the fact that each function f_k satisfies the MoM equation (2.60c), so that the trial sum f_t also satisfies this equation. Using the identity between the right-hand sides of (2.71) yields the error on I as

$$I - I_t = \langle \mathcal{L}(f), w \rangle - \langle \mathcal{L}(f_t), w_t \rangle \qquad (2.72a)$$
$$= \langle \mathcal{L}(f - f_t), w \rangle \qquad (2.72b)$$
$$= \langle \mathcal{L}(\varepsilon_f), w_t \rangle + \langle \mathcal{L}(\varepsilon_f), \mathcal{L}(\varepsilon_w) \rangle \qquad (2.72c)$$

The first term of (2.72c) vanishes, because it can be rewritten as

$$\langle \mathcal{L}(f), w_t \rangle - \langle \mathcal{L}(f_t), w_t \rangle = \langle g, w_t \rangle - \langle \mathcal{L}(f_t), w_t \rangle \qquad (2.73)$$

which forms the equation satisfied by solution f_t of the MoM, since each trial function w_k satisfies (2.49).

Hence, it is demonstrated that the error produced on I by the MoM is second-order, since it is proportional to the product of the errors made on the trial and weighting functions:

$$I - I_t = \langle \mathcal{L}(\varepsilon_f), \mathcal{L}(\varepsilon_w) \rangle \tag{2.74}$$

When the operator is self-adjoint, identities (2.61a,b,c) apply, with $f^a = w$, and the functional is shown to be stationary about any error made on the solution f:

$$I - I_t = \langle \mathcal{L}(\varepsilon_f), \mathcal{L}(\varepsilon_f) \rangle \tag{2.75}$$

This equation is less restrictive than the statement made by others authors that the error made on the functional I (or ρ) *is second-order when the Galerkin method is applied together with a self-adjoint operator* [2.23][2.24].

D. Comments

These developments lead to the following conclusions.

1. We have proven that the functional I (or ρ) is stationary, provided that the trial distribution $\{f_k\}$ and associated adjoint distribution $\{w_k\}$ are solutions of problems (2.66) and (2.69a,b) yielded by the MoM. This means that those distributions possess, among all the possible sets of coefficients $\{\alpha_k\}\{\beta_k\}$, one set of coefficients which minimizes the functional under investigation and is obtained as the solution of the MoM equations (2.52b) and (2.60c). This also ensures that the following global equations, that we used as identities to prove the stationarity, are satisfied:

$$\langle g, w_t \rangle = \langle \mathcal{L}(f_t), w_t \rangle \tag{2.76a}$$

$$\langle f_t, h \rangle = \langle f_t, \mathcal{L}^a(w_t) \rangle \tag{2.76b}$$

2. At this stage of the discussion, we have no idea how the set of functions f_k influences the value of the functional. Stationarity is proven with respect to the coefficients of the serial expansion, and not about the shape of the distribution $\{f_k\}$. One may, however, wonder about the influence of the shape of $\{f_k\}$, as it is assumed that $\{f_k\}$ forms a complete set of solutions. In fact what is important is the influence on the functional ρ of the error made on the global f. This error can indeed be reported on the sole coefficients $\{\alpha_k\}$, if $\{f_k\}$ forms a complete set enabling to describe all possible field distributions.

3. As a conclusion of the discussion, it is clear that we did not prove that the MoM minimizes the error made on the unknown function f solution of (2.46). We only know that system (2.76a) is solved exactly. This does not guarantee that problem (2.46) is solved exactly. Richmond [2.25], however, states that the system (2.76a) will provide the exact solution if, and only if, the set of truncated functions $\{f_k\}$ forming the expansion (2.67a) is able to represent it exactly. This is usually not the case.

2.3.2.3 The MoM and the reaction concept

We are now going to show, keeping the above theory in mind, that the MoM can be considered as a particular way to solve a reaction problem. Remembering the definition (2.31) of Rumsey's reaction, and limiting the developments for convenience to electric sources of current in isotropic media, yields

$$\langle A, b \rangle = \int \{ \overline{E}^a \cdot \overline{dJ}^b \} \tag{2.77a}$$

where the fields $\overline{E}^{a,b}$ and $\overline{H}^{a,b}$ are generated by the sources a and b respectively, satisfying

$$\nabla \times \overline{E}^{a,b} = -j\omega \overline{\overline{\mu}} \cdot \overline{H}^{a,b} \tag{2.77b}$$

$$\nabla \times \overline{H}^{a,b} = j\omega \overline{\overline{\varepsilon}} \cdot \overline{E}^{a,b} + \overline{J}^{a,b} \tag{2.77c}$$

Entering (2.77b) into (2.77c) yields a linear dyadic relation between the current sources and the electric field:

$$\nabla \times \{ \overline{\overline{\mu}}^{-1} \nabla \times \overline{E}^{a,b} \} = \omega^2 \overline{\overline{\varepsilon}} \cdot \overline{E}^{a,b} - j\omega \overline{J}^{a,b} \tag{2.78}$$

which can be rewritten using a linear operator $\mathcal{M}$ as

$$\mathcal{M}(\overline{E}^{a,b}) = -j\omega \overline{J}^{a,b} \tag{2.79}$$

Provided that the inverse of $\mathcal{M}$ exists, one has

$$\overline{E}^{a,b} = -j\omega \mathcal{M}^{-1}(\overline{J}^{a,b}) \triangleq \mathcal{L}(\overline{J}^{a,b}) \tag{2.80}$$

and the reaction (2.77a) has the form

$$\langle A, b \rangle = \int \{ \mathcal{L}(\overline{J}^a) \cdot \overline{dJ}^b \} \tag{2.81a}$$

which is typical of a linear integral operator. Hence, if we want to calculate the reaction (2.77a), subject to the fact that the electric field satisfies specific conditions on a particular boundary, we formulate the problem as

$$\overline{E}^a = \mathcal{L}(\overline{J}^a) = g \tag{2.81b}$$

which forms the moment equation to be solved. In (2.81a), $\overline{J}^a$ is the unknown current distribution to be solved for, while $\overline{J}^b$ is the weighting function. Usually, condition (2.81b) is imposed on one component of the electric field only, and the problem (2.81b) becomes scalar. According to Rumsey, the reaction (2.81a) is variational if (2.38) is satisfied, which implies the following relationship between trial and exact quantities:

$$\langle \mathcal{L}(\overline{J}_t^a), \overline{J}_t^b \rangle = \langle \mathcal{L}(\overline{J}^a), \overline{J}_t^b \rangle = \langle \mathcal{L}(\overline{J}_t^a), \overline{J}^b \rangle \tag{2.82}$$

On the other hand, we know from (2.76a) that the MoM minimizes the functional $\langle \mathcal{L}(f), w \rangle$ and satisfies

$$\langle w_t, \mathcal{L}(f_t) \rangle = \langle w_t, g \rangle = \langle w_t, \mathcal{L}(f) \rangle \tag{2.83a}$$

while identities (2.71a,b) ensure that

$$\langle \mathcal{L}(f_t), w_t \rangle = \langle \mathcal{L}(f_t), w \rangle \tag{2.83b}$$

Hence (2.83b) reproduces condition (2.82), to be satisfied by trial distribution $\overline{J_t}$ for making reaction $\overline{J}$ stationary. The MoM is thus a way to find adequate trial distributions satisfying (2.82). Other trial configurations can, however, be used for computing reactions satisfying (2.82) and (2.83), without necessarily applying the MoM. This will be illustrated in Section 2.5.

As a matter of fact, how to compute a reaction depends on its use. As will be shown in Section 2.5, reaction in electromagnetics can be related to important circuit parameters, such as input impedance and radar cross-section. As we are interested only in those circuit parameters, we do not need to solve exactly the field distribution inside the structure, and the MoM formalism may be avoided. On the other hand, when we need to solve problem (2.46), we obtain the stationarity of product $\langle \mathcal{L}(f_t), w_t \rangle$ as a by-product of the MoM. As we have shown, this inner product is equivalent to the reaction, and the MoM yields trial quantities which match conditions (2.38) for a stationary reaction.

To conclude this section, we will mention that the non-standard eigenvalue formulation and problem introduced by Lindell [2.10] is a particular case of the reaction method presented in Subsection 2.3.1.2. The non-standard eigenvalue problem is formulated by equation

$$\mathcal{L}(\lambda)f = 0 \tag{2.84}$$

meaning that what is looked at is the solution of a problem described by a linear integral operator applied to an unknown distribution f, and involving an unknown parameter λ. Since the value of λ depends on the solution f, the problem is an eigenvalue one. Lindell proposes to take the inner product between the two sides of equation (2.84)

$$\langle f, \mathcal{L}(\lambda)f \rangle = 0 \tag{2.85}$$

When a trial function f is introduced in (2.85), we take advantage of the fact that the right-hand side of (2.84) is known *a priori*. Developing (2.85) around the exact solution $\{\lambda_0, f_0\}$ yields

$$\langle \delta f, \mathcal{L}(\lambda_0)f_0 \rangle + \langle f_0, \mathcal{L}(\lambda_0)\delta f \rangle + \langle f_0, \frac{\partial \mathcal{L}(\lambda)f_0}{\partial \lambda} \rangle \delta\lambda = 0 \tag{2.86}$$

Assuming that the operator is self-adjoint, yields

$$\langle f_0, \frac{\partial \mathcal{L}(\lambda)}{\partial \lambda} f_0 \rangle \delta\lambda = -2\langle \delta f, \mathcal{L}(\lambda_0)f_0 \rangle = 0 \tag{2.87}$$

since the exact solution $\{\lambda_0, f_0\}$ satisfies (2.84). The final result (2.87) is identical to (2.45). However, it has to be noted that the assumptions made are not the same. The non-standard eigenvalue problem starts with a homogeneous equation, yielding directly result (2.87) provided that the operator is self-adjoint. The stationarity of p in (2.44) can be obtained from the assumption that the self-reaction is stationary, regardless of the fact that it vanishes, even if Harrington does assume it.

2.3.2.4   MoM and perturbation

MoM concepts can be used for solving problems in perturbed configurations, because they propose an alternate way to the classical perturbation theory. Harrington proposes a perturbational solution for the MoM as follows [2.18]: solution f_0 of the exact unperturbed configuration (subscript 0) is supposed to be known:

$$\langle \mathcal{L}_0(f_0), w_0 \rangle = \langle g, w_0 \rangle \qquad (2.88a)$$

while the perturbed problem is described by operator $\mathcal{L}$ and associated equation:

$$\langle \mathcal{L}(f), w \rangle = \langle \{\mathcal{L}_0 + \mathcal{M}\}(f), w \rangle = \langle g, w \rangle \qquad (2.88b)$$

Limiting the case to self-adjoint operators and the Galerkin procedure, a first-order perturbation solution is proposed by Harrington as

$$f = \alpha f_0 \qquad (2.89a)$$

$$w = f_0 \qquad (2.89b)$$

which, introduced into (2.88) yields

$$\alpha \langle \{\mathcal{L}_0 + \mathcal{M}\}(f_0), f_0 \rangle = \alpha \{\langle \mathcal{L}_0(f_0), f_0 \rangle + \langle \mathcal{M}(f_0), f_0 \rangle\}$$
$$= \langle g, f_0 \rangle = \langle \mathcal{L}_0(f_0), f_0 \rangle \qquad (2.90)$$

and

$$\alpha = 1 - \frac{\langle \mathcal{M}(f_0), f_0 \rangle}{\langle \mathcal{L}_0(f_0), f_0 \rangle + \langle \mathcal{M}(f_0), f_0 \rangle} \qquad (2.91)$$

If the perturbation is "truly small", then the second inner product in the denominator is small compared to the first one, and the perturbed solution has the (approximate) form

$$f = \left\{ 1 - \frac{\langle \mathcal{M}(f_0), f_0 \rangle}{\langle \mathcal{L}_0(f_0), f_0 \rangle} \right\} f_0 \qquad (2.92)$$

This is indeed a perturbational formula for the unknown f.

2.3.3 Variational principles for solving perturbation problems

Some authors [2.9],[2.26] mention that variational principles are useful for solving perturbation problems. Berk solves the insertion of a small sample into a cavity with the variational principle introduced in Section 2.2. When the distribution of matter within the waveguide is discontinuous, (2.30) can be further modified so that both electric and magnetic trial fields are arbitrary at the boundary. The modification is in adding to the numerator of (2.30) the term

$$\oint_C \overline{n} \cdot (\overline{e} \times \overline{h}^*) \, dC \tag{2.93}$$

where C is the boundary of the waveguide cross-section. Assuming that permeability and/or permittivity are modified in a small area of the waveguide cross-section in turns modifies the field distribution. As (2.30) is variational, however, trial fields for the perturbed case can be taken equal to be to the fields obtained for the unperturbed case, since the formulation has been made insensitive to boundary conditions by adding (2.93). Hence, a formulation correct to second-order is obtained for the shift on the propagation constant:

$$
\begin{aligned}
j(\beta_1 - \beta) \;=\; & \frac{j\omega \int_S \{\overline{e_1}^* \cdot (\overline{\overline{\varepsilon}}_1 \cdot \overline{e_1}) + \overline{h_1}^* \cdot (\overline{\overline{\mu}}_1 \cdot \overline{h_1})\} \, dS}{\int_S \{\overline{h_1}^* \cdot (\overline{a}_z \times \overline{e_1}) - \overline{e_1}^* \cdot (\overline{a}_z \times \overline{e_1})\} \, dS} \\[2ex]
& - \frac{\int_S \{\overline{e_1}^* \cdot (\nabla \times \overline{h_1}) - \overline{h_1}^* \cdot (\nabla \times \overline{e_1})\} \, dS}{\int_S \{\overline{h_1}^* \cdot (\overline{a}_z \times \overline{e_1}) - \overline{e_1}^* \cdot (\overline{a}_z \times \overline{e_1})\} \, dS} \\[2ex]
& - \frac{j\omega \int_S \{\overline{e}^* \cdot (\overline{\overline{\varepsilon}} \cdot \overline{e}) + \overline{h}^* \cdot (\overline{\overline{\mu}} \cdot \overline{h})\} \, dS}{\int_S \{\overline{h}^* \cdot (\overline{a}_z \times \overline{e}) - \overline{e}^* \cdot (\overline{a}_z \times \overline{h})\} \, dS} \\[2ex]
& + \frac{\int_S \{\overline{e}^* \cdot (\nabla \times \overline{h}) - \overline{h}^* \cdot (\nabla \times \overline{e})\} \, dS}{\int_S \{\overline{h}^* \cdot (\overline{a}_z \times \overline{e}) - \overline{e}^* \cdot (\overline{a}_z \times \overline{e})\} \, dS} \\[2ex]
& + \frac{\oint_C \overline{n} \cdot (\overline{e_1} \times \overline{h_1}^*) \, dC}{\int_S \{\overline{h_1}^* \cdot (\overline{a}_z \times \overline{e_1}) - \overline{e_1}^* \cdot (\overline{a}_z \times \overline{e_1})\} \, dS} \\[2ex]
& - \frac{\oint_C \overline{n} \cdot (\overline{e} \times \overline{h}^*) \, dC}{\int_S \{\overline{h}^* \cdot (\overline{a}_z \times \overline{e}) - \overline{e}^* \cdot (\overline{a}_z \times \overline{e})\} \, dS}
\end{aligned}
\tag{2.94a}
$$

$$\approx \frac{jw \int_S \{\overline{e}^* \cdot (\Delta\overline{\overline{\varepsilon}} \cdot \overline{e}) + \overline{h}^* \cdot (\Delta\overline{\overline{\mu}} \cdot \overline{h})\}\, dS}{\int_S \overline{h}^* \cdot (\overline{a}_z \times \overline{e})\, dS - \int_S \overline{e}^* \cdot (\overline{a}_z \times \overline{h})\, dS} \tag{2.94b}$$

$$\text{with } \Delta\overline{\overline{\varepsilon}} = \overline{\overline{\varepsilon}}_1 - \overline{\overline{\varepsilon}}$$
$$\Delta\overline{\overline{\mu}} = \overline{\overline{\mu}}_1 - \overline{\overline{\mu}}$$

The only approximation made is equating the field distribution $\{\overline{e_1}, \overline{h_1}\}$ in the presence of perturbing material $\{\overline{\overline{\varepsilon}}_1, \overline{\overline{\mu}}_1\}$ to the field distribution $\{\overline{e}, \overline{h}\}$ in the absence of perturbation. No assumption is made about the magnitude of the perturbation and hence about the magnitude of the difference $\overline{e_1} - \overline{e}$, and $\overline{h_1} - \overline{h}$. We require these differences to be zero, taking advantage of the fact that formula (2.30) is insensitive to the first-order to any error made on the perturbed field distribution $\{\overline{e_1}, \overline{h_1}\}$. This departs from traditional perturbational solutions.

The efficiency of formula (2.94a) for planar lines will be illustrated in Chapter 5. To conclude this subsection, we note that (2.92) is a perturbational expression for a field distribution solution of (2.46), while (2.94b) provides a variational solution for a transmission line parameter.

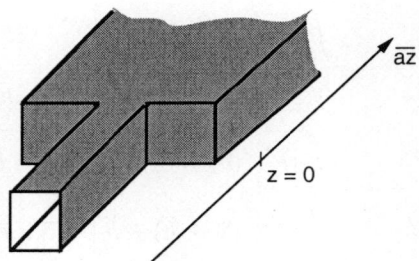

Fig. 2.3 *Step discontinuity in planar waveguide*

2.3.4   Mode-matching technique

The mode-matching technique is an efficient method for describing lumped-circuit elements of discontinuities. It is extensively presented in [2.27] and briefly summarized here in a specific example. The method is typically applied to the problem of scattering into waveguides on both sides of a discontinuity, such as a step, represented in Figure 2.3.

The fields in the access waveguides on both sides of the step are expanded into a set of orthogonal modes of the waveguide, with unknown coefficients. Coefficients A_k are used for the input waveguide, and B_k for the output waveguide. For TE_{n0} excitation for example, we expand the E_y and H_y field in terms of the modal field in each waveguide. In the incident waveguide, we assume a wave of magnitude A_1 incident from the left of the junction. It is scattered at the discontinuity, generating a reflected wave of magnitude $\Gamma_1 A_1$ and an infinite number of higher order modes evanescent towards the left of the discontinuity ($z < 0$):

$$E_{yl} = \left(A_1 e^{-\gamma_1 z} + \Gamma_1 A_1 e^{\gamma_1 z}\right) \Phi_1 + \sum_{n=2}^{\infty} B_n \Phi_n e^{\gamma_n z} \qquad (2.95a)$$

$$H_{yl} = -Y_{l1} \left(A_1 e^{-\gamma_1 z} - \Gamma_1 A_1 e^{\gamma_1 z}\right) \Phi_1 + \sum_{n=2}^{\infty} B_n Y_{ln} \Phi_n e^{\gamma_n z} \qquad (2.95b)$$

while at the right of the discontinuity ($z > 0$), the fields are described by

$$E_{yr} = \left(B_1 e^{-\gamma_1 z}\right) \Psi_1 + \sum_{m=2}^{\infty} B_m \Psi_m e^{-\gamma_m z} \qquad (2.96a)$$

$$H_{yr} = -Y_{r1} \left(B_1 e^{-\gamma_1 z}\right) \Psi_1 - \sum_{m=2}^{\infty} B_m Y_{rm} \Psi_m e^{-\gamma_m z} \qquad (2.96b)$$

Subscripts l and r refer to areas respectively to the left and right of the discontinuity, respectively, while Y_{ln} and Y_{rm} denote the admittance of mode n to the left of the discontinuity and m to the right of the discontinuity, respectively. The next step is to impose the two fundamental continuity equations in the plane of the discontinuity $z = 0$:

$$A_1 \left(1 + \Gamma_1\right) \Phi_1 + \sum_{n=2}^{\infty} B_n \Phi_n = B_1 \Psi_1 + \sum_{m=2}^{\infty} B_m \Psi_m \qquad (2.97a)$$

$$Y_{l1} A_1 \left(1 - \Gamma_1\right) \Phi_1 = \sum_{n=2}^{\infty} B_n Y_{ln} \Phi_n + Y_{r1} B_1 \Psi_1 + \sum_{m=2}^{\infty} B_m Y_{rm} \Psi_m \qquad (2.97b)$$

This is the exact solution. Next, we use the mode orthogonality in the two waveguides. Multiplying the two sides of equations (2.97a,b) by the dominant mode of input waveguide Φ_1, and integrating over the cross-section we obtain

$$A_1 \left(1 + \Gamma_1\right) = B_1 \int_S \Psi_1 \Phi_1 \, dS + \sum_{m=2}^{\infty} B_m \int_S \Psi_m \Phi_1 \, dS \qquad (2.98a)$$

$$Y_{l1} A_1 \left(1 - \Gamma_1\right) = Y_{r1} B_1 \int_S \Psi_1 \Phi_1 \, dS + \sum_{m=2}^{\infty} B_m Y_{rm} \int_S \Psi_m \Phi_1 \, dS \qquad (2.98b)$$

Use has been made of the fact that the functions $\{\Phi_n\}$ form a set of orthogonal functions on the input waveguide aperture. The procedure leading to equations (2.98a,b) is called the mode-matching technique. The problem is now that we have to repeat the procedure several times (theoretically for a double infinity of modes) with all orthogonal modes Φ_n and Ψ_m in the two waveguides, in order to generate a sufficient number of equations relating the B_m and B_n to Γ_1, and then eliminate them to solve the problem for the reflection coefficient Γ_1 of the dominant mode only.

A variational approach may be helpful to solve the problem [2.4]. Assuming that in the junction aperture the electric field has the (unknown) distribution $E(x,y)$, equation (2.97a) can be rewritten as

$$A_1 \left(1 + \Gamma_1\right) \Phi_1 + \sum_{n=2}^{\infty} B_n \Phi_n = B_1 \Psi_1 + \sum_{m=2}^{\infty} B_m \Psi_m = E(x,y) \tag{2.99}$$

This means that first, the unknown distribution can be expanded in terms of the modal solutions in each of the two waveguides, and second, that each coefficient of field expansions (2.95a,b) and (2.96a,b) can be derived from this distribution, by applying the orthogonality relationship between modes:

$$B_n = \int_{S'} \Phi_n E(x', y') \, dS' \tag{2.100a}$$

$$B_m = \int_{S'} \Psi_m E(x', y') \, dS' \tag{2.100b}$$

$$A_1 \left(1 + \Gamma_1\right) = \int_{S'} \Phi_1 E(x', y') \, dS' \tag{2.100c}$$

When doing this, we perform a mode-matching between the waveguide modes in the left and right areas and the unknown field distribution in the aperture. The input admittance in presence of the discontinuity is

$$Y_{in} = \frac{1 - \Gamma_1}{1 + \Gamma_1} Y_{l1} \tag{2.101}$$

Introducing (2.100a,b,c) into (2.97b), and combining the result with (2.100c) yields

$$Y_{l1} A_1 \left(1 - \Gamma_1\right) \Phi_1 \frac{\int_{S'} \Phi_1 E(x', y') \, dS'}{A_1 \left(1 + \Gamma_1\right)}$$
$$= \sum_{n=2}^{\infty} Y_{ln} \Phi_n \int_{S'} \Phi_n E(x', y') \, dS' + \sum_{m=1}^{\infty} Y_{rm} \Psi_m \int_{S'} \Psi_m E(x', y') \, dS' \tag{2.102}$$

which finally provides a relationship for the input admittance of the discontinuity

$$Y_{in}\Phi_1 \int_{S'} \Phi_1 E(x',y')\, dS' = \int_{S'} E(x',y')G(x',y'|x,y)\, dS' \qquad (2.103a)$$

provided that the Green's function (Appendix A) is defined by

$$G(x,y|x',y') = \sum_{n=2}^{\infty} Y_{ln}\Phi_n(x',y')\Phi_n(x,y) + \sum_{m=1}^{\infty} Y_{rm}\Psi_m(x',y')\Psi_m(x,y)$$

$$(2.103b)$$

To obtain a convenient expression for the input admittance, it is now sufficient to take the inner product of both sides of (2.103a) with the trial function $E(x,y)$:

$$Y_{in} \int_S E(x,y)\Phi_1 \int_{S'} \Phi_1 E(x',y')\, dS'\, dS$$
$$= \int_S E(x,y) \int_{S'} E(x',y')G(x',y'|x,y)\, dS'\, dS \qquad (2.104)$$

which finally yields Y_{in} as the ratio of two integrals. We finally obtain for the input admittance of the discontinuity

$$Y_{in} = \frac{\int_S \int_{S'} E(x,y)E(x',y')G(x',y'|x,y)\, dS'\, dS}{\left\{ \int_S \Phi_1 E(x,y)\, dS \right\}^2} \qquad (2.105)$$

Comparing this expression with equation (1.2), it is obvious that it is an eigenvalue problem, with Green's function (2.103b) as linear operator. Hence, the eigenvalue Y_{in} is variational about the field distribution $E(x,y)$. This mode-matching approach has been used by some authors [2.28] to calculate the input admittance of a planar microstrip antenna, fed by a coaxial cable.

2.4 Trial field distribution

2.4.1 Rayleigh-Ritz procedure

For planar transmission lines, the quasi-static analysis is based on the computation of a quasi-TEM lumped-circuit element, capacitance or inductance, noted p. For a number of variational principles found in the literature, the expression obtained for p is explicit, like

$$p = \frac{\mathcal{L}_1[\overline{F}_i(x_1, x_2, \ldots, x_n)\overline{F}_j(x_1, x_2, \ldots, x_n)]}{\mathcal{L}_2[\overline{F}_i(x_1, x_2, \ldots, x_n)\overline{F}_j(x_1, x_2, \ldots, x_n)]} \qquad (2.106)$$

where $x_1, x_2, \ldots, x_n$ are the variables describing the domain of the problem
$\overline{F}_i$ and $\overline{F}_j$ the scalar or vector quantities associated to the problem
$\mathcal{L}_1, \mathcal{L}_2$ denote linear operators.

The variational behavior of the scalar quantity p holds only when specific conditions are verified by trial expressions $\overline{F}_t$. In electromagnetics, these trials are potentials or fields, and they have to satisfy Maxwell's equations and boundary conditions. This is usually not sufficient to determine efficient shapes of the trial field, so that the stationarity of p is frequently used to improve the description of trial quantities $\overline{F}_t$. Methods using the stationarity of p, are referred to as variational methods in [2.29]. In these methods, the trial expressions are expanded into a set of suitable functions weighted by unknown coefficients. These coefficients, called variational parameters, are then found as rendering p extremum. When the expansion is linear, the method is known as the Rayleigh-Ritz method. Assuming that $\overline{F}_t$ is described by

$$\overline{F}_{t\,i,j} = \overline{F}^0_{i,j} + \sum_{k=1}^{N} \alpha^k_{i,j} \overline{F}^k_{i,j} \tag{2.107}$$

the Rayleigh-Ritz method imposes

$$\frac{\partial p}{\partial \alpha^k_{i,j}} = 0 \qquad \text{for } k = 1, \ldots, N \tag{2.108}$$

The coefficient $\alpha^0_{i,j}$ has been normalized to 1, because p is calculated as a ratio. Since only products of coefficients $\alpha^k_{i,j}$ appear in the expression of p, by virtue of (2.106), we obtain from (2.108) a set of N linear equations to be solved for the unknowns coefficients $\alpha^k_{i,j}$. Hence, having obtained the values of the coefficients $\alpha^k_{i,j}$, we find p by using (2.107). It should be underlined that the value obtained is the best one for the number N and set of functions $\overline{F}^k_{i,j}$ that have been chosen. If we change either the number of terms in the serial expansion or the shape of the functions $\overline{F}^k_{i,j}$, we do not know *a priori* if we find a better value for p (higher or lower, depending on whether we are looking for a maximum or a minimum, respectively). So, with a given expansion (2.107) we are never certain to have found the "best" extremum.

Finally, it is now obvious that solving equations (2.52b) and (2.60c) of the MoM presented in Section 2.3, is equivalent to applying the Rayleigh-Ritz method to the functional ρ (2.58) associated with the problem or to the error functional (2.54a).

2.4.2   Accuracy on fields

The fact that some functionals are stationary with respect to some field distributions or parameters gives an insight into the accuracy of the fields. We choose trial functions and adjust their parameters until the value of the functional, expressed in terms of the trial function, is minimized (or maximized).

Actually, by doing this, it is never possible to be sure that the estimate obtained for the functional is the lowest (or the highest) one. All that can be said is that it is the best estimate that can be obtained with such a class of trial functions. It is not possible to determine whether the trial distribution which minimizes (or maximizes) the functional approaches the exact solution for the distribution. This will be true only if the trial distribution forms a complete set of eigenfunctions capable of describing all distributions, and if this is proven, when the functional is extremum, the corresponding trial functions are solutions of the equation describing the problem. What can be said, however, is that the exact value will always be lower (or higher) than the best estimate found, and that, if the estimates of the eigenvalues appear to converge to a limit when improving the trial functions, that limit is probably an eigenvalue of the system [2.26].

Also, this suggests a way to solve for higher-order modes parameters, using trial functions, trial fields and variational principles. Once we have identified one modal solution (dominant mode, for example), with its own set of trial functions obtained from the Rayleigh-Ritz method, it is sufficient, for solving for a higher-order mode, to remove from the set of trial functions used those already involved in the description of the dominant mode. This will be illustrated in an example in the next section.

2.5 Variational principles for circuit parameters

Various applications of variational principles to electromagnetic problems are found in literature. Many authors provide a review of some typical problems solvable with variational principles. They can be classified as follows: variational principles based on energy, variational principles associated with an eigenvalue problem (explicit or implicit), and variational expressions based on a reaction. Some examples are presented in this section, with comments based on the concepts reviewed throughout this chapter.

2.5.1 Based on electromagnetic energy

Electrostatic energy is a stationary quantity. Hence, expressions of circuit parameters depending linearly on energy guarantee the stationarity of those parameters. In Section 2.1.2, this was introduced for the case of the capacitance per unit length of a transmission line. A rigorous proof will be presented in Chapter 3, for the case of the capacitance per unit length, and for its dual parameter, the inductance per unit length, that we will show to be related to the magnetostatic energy. Baldomir and Hammond [2.30] present a geometrical formulation for variational electromagnetics based on energy, for both static and dynamic cases. The stationarity of electrostatic energy is helpful for solving in terms of electrostatic potential. It can be demonstrated (Section 2.1.2) that under specific conditions the potential distribution which minimizes the electrostatic energy is also the distribution which is the solution of Laplace's equation. This feature is similar to that observed for field distributions, where the stationarity of a given functional is often used to derive the solution of a problem, instead of solving exactly the equations of this problem.

2.5.2   Based on eigenvalue approach

Waldron [2.26] proposes a series of variational principles based on an eigenvalue approach for waveguide components, such as cut-off frequency, propagation constant, and resonant frequency of a waveguide resonator. He bases his approach on the fact that "In the case of a system which is carrying waves, the principle tells us that of all conceivable wave functions, the one actually obtained will be the one which minimizes an appropriate eigenvalue". This suggests of course the Rayleigh-Ritz method, which we have seen to be efficient for choosing trial field expressions. Waldron, however, *postulates* that the eigenvalue of a waveguide problem is stationary because he relates it to the principle of least action. With this in mind, the resonant frequency of an oscillating system, for example, will always be as low as possible. We will show in this section that the proof of the stationarity of waveguide eigenvalues is immediate.

For the resonant frequency of a waveguide resonator, Waldron starts from Helmoltz equation:

$$\nabla^2 \Psi + \omega^2 \varepsilon_0 \mu_0 \Psi = 0 \qquad (2.109)$$

where ω is the unknown resonant frequency, and Ψ is the z-component of either the E or H field. Multiplying both sides of (2.109) by Ψ, integrating on the volume of the cavity, and rearranging, yields

$$\omega^2 \varepsilon_0 \mu_0 = -\frac{\displaystyle\int_V \Psi \nabla^2 \Psi \, dV}{\displaystyle\int_V \Psi^2 \, dV} \qquad (2.110)$$

which forms an eigenvalue problem similar to (1.1) and (2.4), with $\lambda = \omega^2 \varepsilon_0 \mu_0$ as eigenfunction and Ψ as eigensolution. Waldron does not prove this stationarity. It is immediate however, since varying λ and Ψ in (2.110) and neglecting second-order variations yields

$$\delta\lambda \int_V \Psi^2 \, dV + 2\lambda \int_V \delta\Psi \, \Psi \, dV + \int_V \delta\Psi \nabla^2 \Psi \, dV$$
$$+ \int_V \Psi \nabla^2 \delta\Psi \, dV + \lambda \int_V \Psi^2 \, dV + \int_V \Psi \nabla^2 \Psi \, dV = 0 \qquad (2.111)$$

The invariant terms are those of equation (2.110), which is satisfied by the exact field. Next, applying the Green's theorem (Appendix B)

$$\int_V A \nabla^2 B \, dV = \int_V B \nabla^2 A \, dV + \int_S A \nabla B \cdot \overline{n} \, dS - \int_S B \nabla A \cdot \overline{n} \, dS \quad (2.112)$$

(2.111) becomes

$$\delta\lambda \int_V \Psi^2 \, dV + 2 \int_V \delta\Psi \{\lambda\Psi + \nabla^2\Psi\} \, dV$$
$$+ \int_S \Psi\nabla\delta\Psi \cdot \overline{n} \, dS - \int_S \delta\Psi\nabla\Psi \cdot \overline{n} \, dS = 0 \tag{2.113}$$

Since (2.109) is satisfied, the only way to cancel the first-order error $\delta\lambda$ is to assume that one has

$$\int_S \Psi\nabla\delta\Psi \cdot \overline{n} \, dS - \int_S \delta\Psi\nabla\Psi \cdot \overline{n} \, dS = 0 \tag{2.114}$$

On the boundary of the cross-section of an empty waveguide resonator, either Ψ or the normal component of $\nabla\Psi$ vanishes. The only way to satisfy (2.114) is to assume that the trial distribution $(\Psi + \delta\Psi)$ satisfies the same boundary conditions on the surface S as the exact solution Ψ does.

For the cut-off frequency of a waveguide, noted ω_c, the formulas proposed are similar to (2.109) and (2.110), but since a waveguide at cut-off can be viewed as a resonator having no variation along the z-axis, the integration is limited to the cross-section S of the waveguide:

$$\omega_c^2 \varepsilon_0 \mu_0 = -\frac{\displaystyle\int_S \Psi\nabla^2\Psi \, dS}{\displaystyle\int_S \varepsilon_r \Psi^2 \, dS} \tag{2.115}$$

Away from cut-off, the wave equation becomes

$$\nabla^2\Psi + \{\omega^2\varepsilon_r\varepsilon_0\mu_0 - \beta^2\}\Psi = 0 \tag{2.116}$$

yielding directly the variational expression for the propagation constant β

$$\beta^2 = \frac{\displaystyle\int_S \Psi\nabla^2\Psi \, dS + \int_S \omega^2\varepsilon_r\varepsilon_0\mu_0\Psi^2 \, dS}{\displaystyle\int_S \Psi^2 \, dS} \tag{2.117}$$

Equations (2.115) and (2.117) are valid even if the permittivity varies with the position in the waveguide: the relative dielectric constant ε_r is left inside the integrals. Applying such a formulation to waveguides is easy, because the wave function Ψ is not affected by medium inhomogeneities in the transverse section: the function is related to the z-component of either the electric or the magnetic field in the waveguide structure, which are not subject to dielectric-dependent boundary conditions in the transverse section. Also, Green's theorem used for the proof does not involve the dielectric constants, so that the derivation of the proof carried out for the resonant frequency of an empty waveguide resonator remains valid for the cut-off frequency of the dielectric-loaded waveguide. It can also be verified that the proof for the propagation constant, even for an inhomogeneous case, yields an equation very similar to (2.113) with the same final constraint (2.114).

To conclude this subsection, we illustrate the efficiency of the Rayleigh-Ritz method, by calculating the cut-off frequency of a hollow rectangular waveguide of width $2a$ ($a = 1$ cm), using variational principle (2.115). We assume as first estimate that the longitudinal component H_z x-dependence has the general polynomial form

$$\Psi = Ax^5 + Bx^4 + Cx^3 + Dx^2 + Ex + F \tag{2.118}$$

From the general theory of rectangular waveguides, we know that the H_z component of the dominant mode has an odd x-dependence, and that all field components have no variation along the y-axis. Also, the boundary conditions require that the H_z component vanishes at the lateral boundaries of the waveguide. Hence, we may a priori simplify the trial function for the dominant mode as

$$\Psi_{dom} = B'x^5 + A'x^3 + x \tag{2.119a}$$

and impose for $x = \pm a$

$$\frac{\partial \Psi_{dom}}{\partial x} = 5B'a^4 + 3A'a^2 + 1 = 0 \tag{2.119b}$$

which yields

$$B' = -\frac{1 + 3A'a^2}{5a^4} \tag{2.119c}$$

and the final form for the trial dominant mode as

$$\Psi_{dom} = -\frac{1 + 3A'a^2}{5a^4}x^5 + A'x^3 + x \tag{2.120}$$

Introducing this trial in the variational principle (2.115) provides an expression for cut-off frequency f_c as a function of the unknown parameter A'. Figure 2.4a shows the dependence of the resulting cut-off frequency on the value of A'. It exhibits a minimum for $A' = -5000$. For this particular value, the resonant frequency yielded by the variational principle is 7.51 GHz, which corresponds closely to the exact theoretical value (dashed line)

$$f_c = \frac{c_0}{2}2a = 7.5 \text{ GHz} \tag{2.121}$$

For this value of A', Figure 2.4b compares the trial Ψ_{dom} obtained from (2.120) to the exact solution $H_z = \sin(\frac{\pi x}{2a})$. The trial solution is a very good

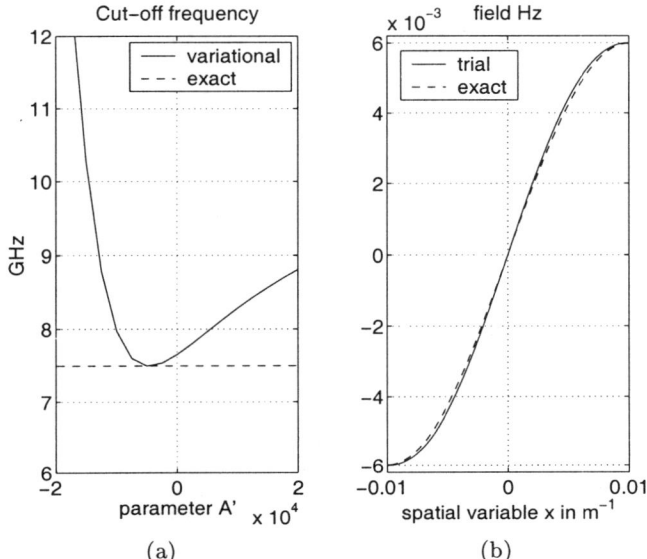

Fig. 2.4 *Rayleigh-Ritz procedure for dominant mode of rectangular waveguide of width $2a = 1$ cm (a) functional (2.115) versus A'; (b) exact component H_z and trial Ψ_{dom} for $A' = -5000$*

approximation of the true field. Hence, applying the Rayleigh-Ritz procedure to variational principle (2.115) expressed as a function of the unknown parameter A' provides the minimum value of 7.51 GHz for the resonant frequency, located at the abscissa $A' = -5000$. The resulting trial distribution associated to this extremum is shown in Figure 2.4b.

A similar reasoning is made for the first higher-order mode, assuming that its H_z x-dependence is even:

$$\Psi_{higher} = B''x^4 + A''x^2 + 1 \tag{2.122a}$$

$$\frac{\partial \Psi_{higher}}{\partial x} = 4B''a^3 + 2A''a = 0 \tag{2.122b}$$

yielding

$$\Psi_{higher} = -\frac{A''}{2a^2}x^4 + A''x^2 + 1 \tag{2.122c}$$

Results similar to Figure 2.4 are obtained in Figure 2.5. The extremum value of the cut-off frequency is now about 15 GHz. It is obtained for $A'' = -42500$, while the associated trial distribution slightly differs from the exact one ($H_z = \cos(\frac{2\pi x}{2a})$). The exact cut-off frequency for this higher-order mode is $f_c = 15$ GHz.

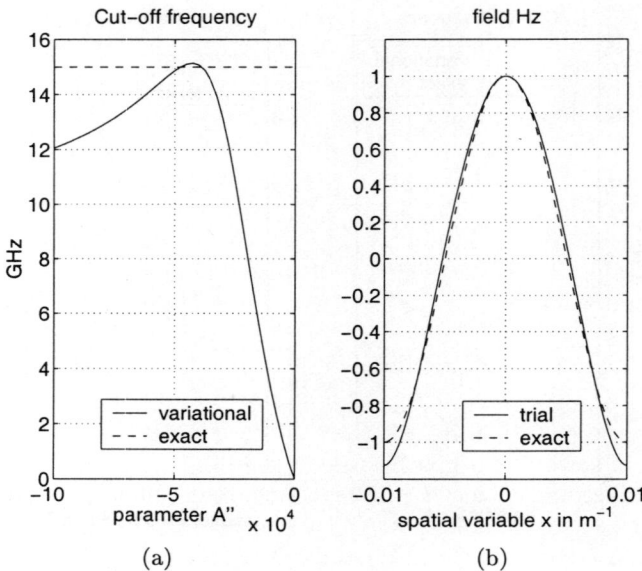

Fig. 2.5 *Rayleigh-Ritz procedure for first-higher order mode of rectangular waveguide of width $2a = 1$ cm (a) functional (2.115) versus A"; (b) exact component H_z and trial Ψ_{higher} for $A" = -42500$*

Both figures illustrate that accurate values of the functional ω_c^2 can be obtained once different distributions are chosen for the dominant (odd x-dependence) and the higher-order modes (even x-dependence). We remove the odd effect of the dominant mode from the general trial function (2.118) when we search for the first-higher order mode, yielding accurate results for this mode. In Chapters 3 and 4, other applications of the Rayleigh-Ritz method will be given.

2.5.3   Based on reaction

There are numerous formulations for circuit and system parameters which are based on reaction in electromagnetics. The reader will find in this section the most popular ones: input impedance of an antenna, echo of an obstacle, and transmission coefficient through an aperture.

2.5.3.1   Input impedance of an antenna

Harrington [2.17] proposes the following formula for the input impedance of an antenna, supposed to be perfectly conducting:

$$Z_{in} = \frac{\langle A, a \rangle}{I^2} = -\frac{1}{I^2} \int_S \overline{E}^a \cdot \overline{K}_s^a \, dS \tag{2.123}$$

where $\overline{K}_s$ is a surface current (A/m). The antenna is fed by a current source I, and the current distributes itself on the surface S in such a manner that the resulting electric field has vanishing tangential components on the conductors of the antenna. Hence, the product $\overline{E}^c \cdot \overline{K}_s^c$ is zero, except at the feeder point on the surface of the antenna, where the total current I incident from the feeding source spreads onto the conductor, while the voltage between the two terminals of the feeder creates the electric field normal to the antenna conductor at the feeding point. Hence, the reaction $\langle A_0, a_0 \rangle$ between correct field and source reduces to the integration of their product between the terminals of the antenna, supposed to be close together, as shown by:

$$Z_{in\,c} = \frac{\langle A_0, a_0 \rangle}{I^2} = -V_0 I \qquad (2.124)$$

where V_0 is the correct voltage across the terminals. When a trial surface current $\overline{K}_s^a$ is assumed, it can be chosen such that its integration gives the same value of feeding current I (this is just a matter of normalization). Hence, the reaction $\langle A_0, a \rangle$ (where a correct current distribution is assumed) is still equal to the product $-V_0 I$, which validates the equality

$$\langle A_0, a \rangle = \langle A_0, a_0 \rangle \qquad (2.125)$$

Next, the nature of the problem (antenna configuration in isotropic medium) ensures that reciprocity occurs :

$$\langle A_0, a \rangle = \langle A, a_0 \rangle \qquad (2.126)$$

Using the notations of Subsection 2.3.1, the reactions in (2.126) are rewritten as

$$\langle A_0, a \rangle = \langle A_0, a_0 \rangle + \langle A_0, \delta a \rangle \qquad (2.127a)$$

$$\langle A, a_0 \rangle = \langle A_0, a_0 \rangle + \langle \delta A, a_0 \rangle \qquad (2.127b)$$

which imposes

$$\langle A_0, \delta a \rangle = \langle \delta A, a_0 \rangle = 0 \qquad (2.128)$$

On the other hand, the trial self-reaction deduced from (2.39) is

$$\langle A, a \rangle = \langle A_0, a_0 \rangle + \langle \delta A, a_0 \rangle + \langle A_0, \delta a \rangle + \langle \delta A, \delta a \rangle \qquad (2.129)$$

which, using (2.128), reduces to

$$\langle A, a \rangle = \langle A_0, a_0 \rangle + \langle \delta A, \delta a \rangle \qquad (2.130)$$

It demonstrates that the impedance of a perfect conducting antenna computed using (2.123) is variational about the current density and associated electric field.

The formula is widely used for antennas calculations. Again, it has to be underlined that the trial current and field are not required to satisfy exact boundary conditions on the conductor of the antenna. Identity (2.126) only imposes that the product of the exact field by the trial current, integrated on the surface of the antenna, is equal to the product of the trial field by the exact current, integrated on the same surface. This indeed does not ensure that the tangential component of the trial field vanishes on each point of the surface, although it is the case for the exact field.

Obviously, it can be easily demonstrated that the following is a variational formula for the mutual impedance between two antennas, supporting surface current densities $\overline{K}_s^a$ and $\overline{K}_s^b$:

$$Z_m = \frac{\langle B, a \rangle}{I^a I^b} = -\frac{1}{I^a I^b} \int_S \overline{E}^b \cdot \overline{K}_s^a \, dS \tag{2.131}$$

2.5.3.2 Stationary formula for scattering

The electric field scattered by an obstacle is obtainable by a variational reaction. We assume that the obstacle is a perfect electric conductor. Assuming that the source is a current element $I\,l$, we denote by $\overline{E}^i$ the field generated by this current, and by $\overline{E}^s$ the field scattered by the obstacle. The sum of those two fields form the total field in the whole space. The reaction of the scattered field on the current element is noted $\langle S, i \rangle$:

$$\langle S, i \rangle = I\,l E_l^s = -I V^s \tag{2.132}$$

where V^s is the voltage appearing across l. The echo is then defined as the ratio between E_l^s and the current element $I\,l$:

$$\text{echo} \triangleq \frac{E_l^s}{I\,l} = \frac{\langle S, i \rangle}{(I\,l)^2} = \frac{\langle I, s \rangle}{(I\,l)^2} \tag{2.133a}$$

$$= \frac{1}{(I\,l)^2} \int_S \overline{E}^i \cdot \overline{K}_s^s \, dS \tag{2.133b}$$

$$= -\frac{1}{(I\,l)^2} \int_S \overline{E}^s \cdot \overline{K}_s^s \, dS = -\frac{\langle S, s \rangle}{(I\,l)^2} \tag{2.133c}$$

where reciprocity has been assumed, and $\overline{K}_s^s$ is the surface current induced on the perfect conducting obstacle. On the surface, the total electric field vanishes, yielding equation (2.133c), relating the echo to the self-reaction $\langle S, s \rangle$. Equations (2.133) are valid for reactions written with the exact fields, so they must be written with the subscript 0. Replacing $\langle S_0, s_0 \rangle$ by the trial $\langle S, s \rangle$ yields a variational formula for the echo, provided that (2.38) is satisfied:

$$\langle S, s \rangle = \langle S_0, s \rangle \tag{2.134a}$$

$$= -\langle I_0, s \rangle \tag{2.134b}$$

To render reaction $\langle S, s \rangle$ insensitive to the unknown magnitude of $\overline{K}^s_s$, the reaction is finally expressed, imposing (2.134b), as

$$\langle S, s \rangle = \frac{\langle I_0, s \rangle^2}{\langle S, s \rangle} \tag{2.135}$$

yielding the final variational form for the echo as

$$\text{echo} = -\frac{\langle I_0, s \rangle^2}{(I\,l)^2 \langle S, s \rangle} = \frac{\left(\int_S \overline{E}^i \cdot \overline{K}^s_s \, dS \right)^2}{(I\,l)^2 \int_S \overline{E}^s \cdot \overline{K}^s_s \, dS} \tag{2.136}$$

It has to be noted that the incident field is assumed to be derived from the current element, so that no error results on I.

2.5.3.3 Stationary formula for transmission through apertures

As shown in [2.31], the direct application of Babinet's principle to the problem of transmission through apertures is illustrated in Figure 2.6: the field transmitted by an aperture in a plane conducting screen is equal to the opposite of the field scattered by the complementary obstacle. The complementary

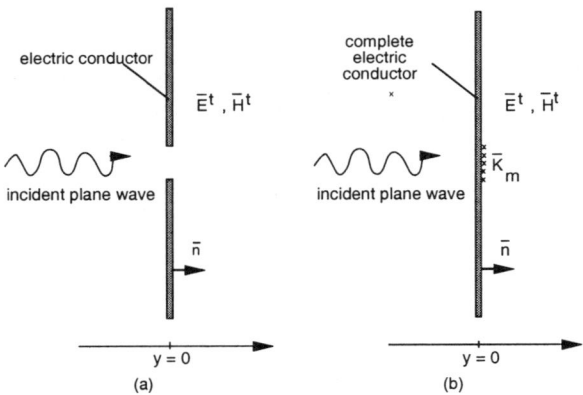

Fig. 2.6 *Transmission through aperture (a) initial configuration; (b) application of Babinet's principle*

obstacle is a sheet of perfect electric conductor, with a perfect surface magnetic current, denoted by $\overline{K}_m$, flowing on it. The transmission coefficient T of the aperture is commonly defined as the ratio between power transmitted through the aperture and power incident at the aperture:

$$T_{aperture} \triangleq \frac{\mathrm{Re}(P_t)}{\mathrm{Re}(P_i)} = \frac{\mathrm{Re}\left\{\int_{aperture} \overline{E}^t \times (\overline{H}^t)^* \, dS\right\}}{\mathrm{Re}\left\{\int_{aperture} \overline{E}^i \times (\overline{H}^i)^* \, dS\right\}} \tag{2.137}$$

The incident power is known, since it depends only on incident source fields. In the aperture, upon application of Babinet's principle, the transmitted magnetic field is related to the incident field by

$$\overline{n} \times \overline{H}^t = \overline{n} \times \overline{H}^i \tag{2.138}$$

Assuming that the magnetic field is adjusted to be real in the aperture, then the conjugate can be omitted in (2.137). The equivalent magnetic current in Figure 2.6b is:

$$\overline{K}_m = -\overline{n} \times \overline{E}^t \tag{2.139}$$

yielding the reaction formulation of the transmitted power:

$$\mathrm{Re}(P_t) = -\mathrm{Re}\left\{\int_{aperture} \overline{K}_m \cdot \overline{H}^i \, dS\right\} = \mathrm{Re}(\langle I, c\rangle) \tag{2.140}$$

where c denotes that the correct magnetic current distribution is assumed. We denote by a the reactions involving trial magnetic currents. By virtue of (2.138), we also have

$$\langle I, a\rangle = \langle A, a\rangle \tag{2.141}$$

and by reciprocity, $\langle I, a\rangle = \langle A, i\rangle$. Hence, we have the conditions for a stationary reaction. As for echo calculations, we normalize the reaction and obtain the final variational principle for the real part of transmitted power as

$$\mathrm{Re}(P_t) = \frac{\langle I, a\rangle^2}{\langle A, a\rangle} = -\frac{\left(\mathrm{Re}\left\{\int_{aperture} \overline{K}_m^a \cdot \overline{H}^i \, dS\right\}\right)^2}{\mathrm{Re}\left\{\int_{aperture} \overline{K}_m^a \cdot \overline{H}^a \, dS\right\}} \tag{2.142}$$

For a plane wave normally incident to the aperture, with a magnetic field of unit magnitude oriented along direction $\overline{u}$ and a wave impedance η, the incident power is given by the product

$$\mathrm{Re}(P_i) = \eta S_{aperture} \tag{2.143}$$

and the final variational principle for the transmission coefficient is

$$\mathrm{Re}(P_t) = \frac{\langle I, a\rangle^2}{\langle A, a\rangle} = \frac{\left(\mathrm{Re}\left\{\int_{aperture} \overline{u} \cdot \overline{n} \times \overline{E}^a \, dS\right\}\right)^2}{\eta S_{aperture} \, \mathrm{Re}\left\{\int_{aperture} \overline{H}^a \cdot \overline{n} \times \overline{E}^a \, dS\right\}} \tag{2.144}$$

2.6 Summary

Variational principles have been introduced in this chapter in general terms to calculate impedances and propagation constants as, in this book, we are essentially interested with transmission lines and resonators. Variational principles have first been developed so that the value of the variational quantity for the exact unknown has a physical significance, like energy and eigenvalues. It has been shown that a method exists for systematically improving upon the trial function by iteration, thus providing an estimate of the error, by giving both an upper and a lower bound to those quantities which are being varied. Then, we developed and illustrated methods for establishing variational principles by adequately re-arranging Maxwell's equations. It has been shown that explicit as well as implicit expressions can be obtained.

The main part of the chapter has been devoted to comparing variational methods with other concepts and methods. The reaction concept has been demonstrated to be efficient in obtaining variational principles for a wide variety of electromagnetic parameters, and self-reaction has been shown to have some additional interesting properties. The method of moments (MoM) has been developed and widely commented upon and it has been shown that it is possible to minimize the error made. Variational functionals associated with the MoM have been discussed. The Galerkin method has also been discussed in detail and we have presented where and when the MoM and Galerkin method are variational or not. Links have been established between the MoM and the reaction concept as well as with the perturbation method. The differences between perturbational and variational methods have been carefully detailed. The mode-matching technique has been briefly summarized, as well as the Rayleigh-Ritz method.

Finally, examples of variational principles for circuit parameters have been presented and concepts reviewed throughout the chapter, such as waveguide cut-off frequencies as eigenvalues, input impedance of an antenna, echo from an obstacle, and transmission coefficient through an aperture.

2.7 References

[2.1] P.M. Morse, H. Feshbach, *Methods of Theoretical Physics*. New York: McGraw-Hill, 1953, pp. 1001-1106.

[2.2] P.M. Morse, H. Feshbach, *Methods of Theoretical Physics*. New York: McGraw-Hill, 1953, pp. 1106-1158.

[2.3] R.E. Collin, *Field Theory of Guided Waves*. New York: McGraw-Hill, 1960, pp. 148-150.

[2.4] R.E. Collin, *Field Theory of Guided Waves*, 2nd ed. New York: IEEE Press, 1991, Ch. 8, pp. 547-552.

[2.5] P.M. Morse, H. Feshbach, *Methods of Theoretical Physics*, pp. 1026-1030 and 1137-1158. New York: McGraw-Hill, 1953.

[2.6]   A. Vander Vorst, R. Govaerts, F. Gardiol, "Application of the variation-iteration method to inhomogeneously loaded waveguides", *Proc. 1st Eur. Microwave Conf.*, pp. 105-109, Sept. 1969.

[2.7]   A. Vander Vorst, R. Govaerts, "Application of a variation-iteration method to inhomogeneously loaded waveguides", *IEEE Trans. Microwave Theory Tech.*, vol. MTT-18, pp. 468-475, Aug. 1970.

[2.8]   A. Laloux, R. Govaerts, A. Vander Vorst, "Application of a variation-iteration method to waveguides with inhomogeneous lossy loads", *IEEE Trans. Microwave Theory Tech.*, vol. MTT-22, pp. 229-236, Mar. 1974.

[2.9]   A.D. Berk, "Variational principles for electromagnetic resonators and waveguides", *IRE Trans. Antennas Propagat.*, vol. AP-4, no. 2, pp. 104-111, Apr. 1956.

[2.10] I.V. Lindell, "Variational methods for nonstandard eigenvalue problems in waveguide and resonator analysis", *IEEE Trans. Microwave Theory Tech.*, vol. MTT-30, pp. 1194-1204, Aug. 1982.

[2.11] M. Oksanen, H. Mäki, I.V. Lindell, "Nonstandard variational method for calculating attenuation in optical fibers", *Microwave Opt. Technol. Lett.*, vol. 3, pp. 160-164, May 1990.

[2.12] W. English, F. Young, "An E vector variational formulation of the Maxwell equations for cylindrical waveguide problems", *IEEE Trans. Microwave Theory Tech.*, vol. MTT-19, pp. 40-46, Jan. 1971.

[2.13] A. Konrad, "Vector variational formulation of electromagnetic fields in anisotropic media", *IEEE Trans. Microwave Theory Tech.*, vol. MTT-24, pp. 553-559, Sept. 1976.

[2.14] K. Morishita, N. Kumagai, "Unified approach to the derivation of variational expressions for electromagnetic fields ", *IEEE Trans. Microwave Theory Tech.*, vol. MTT-25, pp. 34-40, Jan. 1977.

[2.15] V.H. Rumsey, "Reaction concept in electromagnetic theory", *Phys.Rev.*, vol. 94, pp. 1483-1491, June 1954, and vol. 95, p. 1705, Sept. 1954.

[2.16] R.F. Harrington, A. T. Villeneuve, "Reciprocity relationships for gyrotropic media", *IRE Trans. Microwave Theory Tech.*, pp. 308-310, July 1958.

[2.17] R.F. Harrington, *Time-Harmonic Electromagnetic Fields*. New York : McGraw-Hill, 1961, Ch. 7, p. 341.

[2.18] R.F. Harrington, *Field Computation by Moment Methods*. New York: Macmillan, 1968.

[2.19] E.K. Miller, L. Medguesi-Mitschang, E.H. Newman (Editors), *Computational Electromagnetics - Frequency-Domain Method of Moments*. New York: IEEE Press, 1992.

[2.20] I. Huynen, *Modelling planar circuits at centimeter and millimeter wavelengths by using a new variational principle*, Ph.D. dissertation. Louvain-la-Neuve, Belgium: Microwaves UCL, 1994.

[2.21] R.F. Harrington, "Origin and development of the method of moments for field computations", *in Computational Electromagnetics - Frequency-Domain Method of Moments* (edited by E. K. Miller, L. Medguesi-Mitschang, E.H. Newman). New York: IEEE Press, 1992, pp. 43-47.

[2.22] A.F. Peterson, D.R. Wilton, R.E. Jorgenson, "Variational Nature of Galerkin and Non-Galerkin Moment Method Solutions", *IEEE Trans. Antennas Propagat.*, vol. 44, no. 4, pp. 500-503, Apr. 1996.

[2.23] R.H. Jansen, "The spectral domain approach for microwave integrated circuits", *IEEE Trans. Microwave Theory Tech.*, vol. MTT-33, no. 10, pp. 1043-1056, Oct. 1985.

[2.24] T. Itoh, "An overview on numerical techniques for modelling miniaturized passive components", *Ann. Télécommun.*, vol. 41, pp. 449-462, Sept.-Oct. 1986.

[2.25] J.H. Richmond, "On the variational aspects of the Moment Method", *IEEE Trans. Antennas Propagat.*, vol. 39, no. 4, pp. 473-479, Apr. 1991.

[2.26] R.A. Waldron, *Theory of Guided Electromagnetic Waves*. London: Van Nostrand Reinhold Company, 1969.

[2.27] T. Itoh, *Numerical Methods for Microwave and Millimeter Wave Passive structures*. New York: John Wiley&Sons, 1989, Ch. 5.

[2.28] W. C. Chew, Z. Nie, and T.T. Lo, "The effect of feed on the input impedance of a microstrip antenna", *Microwave Opt. Technol. Lett.*, vol. 3, no. 3, March 1990, pp. 79-82.

[2.29] P.M. Morse, H. Feshbach, *Methods of Theoretical Physics*, Volume II. New York: McGraw-Hill, 1953, Ch. 11.

[2.30] D. Baldomir, P. Hammond, "Geometrical formulation for variational electromagnetics", *IEE Proceedings*, vol. 137, Pt. A, no. 6, pp. 321-330, Nov. 1990.

[2.31] R.F. Harrington, *Time-Harmonic Electromagnetic Fields*. New York : McGraw-Hill, 1961, Ch. 7, p. 365.

CHAPTER 3

Analysis of planar transmission lines

This chapter details the various forms of variational principles presented in Chapter 2 for the analysis of planar transmission lines and circuits. The formalism is based on the use of Green's functions and Fourier transforms. After a brief overview of the various analysis techniques, the emphasis is put on the variational methods and their variants. They will be compared with a general variational model in Chapter 4. First the three basic line topologies used in microwave integrated circuits (MICs) are presented, namely the microstrip line, the slot-line and the coplanar waveguide, as well as their shielded variants. Secondly, the various analysis methods of planar lines are described. They are divided into two classes: approximate quasi-static methods and full-wave methods. The advantages and drawbacks of each method are highlighted. Thirdly, an in-depth analysis of the variational behavior underlying each method is proposed.

3.1 Topologies used in MICs

The analysis methods for planar lines at microwave and millimeter wave frequencies can be divided into two classes: the approximate quasi-static methods and the full-wave dynamic methods. The first category is based on the assumption that the dominant mode, propagating along the z-axis of the line, is well approximated by a TEM description. This is the case when the actual mode is of a TM kind, with the TEM mode being a particular case of a TM mode. The second category uses a hybrid TE+TM representation of the fields in the guiding structure [3.1], yielding an accurate frequency-dependent description of the propagation characteristics. The basic line topologies used in microwave integrated circuits are represented in Figure 3.1. They are respectively: the microstrip line (a), the slot-line (c), and the coplanar waveguide (e). Various combinations of these lines may be considered with a view to obtaining coupled structures. The basic configurations for coupled lines are depicted in Figure 3.1b, d, and f. It should be noted that the coplanar waveguide topology is in fact the odd-coupled version of the slot-line (with a change of sign of the electric field component across the slot), while the even-coupled version of the slot-line is referred to as coupled slots in the literature (Fig. 3.1f). The basic line topologies of Figure 3.1 are open. The shielded versions are used at high frequencies to avoid radiation losses: they consist in inserting the planar line into a metallic enclosure. In particular, the shielded version of the slot-line, more commonly denoted finline, is widely used for applications at millimeter waves.

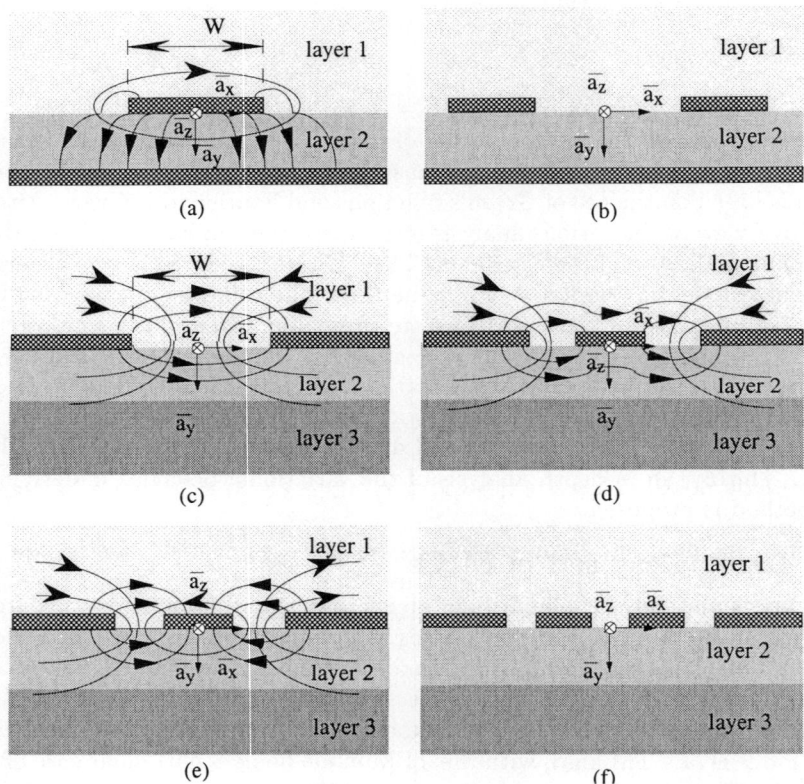

Fig. 3.1 *Basic line topologies for MICs (a) open microstrip; (b) coupled microstrips; (c) open slot-line; (d) coupled slot-lines; (e) open coplanar waveguide; (f) coupled coplanar waveguides*

From these basic topologies a number of combinations are possible, involving structures having metallizations in different planes and layers of different width. Two particular structures are shown in Figure 3.2a, b [3.2]. They are the antipodal finline and the p-i-n travelling wave photodetector respectively. These two structures act like transmission lines at microwave and millimeter wave frequencies. Hence, they can be analysed using conventional transmission lines analysis techniques.

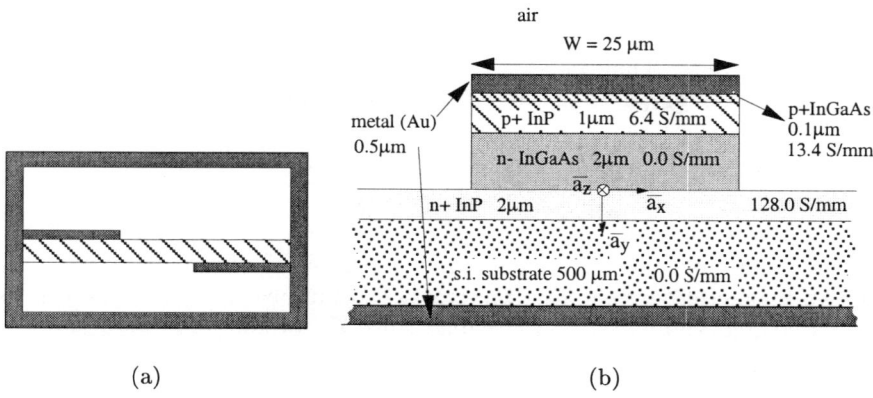

Fig. 3.2 *Examples of combinations of basic line topologies (a) antipodal fin-line; (b) p-i-n photodetector*

3.2 Quasi-static methods

The quasi-static methods are based on the fact that the TEM approximation holds. This is the case when the line has at least two conductors. Under the TEM assumption, the transverse fields are very close to the static electric and magnetic fields. They are derived from the electrostatic scalar or magnetostatic vector potentials respectively, which are both solution of Laplace's equation. The main advantage of the quasi-static approach is that the equations for the electric and magnetic fields are totally decoupled and hence much simpler to handle: we are left with two static problems, electric and magnetic respectively. The propagation constant and the characteristic impedance of the line are expressed as [3.3][3.4]

$$\beta = \omega\sqrt{LC} \tag{3.1a}$$

$$Z_c = \sqrt{\frac{L}{C}} \tag{3.1b}$$

where L and C are the inductance and capacitance per unit length respectively, deduced from the static fields. When the two conductors of the line are surrounded by a homogeneous medium having the constitutive parameters ε, μ, the phase velocity on the line is given by

$$v_{ph} = \frac{1}{\sqrt{LC}} = \frac{1}{\sqrt{\varepsilon\mu}} = \frac{c_0}{\sqrt{\varepsilon_r\mu_r}} \tag{3.2}$$

with c_0 phase velocity in vacuum

ε_r, μ_r relative permittivity and permeability respectively.

When the structure is multilayered, relations (3.2) are no longer valid, because the constitutive parameters are different in the two material layers. The main feature of the quasi-static method is that one can derive the equivalent parameters of a unique effective homogeneous medium from the parameters of the two layers. They provide a single effective phase velocity, which is computed from the value of the effective static parameters L or C of the multilayered structure:

$$v_{ph} = \frac{1}{\sqrt{LC}} = \frac{1}{\sqrt{\varepsilon_{eff}\mu_{eff}}} = \frac{c_0}{\sqrt{\varepsilon_{reff}\mu_{reff}}} \tag{3.3}$$

For multilayered dielectric structures, where all the layers have the same relative magnetic permeability, the static capacitance C of the structure is calculated and compared to the capacitance C_0 of the same structure for which all layers are assumed to have the parameters ε_0, μ_0 of vacuum. This yields an effective relative permittivity

$$\varepsilon_{reff} = \frac{C}{C_0} \tag{3.4a}$$

L_0 is obtained from C_0 as

$$L_0 = \frac{1}{C_0 c_0^2} \tag{3.4b}$$

and the line is described as an equivalent TEM-line having the parameters L_0, C. For multilayered magnetic structures, where all layers have the same relative dielectric permittivity, the inductance L of the structure is calculated and compared to the inductance L_0 - the same structure for which all the layers are assumed to have the parameters ε_0, μ_0 of vacuum. This yields an effective relative permeability

$$\mu_{reff} = \frac{L}{L_0} \tag{3.5a}$$

C_0 is obtained from L_0 as

$$C_0 = \frac{1}{L_0 c_0^2} \tag{3.5b}$$

and the line is described as an equivalent TEM-line with parameters L, C_0. When the layers are both electrically and magnetically inhomogeneous, the two previous methods are combined in the following manner: the inductance L is computed for an electrically homogeneous medium using (3.5a) and the capacitance C is computed for a magnetically homogeneous medium using (3.4b). Those two static parameters are combined to form the quasi-TEM transmission line parameters (3.1b). The quasi-static methods only differ on how the static capacitance C or inductance L are obtained. This is explained below. It should be noted that quasi-static methods cannot be applied to slot lines, because these support a dominant mode which is fundamentally TE.

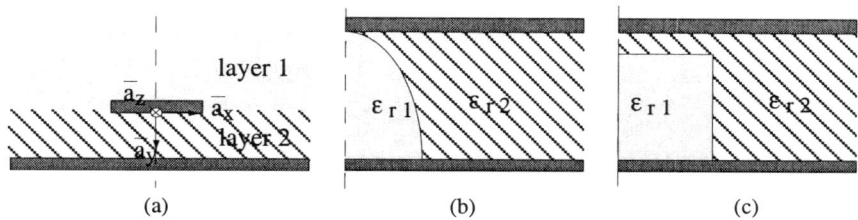

Fig. 3.3 *Conformal mapping for microstrip line (a) original structure; (b) rigorous mapping of right-hand side of original structure; (c) approximate capacitive structure*

3.2.1 Modified conformal mapping

The modified conformal mapping method has been widely used to calculate the equivalent static capacitance of microstrip and coplanar waveguide lines. It consists of a transformation of the original structure into a parallel plate capacitor in a transformed domain, for which the calculation of the capacitance or inductance is straightforward. Conformal mapping is applied first to the homogeneous line, and then to the multilayered line. The application of this method is illustrated in Figure 3.3 for microstrip lines and in Figure 3.4 for coplanar waveguides respectively.

In the case of original microstrip topologies (Fig. 3.3a), an exact and simple analytical expression can always be found for the capacitance C_0 ($\varepsilon_{r1} = \varepsilon_{r2} = 1$). This is not true for the capacitance C, because the interface $y = 0$ is transformed into an elliptical-looking curve (Fig. 3.3b). Wheeler [3.5][3.6] proposed to approximate this transformed boundary by a rectangular boundary, as shown at Figure 3.3c. Hence, an equivalent filling factor is defined for the parallel plate capacitance, and an equivalent effective dielectric constant is obtained for the line. Unfortunately, the transformation of the dielectric-air boundary results in simple expressions only when this boundary corresponds to a line of electric field, which is not the case for the microstrip introduced in Figure 3.1a. Moreover the calculation of the conformal mapping requires the evaluation of elliptic functions, which converge slowly.

In the case of coplanar lines (Fig. 3.1e), the conformal mapping is more rigorous, since the electric field across the slot is tangential to the dielectric interface. Hence, when layer 2 is of infinite thickness, an exact solution for the capacitance of the transformed structure of Figure 3.4b is obtained using a Schwarz-Christoffel conformal mapping [3.7]. Due to the obvious symmetry of the structure, the effective dielectric constant is the mean value of the dielectric constants of the two layers as shown by:

$$\varepsilon_{eff} = \frac{\varepsilon_1 + \varepsilon_2}{2} \tag{3.6}$$

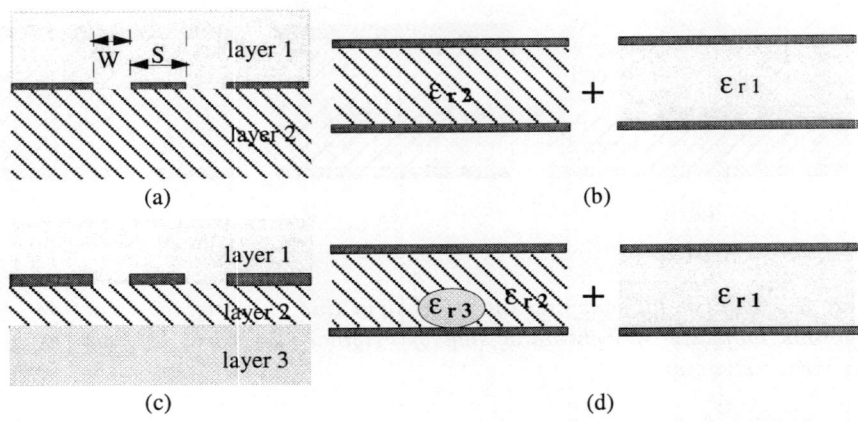

Fig. 3.4 *Conformal mapping for coplanar waveguide (a) original two-layered structure; (b) partial capacitances corresponding to two-layered structure; (c) original three-layered structure; (d) partial capacitances corresponding to three-layered structure*

while the total capacitance C_0 is obtained as a function of the complete elliptic integral of the first kind $K(k)$:

$$C_0 = 4\varepsilon_0 \frac{K(k)}{K(k')} \tag{3.7}$$

with $k = S/(S+2W)$ $k' = \sqrt{1-k^2}$

When layer 2 has a finite thickness (Fig. 3.4c), the structure in the transformed domain has to be modified as shown in Figure 3.4d: layer 3 is mapped into an area of elliptic shape. In this case the capacitance has to be computed via a distributed shunt-series capacitances modeling, as done by Davis *et al.* [3.8].

In addition to the static assumption made for the calculation of the fields, the conformal mapping method has important limitations. First, it is only applicable to open structures (Fig. 3.1a, c and e). Secondly, the effect of the finite thickness of the metallization is never taken into account. Thirdly, the finite thickness of the layers has to be evaluated by numerical discretizations or curve fittings on a presumed transformed boundary, without any *a priori* knowledge of the influence of the approximation on the result.

3.2.2   Planar waveguide model

The planar waveguide model is applicable to open microstrip lines. It has been developed by Kompa [3.9] and Mehran [3.10]. It consists of a waveguide having perfect conductive top and bottom shieldings, while lateral limits are perfect

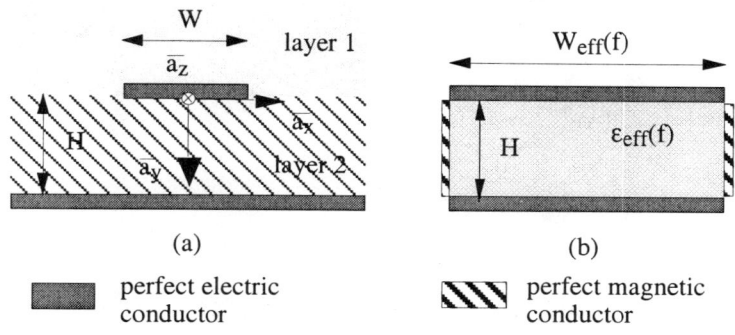

Fig. 3.5 *Planar waveguide model (a) original microstrip structure; (b) equivalent waveguide*

magnetic walls. The guide is filled by the effective dielectric constant of the microstrip line (Fig. 3.5), calculated by another method [3.11]. The effective dielectric constant and the width of the waveguide are frequency-dependent, to ensure that the impedance of the TEM mode supported by this waveguide is a good representation of the actual impedance of the dominant mode on the microstrip. By extrapolation, this structure is expected to be efficient for the calculation of the parameters of higher-order modes supported by the microstrip line. The model is mainly applied to model planar microstrip discontinuities. In this case, mode matching requires the knowledge of the higher-order modes supported by the microstrip line.

3.2.3 Finite-difference method for solving Laplace's equation

The transverse dependence of the electric field of the TEM mode is derived from the static potential Φ, which has to satisfy the two-dimensional Laplace's equation

$$\nabla^2 \Phi(x,y) = \frac{\partial^2 \Phi}{\partial x^2} + \frac{\partial^2 \Phi}{\partial y^2} = 0 \tag{3.8}$$

The equation is discretized by the finite-difference method. Hence, the method is only applicable to shielded lines [3.11], as illustrated in Figure 3.6. The potential is assumed to be zero on one conductor and V_0 on the other. The application of the finite-difference method, taking into account the boundary condition at the dielectric interface, is detailed in [3.12]. Basically, the solution for the potential in the neighborhood of point A (points B, C, D, E) is expanded into a Taylor's series around the solution at point A, noted Φ_A:

$$\Phi_B = \Phi_A + h \left.\frac{\partial \Phi}{\partial x}\right|_A + \frac{h^2}{2!} \left.\frac{\partial^2 \Phi}{\partial x^2}\right|_A + \frac{h^3}{3!} \left.\frac{\partial^3 \Phi}{\partial x^3}\right|_A + O(h^4) \tag{3.9a}$$

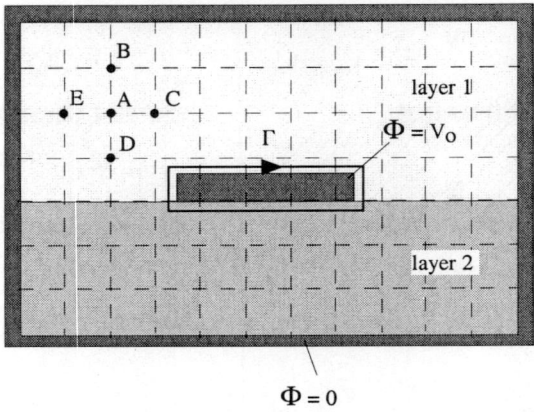

Fig. 3.6 *Discretized domain for analysis of shielded microstrip line by using the finite-difference method*

$$\Phi_D = \Phi_A - h \left.\frac{\partial \Phi}{\partial x}\right|_A + \frac{h^2}{2!} \left.\frac{\partial^2 \Phi}{\partial x^2}\right|_A - \frac{h^3}{3!} \left.\frac{\partial^3 \Phi}{\partial x^3}\right|_A + O(h^4) \tag{3.9b}$$

$$\Phi_C = \Phi_A + h \left.\frac{\partial \Phi}{\partial y}\right|_A + \frac{h^2}{2!} \left.\frac{\partial^2 \Phi}{\partial y^2}\right|_A + \frac{h^3}{3!} \left.\frac{\partial^3 \Phi}{\partial y^3}\right|_A + O(h^4) \tag{3.9c}$$

$$\Phi_E = \Phi_A - h \left.\frac{\partial \Phi}{\partial y}\right|_A + \frac{h^2}{2!} \left.\frac{\partial^2 \Phi}{\partial y^2}\right|_A - \frac{h^3}{3!} \left.\frac{\partial^3 \Phi}{\partial y^3}\right|_A + O(h^4) \tag{3.9d}$$

where subscript A means that quantities are evaluated at point A and where $O(h^4)$ is a quantity proportional to h^4. Adding these four equations yields:

$$\Phi_B + \Phi_C + \Phi_D + \Phi_E = 4\Phi_A + h^2 \left.\left(\frac{\partial^2 \Phi}{\partial x^2} + \frac{\partial^2 \Phi}{\partial y^2}\right)\right|_A + O(h^4) \tag{3.10}$$

Since Φ has to satisfy (3.8) everywhere, the following is a good approximation for the potential solution of (3.8):

$$\Phi_A = \frac{\Phi_B + \Phi_C + \Phi_D + \Phi_E}{4} \tag{3.11}$$

which can be repeated at each mesh point. It has a somewhat different form if point A is located at the interface between two different media. At the boundary points, Φ or its derivative, or a linear combination of both, is specified. Hence, (3.11) written for each mesh point plus the set of boundary conditions may be rewritten using a matrix formalism as

$$\overline{\overline{M}} \, \overline{\Phi} = \overline{b} \tag{3.12}$$

where the right-hand side is a column vector containing information provided by the boundary points, and $\overline{\Phi}$ is a column vector for the unknown values of potential at the various mesh points. Solving (3.12) for $\overline{\Phi}$ is equivalent, to the h^4 order, to obtaining a solution for (3.8) satisfying some imposed boundary conditions. Hence, the method is efficient for small values of h, which implies that the mesh has to be sufficiently small.

Once the potential is obtained, the electric field normal to the strip conductor of Figure 3.6 is calculated and the total charge Q_l per unit length on the strip is obtained as a line integral

$$Q_l = \oint_{\Gamma} \varepsilon_i \overline{E}_i \cdot \overline{dn} \tag{3.13}$$

where Γ is the contour of the strip. Hence, the capacitance per unit length of the line is expressed as

$$C = \frac{Q_l}{V_0} \tag{3.14}$$

The mathematical preprocessing is minimal, and the method can be applied to a wide range of structures but a price to pay is its numerical inefficiency. Open structures have to be truncated to a finite size, and the method requires that mesh points lie on the boundary. The main drawback of the method is that an accurate expression of the capacitance is obtained only when the discretization of the domain of interest is small enough. This may require considerable computation time and memory space. The method, however, is used for characterizing transmission lines containing layers with highly inhomogeneous constitutive parameters as is the case for interconnections on doped semiconducting layers. The finite-difference method is also used in the time domain as a full-wave dynamic method.

What has been described is the original procedure for discretizing the structure, where square - or cubic - mesh elements are used, with a uniform mesh parameter throughout. This procedure has been modified over the years with the aim to reduce errors, memory requirements, and numerical expenditure, as well as toward improving the efficiency of programming techniques. Regular and irregular graded meshes have been used, for adapting the density of the network to local non-uniformities of the fields, resulting from sharp corners of fins, for instance. If different mesh sizes are used along the different coordinate axes, then the inductance and capacitance per unit length as well as the length of the branches, are different according to the directions, and the equations have to take these differences into account. So-called condensed node schemes have led to considerable savings in computer resources, particularly when they are combined with a graded mesh technique in which the density of the mesh can be adapted to the degree of field non-uniformity.

However, it is not our intent to give a thorough presentation of the numerical methods used for solving electromagnetic problems. The interested reader will find a number of good books on the subject.

3.2.4   Integral formulations using Green's functions

The Green's formalism (Appendix A) is interesting for quasi-static formulations, because it offers a elegant way to obtain solutions for Poisson's equation

$$\nabla^2 \Phi(x, y, z) = -\rho(x, y, z) \tag{A.23}$$

This equation is the basis of most electrostatic problems, which usually require finding the charge and potential distributions present in a given structure in order to calculate the equivalent capacitance of the system. A previous section illustrated this approach, when using Laplace's equation. Laplace's equation is a particular case of the Poisson's equation, where the source charge density ρ is assumed to be zero.

A. Equation (3.8) may also be solved using a Green's formalism which imposes boundary conditions on the two line contours modeling the surface of the two conductors in the transverse section of the line (Fig. 3.6). The general Green's function associated with Dirichlet boundary conditions on a surface expressed as

$$\overline{P}(\overline{r}, t) = \int_{\Omega_x} \int_{T_x} \frac{\partial \overline{\overline{G}}(\overline{r}, t | \overline{r}', t')}{\partial \overline{n}} \cdot \{\overline{P}(\overline{r}', t')\}|_{\Omega_x} \, dr' dt' \tag{A.20d}$$

is formulated in two dimensions. The potential is the solution of a homogeneous Poisson's equation which satisfies inhomogeneous boundary conditions on the contours limiting the two conductors in the transverse section of the transmission line (Fig. 3.6):

$$\Phi(\overline{r}) = \int_{\Gamma_1} \frac{\partial G_P}{\partial \overline{n}}(\overline{r}|\overline{r}_1)\Phi(\overline{r}_1)d\Gamma_1 + \int_{\Gamma_2} \frac{\partial G_P}{\partial \overline{n}}(\overline{r}|\overline{r}_2)\Phi(\overline{r}_2)d\Gamma_2 \tag{3.15}$$

where the space vectors are two-dimensional, G_P is the scalar Green's function associated to the two-dimensional Poisson's equation derived from (A.23), and $\overline{n}$ is the normal to the line contour considered. Assuming that $\Phi(\overline{r}_1) = 0$ on Γ_1 and $\Phi(\overline{r}_2) = V_0$ on Γ_2, the first term of the sum vanishes. Then, equations (3.13) and (3.14) with $\Gamma = \Gamma_2$ yield the capacitance C.

B. Collin [3.13] has given another simple expression for the capacitance of a TEM transmission line, also based on a Green's function. The formulation relates the capacitance to the inner product between a charge distribution in the cross-section of the line and the corresponding electrostatic potential:

$$\frac{1}{C} = \frac{1}{Q_i^2} \oint_\Gamma \rho(x, y)\Phi(x, y)d\Gamma \tag{3.16}$$

where the line path Γ is taken on the surface of the inner conductor in the transverse section

$\Phi(x, y)$ is the static potential solution of the two-dimensional Poisson's equation

$\rho(x, y)$ is the distribution of charge density on the surface of the conductor

Q_l is the integral of the charge density on the inner conductor.

The potential is then related to the charge density through the two-dimensional scalar Green's function associated to Poisson's equation, S being the cross-section of the transmission line:

$$\Phi(x, y) = \int_S G_P(x, y | x', y') \rho(x', y') dx' dy' \tag{3.17}$$

This equation simplifies when the inner conductor is assumed to be of zero thickness in plane $y = 0$ and the charge density ρ is localized at its surface. There is then a surface charge density $\rho_s(x)$ distributed on the conducting strip only. The total charge on the inner conductor per unit length is given by the line integral over the width W_c of the inner conductor

$$Q_l = \int_{-\frac{W_c}{2}}^{\frac{W_c}{2}} \rho_s(x) dx \tag{3.18}$$

Since the normal derivative of the potential is proportional to the charge density on the conductor, the Green's formalism associated to Neumann boundary conditions is applicable:

$$\overline{P}(\overline{r}, t) = \int_{\Omega_x} \int_{T_x} \overline{\overline{G}}(\overline{r}, t | \overline{r}', t') \cdot \left. \frac{\partial \overline{P}(\overline{r}', t')}{\partial \overline{n}} \right|_{\Omega_x} dr' dt' \tag{A.20b}$$

The potential can be rewritten as a function of $\rho_s(x)$ using the same Green's function associated with Poisson's equation:

$$\Phi(x, y) = \int_{-\frac{W_c}{2}}^{\frac{W_c}{2}} G_P(x, y | x', 0) \rho_s(x') dx' \tag{3.19}$$

Equation (3.17) illustrates the practical use of two-dimensional Green's functions. Since the transverse section of the line is not homogeneous, the correct continuity has to be imposed between the solutions found for each layer. This can be done either in the space or in the spectral domain along the x-variable. The choice will depend on the complexity of the geometry and, in particular, the location of the interfaces between the different layers, as explained in Section 3.4.

The main problem in using (3.17) and (3.19) is that there is no *a priori* knowledge of the charge distribution. The previous formalism requires the knowledge of the source of excitation. The actual charge distribution, however, results from the boundary conditions for the potential on the strip and at the interfaces between the layers. Hence, the problem seems to have no practical solution. The first way to solve this difficulty is to assume a reasonably adequate distribution for the surface charge ρ_s. Intuitively, the distribution will be relevant if the variational behavior of expression (3.16) about ρ_s has been proven. The second way is to solve the integral equations (3.17) or (3.19) using the moment method, applying a known boundary condition. This is in fact an alternate way to solve an inhomogeneous problem subject to inhomogeneous boundary conditions. For our particular case, the linear integral equation to be considered is

$$
\begin{aligned}
\Phi(x,y)|_{(|x|\leq\frac{W_c}{2},y=0)} &= V_0 \\
&= \int_{-\frac{W_c}{2}}^{\frac{W_c}{2}} G_P(x,y|x',0)\rho_s(x')dx' \Bigg|_{(|x|\leq\frac{W_c}{2},y=0)}
\end{aligned}
\tag{3.20}
$$

It has to be solved for the line path describing the conducting strip, supposed to be at potential V_0. Once the Green's function has been found, it can be considered as a linear integral operator for the boundary condition problem (3.20). Using the notations of Chapter 2, the problem is formulated as

$$
L(f) = g \text{ on } \Gamma \tag{3.21a}
$$

with

$$
f(x,y) = \rho(x,y) \tag{3.21b}
$$

$$
L(f) = \int_S G_P(x,y|x',y')\rho(x',y')dx'dy' \tag{3.21c}
$$

$$
g(x,y) = \Phi(x,y)|_{(x,y)\in\Gamma} = V_0 \tag{3.21d}
$$

which reduces in this case to a one-dimensional problem:

$$
f(x) = \rho_s(x) \tag{3.22a}
$$

$$
L(f) = \int_{-\frac{W_c}{2}}^{\frac{W_c}{2}} G_P(x,0|x',0)\rho_s(x')dx' \tag{3.22b}
$$

$$
g(x) = \Phi(x,0)|_{|x|\leq\frac{W_c}{2}} = V_0 \tag{3.22c}
$$

The charge density is then expanded into a set of basis functions ρ_{sn} weighted by unknown coefficients a_n:

$$
\rho_s(x) = \sum_{n=1}^{N} a_n \rho_{sn} \tag{3.23}
$$

The basis functions are taken such as

$$
\rho_{sn}(x) = 0 \text{ for } |x| > \frac{W}{2} \tag{3.24}
$$

Applying Galerkin's procedure, the coefficients can be determined. Equation (3.18) then yields the total charge Q_l on the inner conductor, and the capacitance is obtained from definition (3.14).

A third procedure has been presented by Gupta *et al.* [3.11]. Equations (3.22) are discretized in order to provide the corresponding matrix formulation:

$$\overline{v} = \overline{\overline{p}}\,\overline{q} \tag{3.25}$$

where $\overline{v}$ and $\overline{q}$ are column matrices representing the potential and the density of surface charge on the conductors respectively, while $\overline{\overline{p}}$ is a matrix containing the effect of the Green's function. Since the potential is usually known on the conductor, system (3.25) may be inverted to find the charge vector $\overline{q}$. The capacitance is then obtained using (3.14) as the integral of the charge density on the conductor:

$$C = \sum_j \sum_k p_{jk}^{-1} \tag{3.26}$$

where p_{jk}^{-1} is the jkth-component of the inverse of matrix $\overline{\overline{p}}$.

To conclude this subsection, it has to be pointed out that the main advantage of the combination Green's function - integral formulation is that the integration is usually performed on a reduced domain of the cross-section, as is illustrated by the transformation of (3.21) into (3.22). Since the Green's function contains information about the entire structure, the potential satisfies the physical laws describing the system (Poisson's equation, for instance). Particular boundary conditions, necessary to ensure a complete determination of the solution, are applied at specific interfaces only (surfaces or lines).

3.3 Full-wave dynamic methods

At high frequencies the quasi-static approximation is no longer valid for multi-layered structures, because the electromagnetic fields supported by the structure have longitudinal components which are no longer negligible. These components are necessary to ensure that the boundary conditions are satisfied at the interfaces. The fields cannot be derived anymore from static potential solutions of Laplace's or Poisson's equation. The fields must be expressed as a combination of TE and TM modes whose parameters differ from one layer to the other, each mode being associated with a scalar potential solution of Helmholtz equation [3.4]:

$$\nabla^2 \Phi_i^{\text{TE}} + (k_i^2 + \gamma^2)\Phi_i^{\text{TE}} = 0 \tag{3.27a}$$

$$\nabla^2 \Phi_i^{\text{TM}} + (k_i^2 + \gamma^2)\Phi_i^{\text{TM}} = 0 \tag{3.27b}$$

with $k_i = \omega\sqrt{\varepsilon_i\mu_i}$ and where subscript i refers to layer i.

A complex $e^{-\gamma z}$ dependence is assumed along the propagation axis. The propagation constant γ is unknown and has to be calculated. Since the physical fields in the structure are combinations of TE and TM modes, they are usually associated with the concept of "hybrid modes".

The most popular method for the full-wave analysis of planar transmission lines is the integral equation expressed as a line boundary condition and solved by Galerkin's procedure. It has been widely used for many years and almost all the methods found in the literature derive from it. It is described in the next subsection. Another method is also presently used for the dynamic analysis of planar lines: the transmission line matrix method (TLM), combined with the finite-difference time domain method (FDTD). It should be mentioned, however, that a particular method has been developed by Cohn for slot-lines [3.14], in order to obtain simple expressions for the parameters of these lines, as has been done for microstrips and coplanar waveguides in the static case. It will be presented in Section 3.3.6.

3.3.1   Formulation of integral equation for boundary conditions

As for the quasi-static applications based on an integral equation, the formulation starts from a relationship involving a Green's function. The Green's function may be derived in two ways, as mentioned by Itoh [3.15]. First it can be derived from the Green's function established for the vector potential solution of the vector wave equation introduced in Appendix A in the following way. For strip-like problems, the strip is replaced by an infinitely thin sheet of current, hence removing the metallization. In each layer, the electric field is derived from a vector wave potential satisfying the vector wave equation

$$\nabla^2 \overline{A}(\overline{r}) + \omega^2 \varepsilon \mu \overline{A}(\overline{r}) = -\mu \overline{J}(\overline{r}) \qquad (A.27)$$

The Green's function associated with this equation is

$$\overline{\overline{G}}_0(\overline{r}|\overline{r}_0) = \overline{\overline{I}} \frac{e^{-j|\overline{k}|\cdot|\overline{r}-\overline{r}_0|}}{|\overline{r}-\overline{r}_0|} \qquad (3.28)$$

It can be shown [3.16] that equation (A.27) for the potential has an equivalent form for the electric field $\overline{E}$:

$$\nabla \times \nabla \times \overline{E} - |\overline{k}|^2 \overline{E} = -j\omega\mu_0 \overline{J} \qquad (3.29)$$

A Green's function $\overline{\overline{G}}_E$ may be directly associated with the electric field solution of this equation, by replacing $\overline{E}$ by $\overline{\overline{G}}_E$ and $-j\omega\mu_0\overline{J}$ by the space impulse unit function

$$\delta(\overline{r}' - \overline{r}_0) = \delta(x' - x_0)\delta(y' - y_0)\delta(z' - z_0) \qquad (A.18d)$$

The solution of (3.29) then becomes [3.16]

$$\overline{\overline{G}}_E(\overline{r}|\overline{r}_0) = (\overline{\overline{I}} + \frac{1}{k_i^2}\nabla\nabla)\frac{e^{-j|\overline{k}|\cdot|\overline{r}-\overline{r}_0|}}{|\overline{r}-\overline{r}_0|} \tag{3.30}$$

from which a Green's relationship between $\overline{E}$ and $\overline{J}$ applies:

$$\overline{E}(\overline{r}) = \int_V \overline{\overline{G}}_E(\overline{r}|\overline{r}')\cdot\overline{J}(\overline{r}')dr' \tag{3.31}$$

where V is the volume wherein the electric field has to be determined. It should be noted that expression (3.30) has to be modified when the medium is finite; thus when specific boundary conditions apply to the electric field at the boundary surface S enclosing the volume V. It has been demonstrated that in this case a second term must be added to the solution [3.17]:

$$\begin{aligned}\overline{E}(\overline{r}) &= \int_V \overline{\overline{G}}_E(\overline{r}|\overline{r}')\cdot\overline{J}(\overline{r}')dr' \\ &+ \int_S (\overline{n}\times\nabla'\times\overline{E})\cdot\overline{\overline{G}}_E dS' + \int_S (\nabla'\times\overline{\overline{G}}_E)\cdot(\overline{n}\times\overline{E})dS'\end{aligned} \tag{3.32}$$

where ∇' means derivatives with respect to the source coordinates $\overline{r}'$. The two surface integrals can be related to equivalent electric and magnetic "polarization" currents flowing on the interface planes, taking into account the boundary conditions at those interfaces. When the medium is multilayered, particular solutions must be added to $\overline{\overline{G}}_E$, so that the boundary conditions at each interface are satisfied [3.18]. For a general case, the solution becomes quite intricate, because the spatial form of Green's functions for a stratified medium is obtained as a serial expansion involving residues integrals. The method is detailed in Wait [3.19].

The second way to derive the Green's function is more convenient to use. It starts from the source-free Helmholtz equations (3.33a,b), for which a general solution may be found in the spectral domain for both TE and TM source-free Hertzian potentials:

$$\nabla^2\overline{\Pi}_i^E + (k_i^2 + \gamma^2)\overline{\Pi}_i^E = 0 \tag{3.33a}$$

$$\nabla^2\overline{\Pi}_i^M + (k_i^2 + \gamma^2)\overline{\Pi}_i^M = 0 \tag{3.33b}$$

Applying Maxwell's equations, the electric and magnetic fields in each layer of the structure (Fig. 3.1) are derived in the spectral domain from the general solution of the respective potentials (3.33a,b). From the derivation of the spectral form of Green's functions (Appendix A), it follows that the partial derivatives in the spatial form of Maxwell's equations are replaced by a product of the corresponding spectral components in the spectral domain. As a result, simple algebraic expressions are obtained between fields and potentials in the spectral domain. The details of the fields derivation will be illustrated in Chapter 4. As another result, electric and magnetic fields do depend upon each other since they derive from the same potentials. An equivalent tangential current density $\overline{J}_t$ is defined as the difference between the tangential magnetic fields at the interface between two layers where there is a planar conductor (plane $y = 0$ in Figure 3.1). Using the dependence between the electric and the magnetic fields, the following equations are obtained in this plane (subscript t holds for tangential), in spectral form:

$$\tilde{\overline{E}}_t(k_x, 0, z) = \tilde{\overline{\overline{G}}}_{st}(k_x, 0; \gamma) \cdot \tilde{\overline{J}}_t(k_x, 0, z) \quad \text{for strip-like problems} \quad (3.34a)$$

$$\tilde{\overline{J}}_t(k_x, 0, z) = \tilde{\overline{\overline{G}}}_{sl}(k_x, 0; \gamma) \cdot \tilde{\overline{E}}_t(k_x, 0, z) \quad \text{for slot-like problems} \quad (3.34b)$$

or, after inversion, in the space domain:

$$\overline{E}_t(x, 0, z) = \int_{-\infty}^{\infty} \overline{\overline{G}}_{st}(x, 0|x', 0; \gamma) \cdot \overline{J}_t(x', 0, z')dx'$$
$$= \mathcal{L}_{st}(\overline{J}_t) \qquad \text{for strip-like problems} \qquad (3.35a)$$

$$\overline{J}_t(x, 0, z) = \int_{-\infty}^{\infty} \overline{\overline{G}}_{sl}(x, 0|x', 0; \gamma) \cdot \overline{E}_t(x', 0, z')dx'$$
$$= \mathcal{L}_{sl}(\overline{E}_t) \qquad \text{for slot-like problems} \qquad (3.35b)$$

In these expressions, $\overline{\overline{G}}$ is the inverse Fourier-transform along the x-variable in the space domain of the spectral dependence $\tilde{\overline{\overline{G}}}$. The spectral quantities are defined by the x-Fourier-transform (Appendix C). It should now be obvious that the $\overline{\overline{G}}$ functions are Green's functions in the sense of (3.31) and (3.32). First, there is a correspondence between the algebraic dyadic relation (3.34) and the spectral definition of the Green's function:

$$\tilde{\overline{P}}(k_x, k_y, k_z, \omega) = \tilde{\overline{\overline{G}}}(k_x, k_y, k_z, \omega) \cdot \tilde{\overline{X}}(k_x, k_y, k_z)$$

Secondly, the potential forms satisfying (A.27) or (3.33a,b) are equivalent since they both derive from Maxwell's equations. Thirdly, the linearity of the Fourier-transform ensures that the homogeneous problem with inhomogeneous boundary conditions solved in the spectral domain for each layer is equivalent to the spatial form (3.32). Equations (3.32) and (3.34a,b) derive indeed from (A.27) and (3.33a,b), respectively.

The potentials are a function of the propagation constant by virtue of (3.27) and (3.33a,b). Hence, the spectral and spatial notations of the Green's functions involve the unknown γ propagation constant along the z-axis. This is in accordance with the general form for fields and potentials having a propagation dependence $e^{-jk_{z0}z'}$ along the z'-direction:

$$\overline{P}(\overline{r},t) = \int_{V_x} \int_{T_x} \overset{\approx}{\overline{G}}(\overline{r},t|\overline{r}',t';k_{z0}) \cdot \overline{X}(\overline{r}',t')dr'dt' \qquad \text{(A.40a)}$$

with

$$\overset{\approx}{\overline{G}}(\overline{r},t|\overline{r}',t';k_{z0}) = \int_{V_x} \int_{T_x} \overline{\overline{G}}(x,y,z,t|x',y',z',t')e^{-jk_{z0}(z'-z)}d(z'-z)$$

For a distributed structure, the z-dependence can be dropped in equations (3.34a,b) and (3.35a,b), by defining the fields and currents as

$$\overline{E}_t(x,y,z) \overset{\Delta}{=} \overline{e}_t(x,y)e^{-\gamma z} \qquad (3.36a)$$

$$\overline{J}_t(x,y,z) \overset{\Delta}{=} \overline{j}_t(x,y)e^{-\gamma z} \qquad (3.36b)$$

Introducing (3.36) into (3.34) and (3.35) and dividing both sides of each equation by $e^{-\gamma z}$ yields

$$\tilde{e}_t(k_x,0) = \overset{\approx}{\overline{G}}_{st}(k_x,0;\gamma) \cdot \tilde{j}_t(k_x,0) \quad \text{for strip-like problems} \qquad (3.37a)$$

$$\tilde{j}_t(k_x,0) = \overset{\approx}{\overline{G}}_{sl}(k_x,0;\gamma) \cdot \tilde{e}_t(k_x,0) \quad \text{for slot-like problems} \qquad (3.37b)$$

or, in the space domain,

$$\overline{e}_t(x,0) = \int_{-\infty}^{\infty} \overline{\overline{G}}_{st}(x,0|x',0;\gamma) \cdot \overline{j}_t(x',0)dx'$$
$$= \mathcal{L}_{st}(\overline{j}_t) \qquad \text{for strip-like problems} \qquad (3.38a)$$

$$\overline{j}_t(x,0) = \int_{-\infty}^{\infty} \overline{\overline{G}}_{sl}(x,0|x',0;\gamma) \cdot \overline{e}_t(x',0)dx'$$
$$= \mathcal{L}_{sl}(\overline{e}_t) \qquad \text{for slot-like problems} \qquad (3.38b)$$

Expressions (3.35a,b) and (3.38a,b) are integral equations where the Green's function is a linear integral operator $\mathcal{L}$. They are basically obtained in a similar way, only the choice of the final expression changes: for strip-like problems, the Green's formulation yields the tangential electric field at the interface of the strip as a function of the current on the strip, while for slot-like problems it yields the current distribution at the interface of the metallization as a function of the electric field in the slot. One may also say that the source for strip-like problems is the current density flowing on the strip, while for slot-like problems it is the electric field in the plane of the slot. Throughout this section, the term "source" will now be used for either $\overline{E}_t$ or $\overline{J}_t$. Using equations (3.33a,b) takes advantage from the fact that fields are calculated in the spectral domain for a homogeneous source-free equation (3.33a,b), and that the integral equation involves the source at the interface only. This is not the case when using (3.32), because sources have to be integrated on the surface limiting the volume of interest for adequately taking into account the boundary conditions. One may also say that, when using equation (3.32), the integral equation has to be applied to the whole space while, when using equations (3.33a,b), source effects apply only in the metallic interface.

Formulation (3.35a,b) presents the same difficulty as the quasi-static problem (3.19), since the source distribution is unknown. If the conductors are perfect, however, it is known that the source has to vanish on the whole interface $y = 0$, except on the line path describing the strip for strip-like problems or the slot for slot-like problems. Hence, formulations (3.35a,b) are most appropriate for an integral boundary condition formulation. The equivalent source in the right-hand side of (3.35a,b) is indeed located on a reduced segment of the interface, which limits the integration on the strip or slot area (Fig. 3.7):

$$\overline{E}_t(x) = 0 \qquad \text{for } |x| > \frac{W}{2} \qquad \text{for slot-like problems} \tag{3.39a}$$

$$\overline{J}_t(x) = 0 \qquad \text{for } |x| > \frac{W}{2} \qquad \text{for strip-like problems} \tag{3.39b}$$

As for the quasi-static boundary condition problem (3.19), conditions (3.39a,b) can be used for selecting either an approximate expression for the equivalent source, or some basis functions for the method of moment, solving for the unknown source distribution while imposing the boundary condition

$$\overline{J}_t(x) = 0 \qquad \text{for } |x| < \frac{W}{2} \qquad \text{for slot-like problems} \tag{3.40a}$$

$$\overline{E}_t(x) = 0 \qquad \text{for } |x| < \frac{W}{2} \qquad \text{for strip-like problems} \tag{3.40b}$$

The user now either inverts the spectral Green's function in the right-hand side of (3.34a,b) and solves in space domain, or inverts the obtained spectral solution for the left-hand side of (3.34a,b). The two approaches are illustrated in the next two sections.

3.3.2   Solution of the integral equation in space domain

Applying an inverse Fourier transform to the right-hand side of equations (3.34a,b), Denlinger [3.20] obtains a pair of coupled integral equations to be solved in space domain on the strip (slot) area. Reasonable expressions for the x and z components of the sources at the interface are expressed as functions of the x variable satisfying (3.39b):

$$J_u(x, 0, z) = I_u j_u(x) e^{-\gamma z} \qquad \text{for strip-like problems} \tag{3.41a}$$

$$E_u(x, 0, z) = V_u e_u(x) e^{-\gamma z} \qquad \text{for slot-like problems} \tag{3.41b}$$

with $u = x, z$ and where I_u and V_u are unknown constants. Their x-Fourier transform are then calculated, while keeping the amplitudes unknown, which

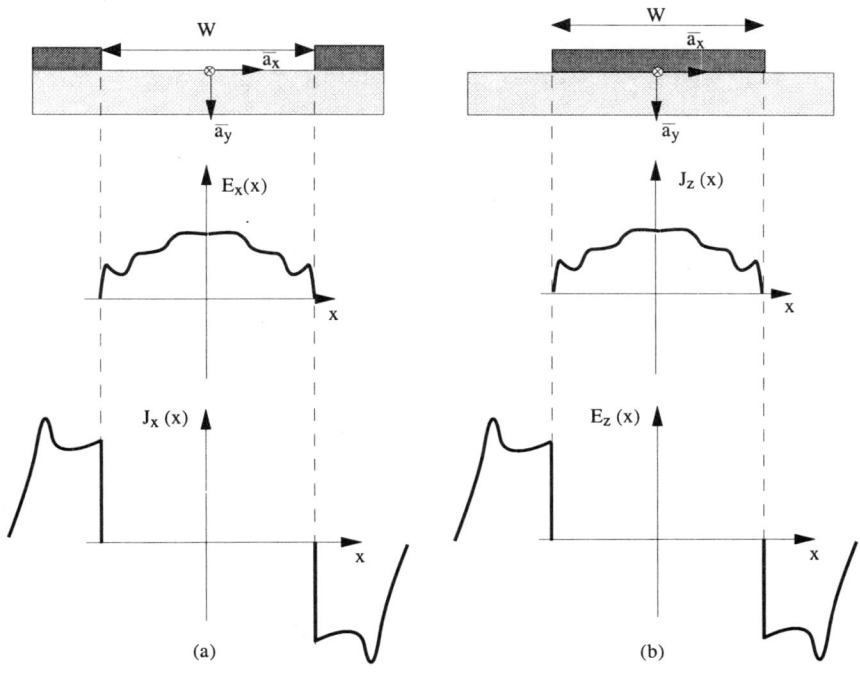

Fig. 3.7 *Shape of source and field quantities for integral formulation of boundary conditions (a) slot case; (b) strip case*

yields the Fourier transforms of the sources in the interface:

$$\tilde{\overline{J}}_t(k_x, 0, \gamma) = \left\{ I_x \tilde{j}_x(k_x)\overline{a}_x + I_z \tilde{j}_z(k_x)\overline{a}_z \right\} e^{-\gamma z} \quad \text{for strip-like problems} \tag{3.42a}$$

$$\tilde{\overline{E}}_t(k_x, 0, \gamma) = \left\{ V_x \tilde{e}_x(k_x)\overline{a}_x + V_z \tilde{e}_z(k_x)\overline{a}_z \right\} e^{-\gamma z} \quad \text{for slot-like problems} \tag{3.42b}$$

The inverse Fourier-transform of the right-hand side of expressions (3.34a,b) yields:

$$\overline{E}_t(x, 0, z) = \int_{-\infty}^{\infty} \tilde{\overline{\overline{G}}}_{st}(k_x, 0; \gamma) \cdot \tilde{\overline{J}}_t(k_x, 0; \gamma) e^{jk_x x} dk_x$$

$$\triangleq \overline{e}_t(x, 0)e^{-\gamma z} \quad \text{for strip-like problems} \tag{3.43a}$$

$$\overline{J}_t(x,0,z) = \int_{-\infty}^{\infty} \overline{\overline{\tilde{G}}}_{sl}(k_x,0;\gamma) \cdot \overline{\tilde{E}}_t(k_x,0;\gamma)e^{jk_xx}dk_x$$

$$\triangleq \overline{\tilde{j}}_t(x,0)e^{-\gamma z} \quad \text{for slot-like problems} \tag{3.43b}$$

Introducing (3.42a,b) into (3.43a,b) and using (3.36a,b) yields the coupled pairs of equations

$$e_x(x,0) = \int_{-\infty}^{\infty} \tilde{G}_{st,xx}(k_x,0;\gamma)I_x\tilde{j}_x(k_x)e^{jk_xx}dk_x$$

$$+ \int_{-\infty}^{\infty} \tilde{G}_{st,xz}(k_x,0;\gamma)I_z\tilde{j}_z(k_x)e^{jk_xx}dk_x \tag{3.44a}$$

$$= 0 \quad \text{for } |x| < \frac{W}{2}$$

$$e_z(x,0) = \int_{-\infty}^{\infty} \tilde{G}_{st,zx}(k_x,0;\gamma)I_x\tilde{j}_x(k_x)e^{jk_xx}dk_x$$

$$+ \int_{-\infty}^{\infty} \tilde{G}_{st,zz}(k_x,0;\gamma)I_z\tilde{j}_z(k_x)e^{jk_xx}dk_x \tag{3.44b}$$

$$= 0 \quad \text{for } |x| < \frac{W}{2}$$

for strip-like problems, while for slot-like problems one has

$$j_x(x,0) = \int_{-\infty}^{\infty} \tilde{G}_{sl,xx}(k_x,0;\gamma)V_x\tilde{e}_x(k_x)e^{jk_xx}dk_x$$

$$+ \int_{-\infty}^{\infty} \tilde{G}_{sl,xz}(k_x,0;\gamma)V_z\tilde{e}_z(k_x)e^{jk_xx}dk_x \tag{3.45a}$$

$$= 0 \quad \text{for } |x| < \frac{W}{2}$$

$$j_z(x,0) = \int_{-\infty}^{\infty} \tilde{G}_{sl,zx}(k_x,0;\gamma)V_x\tilde{e}_x(k_x)e^{jk_xx}dk_x$$

$$+ \int_{-\infty}^{\infty} \tilde{G}_{sl,zz}(k_x,0;\gamma)V_z\tilde{e}_z(k_x)e^{jk_xx}dk_x \tag{3.45b}$$

$$= 0 \quad \text{for } |x| < \frac{W}{2}$$

In (3.44) and (3.45), the unknowns are the amplitudes I_x, I_z for strips and V_x, V_z for slots. Functions $\tilde{j}_{x,k}(k_x)$ and $\tilde{e}_{x,k}(k_x)$ are the Fourier-transforms of the assumed source functions for strip- and slot-like problems, respectively. A more convenient notation for each system of equations for strip-like lines is

$$\overline{\overline{\mathcal{L}}}_{st}(x;\gamma) \begin{bmatrix} I_x \\ I_z \end{bmatrix} = \overline{\overline{0}} \quad \text{for } |x| < \frac{W}{2} \tag{3.46a}$$

with

$$\mathcal{L}_{st,uv}(x;\gamma) \triangleq \int_{-\infty}^{\infty} \tilde{G}_{st,uv}(k_x,0;\gamma)\tilde{j}_v(k_x)e^{jk_x x}dk_x \quad \text{for} \quad u,v = x,z \quad (3.46b)$$

and for slot-like lines

$$\overline{\overline{\mathcal{L}}}_{sl}(x;\gamma)\begin{bmatrix} V_x \\ V_z \end{bmatrix} = \overline{\overline{0}} \qquad \text{for } |x| < \frac{W}{2} \tag{3.47a}$$

with

$$\mathcal{L}_{sl,uv}(x;\gamma) \triangleq \int_{-\infty}^{\infty} \tilde{G}_{sl,uv}(k_x,0;\gamma)\tilde{j}_v(k_x)e^{jk_x x}dk_x \quad \text{for} \quad u,v = x,z \quad (3.47b)$$

Denlinger solved (3.46a) for microstrips [3.20]. The systems (3.46a) and (3.47a) have a non-trivial solution when the determinant of the $\overline{\overline{\mathcal{L}}}$ matrix vanishes. Setting the determinant equal to zero yields an implicit equation to be solved for the unknown propagation constant γ. Following Denlinger, a good choice is to force the determinant to vanish for the value $x = 0$:

$$\det(\overline{\overline{\mathcal{L}}}_{st}(x;\gamma)) = 0 \tag{3.48}$$

If the assumed sources are the actual ones, the characteristic implicit equation provides the same solution for any particular value of x. If an error on γ, however, originates from the approximation made on the sources, it is not know how the choice of the x value influences this error. One may expect that imposing the boundary condition at $x = 0$ is the "best choice", because of the symmetry of the structure. It corresponds to the case of a source current reduced to a filament located in $x = 0$.

3.3.3 Moment method

Since the integral formulations (3.35a,b) and (3.38a,b) act as linear operators for the problem, the method of moments can be used to solve the problem for the unknown source term. Using the notations of Chapter 2 for the moment method, the functional $\overline{f}$ is indeed the source term for each problem:

$$\overline{f} = \overline{j}_t \quad \text{and} \quad \overline{g} = \overline{e}_t \qquad \text{for strip-like problems} \tag{3.49a}$$

$$\overline{f} = \overline{e}_t \quad \text{and} \quad \overline{g} = \overline{j}_t \qquad \text{for slot-like problems} \tag{3.49b}$$

Notation (3.36a,b) for distributed structures is used from now on. As for the scalar static case, the unknown is expanded into a series of functions. Since the actual source only exists in the strip or slot area (3.39a,b), each component of source $\overline{j}_t$ is expanded for strip-like problems into the series

$$\overline{j}_t(x,0) = \sum_{n=1}^{N} a_{nx}j_{nx}(x)\overline{a}_x + \sum_{n=1}^{N} a_{nz}j_{nz}(x)\overline{a}_z \tag{3.50a}$$

where the basis functions are chosen such that $\overline{j}_t$ satisfies (3.39b):

$$j_{nu}(x) = 0 \quad \text{for} \quad |x| > \frac{W}{2} \quad \text{and} \quad u = x, z \tag{3.50b}$$

For slot-like problems, a similar formulation is applied to the source $\overline{e}_t$:

$$\overline{e}_t(x, 0) = \sum_{n=1}^{N} a_{nx}e_{nx}(x)\overline{a}_x + \sum_{n=1}^{N} a_{nz}e_{nz}(x)\overline{a}_z \tag{3.51a}$$

with

$$e_{nu}(x) = 0 \quad \text{for} \quad |x| > \frac{W}{2} \quad \text{and} \quad u = x, z \tag{3.51b}$$

Solutions $\overline{e}_t$ provided by (3.38a) and $\overline{j}_t$ provided by (3.38b) have to satisfy (3.40b) and (3.40a), respectively. This means that, for a given strip or slot topology, $\overline{j}_t$ and $\overline{e}_t$ at interface $y = 0$ may exist only on complementary areas of the domain x of the structure. The pair of equations (3.50b) and (3.40b) indeed have to be satisfied simultaneously for strips, and the pair of equations (3.51b) and (3.40a) for slots. Then the inner product $\langle \overline{f}, \overline{g} \rangle$ between $\overline{j}_t(x, 0)$ and $\overline{e}_t(x, 0)$ is taken. The main feature of the application of the moment method in the case of planar circuits is that a *zero boundary condition is applied a priori*, by virtue of the existing spatial complementarity of actual tangential fields at the partially metallized interface (Fig. 3.7). Instead of imposing that the electric field vanishes on the strip for strip-like problems, or the tangential magnetic field to be continuous on the slot aperture (which is equivalent to saying that the current in the slot aperture is zero), the zero condition is imposed on the inner product $\langle \overline{f}, \overline{g} \rangle$ and not on $\overline{g}$ itself. This yields the final integral formulation of the boundary conditions:

$$\langle \overline{f}, \overline{g} \rangle = \langle \overline{j}_t(x, 0), \overline{e}_t(x, 0) \rangle \triangleq \int_{-\infty}^{\infty} \overline{j}_t(x, 0) \cdot \overline{e}_t(x, 0) dx$$

$$= \int_{-\infty}^{\infty} \overline{j}_t(x, 0) \cdot \mathcal{L}_{st}(\overline{j}_t(x, 0)) dx$$

$$= \int_{-\infty}^{\infty} \overline{j}_t(x, 0) \cdot \{ \int_{-\infty}^{\infty} \overline{\overline{G}}_{st}(x, 0|x', 0; \gamma) \cdot \overline{j}_t(x', 0) dx' \} dx \tag{3.52a}$$

$$= 0 \qquad\qquad \text{for strip-like problems}$$

$$\langle \overline{f}, \overline{g} \rangle = \int_{-\infty}^{\infty} \mathcal{L}_{sl}(\overline{e}_t(x, 0)) \cdot \overline{e}_t(x, 0) dx$$

$$= \int_{-\infty}^{\infty} \{ \int_{-\infty}^{\infty} \overline{\overline{G}}_{sl}(x, 0|x', 0; \gamma) \cdot \overline{e}_t(x', 0) dx' \} \cdot \overline{e}_t(x, 0) dx \tag{3.52b}$$

$$= 0 \qquad\qquad \text{for slot-line problems}$$

Because of (3.50b) or (3.51b), it is expected that setting the inner product (3.52a,b) equal to zero ensures that the solution satisfies (3.40b) or (3.40a), respectively. Equation (3.52a,b) is then solved by the moment method as detailed in Chapter 2. Introducing (3.50a) or (3.51a) into (3.52a,b) and rearranging, yields

$$
\sum_{k=1}^{N} \sum_{n=1}^{N} a_{kx} a_{nx} \langle j_{kx}, \{\mathcal{L}_{st}(j_{nx}\overline{a}_x) \cdot \overline{a}_x\} \rangle
$$

$$
+ \quad \sum_{k=1}^{N} \sum_{n=1}^{N} a_{kx} a_{nz} \langle j_{kx}, \{\mathcal{L}_{st}(j_{nz}\overline{a}_z) \cdot \overline{a}_x\} \rangle
$$

$$
+ \quad \sum_{k=1}^{N} \sum_{n=1}^{N} a_{kz} a_{nx} \langle j_{kz}, \{\mathcal{L}_{st}(j_{nx}\overline{a}_x) \cdot \overline{a}_z\} \rangle
$$

$$
+ \quad \sum_{k=1}^{N} \sum_{n=1}^{N} a_{kz} a_{nz} \langle j_{kz}, \{\mathcal{L}_{st}(j_{nz}\overline{a}_z) \cdot \overline{a}_z\} \rangle
$$

$$
= \quad 0 \qquad \text{for strip-like problems} \qquad (3.53a)
$$

$$
\sum_{k=1}^{N} \sum_{n=1}^{N} a_{kx} a_{nx} \langle e_{kx}, \{\mathcal{L}_{sl}(e_{nx}\overline{a}_x) \cdot \overline{a}_x\} \rangle
$$

$$
+ \quad \sum_{k=1}^{N} \sum_{n=1}^{N} a_{kx} a_{nz} \langle e_{kx}, \{\mathcal{L}_{sl}(e_{nz}\overline{a}_z) \cdot \overline{a}_x\} \rangle
$$

$$
+ \quad \sum_{k=1}^{N} \sum_{n=1}^{N} a_{kz} a_{nx} \langle e_{kz}, \{\mathcal{L}_{sl}(e_{nx}\overline{a}_x) \cdot \overline{a}_z\} \rangle
$$

$$
+ \quad \sum_{k=1}^{N} \sum_{n=1}^{N} a_{kz} a_{nz} \langle e_{kz}, \{\mathcal{L}_{sl}(e_{nz}\overline{a}_z) \cdot \overline{a}_z\} \rangle
$$

$$
= \quad 0 \qquad \text{for slot-like problems} \qquad (3.53b)
$$

In Chapter 2, it has been shown that this solution is equivalent to the following systems, resulting from Galerkin's procedure:

$$
\sum_{n=1}^{N} a_{nx} \langle j_{kx}, \{\mathcal{L}_{st}(j_{nx}\overline{a}_x) \cdot \overline{a}_x\} \rangle + \sum_{n=1}^{N} a_{nz} \langle j_{kx}, \{\mathcal{L}_{st}(j_{nz}\overline{a}_z) \cdot \overline{a}_x\} \rangle = 0
$$

$$
\sum_{n=1}^{N} a_{nx} \langle j_{kz}, \{\mathcal{L}_{st}(j_{nx}\overline{a}_x) \cdot \overline{a}_z\} \rangle + \sum_{n=1}^{N} a_{nz} \langle j_{kz}, \{\mathcal{L}_{st}(j_{nz}\overline{a}_z) \cdot \overline{a}_z\} \rangle = 0
$$

$$
(3.54a)
$$

with $k = 1, ..., N$ for strip-like problems

and

$$\sum_{n=1}^{N} a_{nx} \langle e_{kx}, \{\mathcal{L}_{sl}(e_{nx}\overline{a}_x) \cdot \overline{a}_x\} \rangle + \sum_{n=1}^{N} a_{nz} \langle e_{kx}, \{\mathcal{L}_{sl}(e_{nz}\overline{a}_z) \cdot \overline{a}_x\} \rangle = 0$$

$$\sum_{n=1}^{N} a_{nx} \langle e_{kz}, \{\mathcal{L}_{sl}(e_{nx}\overline{a}_x) \cdot \overline{a}_z\} \rangle + \sum_{n=1}^{N} a_{nz} \langle e_{kz}, \{\mathcal{L}_{sl}(e_{nz}\overline{a}_z) \cdot \overline{a}_z\} \rangle = 0$$

$$(3.54b)$$

with $k = 1, ..., N$ for slot-like problems.

These systems can be formalized as

$$\overline{\overline{\mathcal{L}}}_{st}(\gamma) \begin{bmatrix} \overline{a}_{Nx} \\ \overline{a}_{Nz} \end{bmatrix} = 0 \qquad \text{for strip-like problems} \qquad (3.55a)$$

$$\overline{\overline{\mathcal{L}}}_{sl}(\gamma) \begin{bmatrix} \overline{a}_{Nx} \\ \overline{a}_{Nz} \end{bmatrix} = 0 \qquad \text{for slot-like problems} \qquad (3.55b)$$

where $\overline{a}_{Nx}$ and $\overline{a}_{Nz}$ are column vectors of size N having coefficients a_{nx} and a_{nz} of the serial expansions as nth-component respectively, or

$$\overline{\overline{\mathcal{L}}}_{st,sl}(\gamma) = \begin{bmatrix} \overline{\overline{\mathcal{L}}}_{st,sl}(\gamma)_{xx} & \overline{\overline{\mathcal{L}}}_{st,sl}(\gamma)_{xz} \\ \overline{\overline{\mathcal{L}}}_{st,sl}(\gamma)_{zx} & \overline{\overline{\mathcal{L}}}_{st,sl}(\gamma)_{zz} \end{bmatrix} \qquad (3.56)$$

where $\overline{\overline{\mathcal{L}}}_{st,sl}(\gamma)_{uv}$ are $N \times N$ matrices having the following uvth-component:

$$\mathcal{L}_{st}(\gamma)_{uv,kn} = \langle j_{ku}, \{\mathcal{L}_{st}(j_{nv}\overline{a}_v) \cdot \overline{a}_u\} \rangle$$
$$\text{for strip-like problems with } k, n = 1, ..., N$$

$$\mathcal{L}_{sl}(\gamma)_{uv,kn} = \langle e_{ku}, \{\mathcal{L}_{sl}(e_{nv}\overline{a}_v) \cdot \overline{a}_u\} \rangle$$
$$\text{for slot-like problems with } k, n = 1, ..., N$$

They have a non-trivial solution if their determinant vanishes. Remembering that $\mathcal{L}_{st}$ and $\mathcal{L}_{sl}$ are functions of the unknown γ, making the determinantal equation to vanish provides a characteristic equation for the propagation constant of planar lines:

$$\det(\overline{\overline{\mathcal{L}}}_{st}(\gamma)) = 0 \qquad \text{for strip-like problems} \qquad (3.57a)$$

$$\det(\overline{\overline{\mathcal{L}}}_{sl}(\gamma)) = 0 \qquad \text{for slot-like problems} \qquad (3.57b)$$

These are the characteristic equations usually referred to in literature as "solving the determinantal equation associated with Galerkin's procedure". In fact equations (3.57a,b) are equivalent to the transcendental equations

$$\langle \overline{j}_t(x, 0), \mathcal{L}_{st}(\overline{j}_t(x, 0); \gamma) \rangle = 0 \qquad (3.58a)$$

$$\langle \mathcal{L}_{sl}(\overline{e}_t(x,0);\gamma), \overline{e}_t(x,0)\rangle = 0 \tag{3.58b}$$

which are only more precise notations for (3.52a,b). An implicit equation is thus obtained for the propagation constant, which is equivalent to an implicit eigenvalue problem when using the moment method and Galerkin's procedure: solving for the unknown coefficients of the basis functions provides solutions for those coefficients which are only valid for a particular value of the propagation constant. It is similar to an eigenvalue problem, except that the characteristic equation for the eigenfunctions does not contain explicitly the eigenvalue. Lindell introduces this concept as a "non-standard eigenvalue problem" [3.21]. Galerkin's method has then been formulated in the spectral domain, which has the advantage of replacing the integrals involving Green's functions expressed by the integral operator $\mathcal{L}_{st,sl}$ in (3.52a,b) to (3.58a,b) by algebraic relations.

3.3 4   Spectral domain Galerkin's method

The spectral formulation is based on the well-known Parseval's theorem (Appendix C), relating the product of two functions in the space domain to the product of their corresponding transforms in the spectral domain. Hence, the inner product (3.52a) satisfies

$$\langle \overline{j}_t(x,0), \overline{e}_t(x,0)\rangle = \int_{-\infty}^{\infty} \overline{j}_t(x,0)\cdot\overline{e}_t(x,0)dx = \int_{-\infty}^{\infty} \tilde{\overline{j}}_t(k_x)\cdot\tilde{\overline{e}}_t(k_x)dk_x \tag{3.59a}$$

$$= \int_{-\infty}^{\infty} \tilde{\overline{j}}_t(k_x)\cdot\{\tilde{\overline{\overline{G}}}_{st}(k_x,0;\gamma)\cdot\tilde{\overline{j}}_t(k_x)\}dk_x \tag{3.59b}$$

for strip-like problems

$$= \int_{-\infty}^{\infty} \tilde{\overline{e}}_t(k_x)\cdot\{\tilde{\overline{\overline{G}}}_{sl}(k_x,0;\gamma)\cdot\tilde{\overline{e}}_t(k_x)\}dk_x \tag{3.59c}$$

for slot-like problems

Equation (3.59a) results from the application of Parseval's theorem. Expressions (3.37a,b) have been introduced in relations (3.59b,c). Hence, the characteristic implicit equation (3.57a,b) to be solved becomes:

$$\det(\tilde{\overline{\overline{\mathcal{L}}}}_{st}(\gamma)) = 0 \qquad \text{for strip-like problems} \tag{3.60a}$$

$$\det(\tilde{\overline{\overline{\mathcal{L}}}}_{sl}(\gamma)) = 0 \qquad \text{for slot-like problems} \tag{3.60b}$$

with

$$\tilde{\overline{\overline{\mathcal{L}}}}_{st,sl}(\gamma) = \begin{bmatrix} \tilde{\overline{\overline{\mathcal{L}}}}_{st,sl}(\gamma)_{xx} & \tilde{\overline{\overline{\mathcal{L}}}}_{st,sl}(\gamma)_{xz} \\ \tilde{\overline{\overline{\mathcal{L}}}}_{st,sl}(\gamma)_{zx} & \tilde{\overline{\overline{\mathcal{L}}}}_{st,sl}(\gamma)_{zz} \end{bmatrix} \tag{3.60c}$$

where $\tilde{\mathcal{L}}_{st,sl}(\gamma)_{uv}$ are $N \times N$ matrices having the following uv^{th} component:

$$\tilde{\mathcal{L}}_{st}(\gamma)_{uv,kn} = \int_{-\infty}^{\infty} \tilde{j}_{ku}(k_x)\tilde{G}_{st,uv}(k_x,0;\gamma)\tilde{j}_{nv}(k_x)dk_x \tag{3.60d}$$

$$\text{for strip-like problems}$$

$$\tilde{\mathcal{L}}_{sl}(\gamma)_{uv,kn} = \int_{-\infty}^{\infty} \tilde{e}_{ku}(k_x)\tilde{G}_{sl,uv}(k_x,0;\gamma)\tilde{e}_{nv}(k_x)dk_x \tag{3.60e}$$

$$\text{for slot-like problems}$$

with $k,n = 1,...,N$ and $u,v = x,z$. Each coefficient of the matrix is formulated in the spectral domain as a single integral involving the Fourier-transforms of the basis functions and of the Green's dyadic components.

The spectral domain use of Galerkin's procedure was first developed for microstrip lines by Itoh and Mittra [3.22], and later extended to slot-lines and coplanar waveguides by Knorr and Kuchler [3.23], Itoh [3.24], and Janaswamy and Schaubert [3.25][3.26], and to shielded lines [3.27]-[3.29]. Galerkin's method in the spectral domain is also applicable to generalized transmission lines containing several dielectric layers and conductors located at several dielectric interfaces. This generalized method, known as the spectral immittance approach, was developed by Itoh [3.30][3.31]. Using a suitable coordinate transformation, TE and TM waves are separated in such a way that the Green's function in the spectral domain is derived by simple transmission-line theory. Using this approach, the spectral-domain analysis of generalized printed transmission lines with conductors placed on different layers requires much less analytical effort and has therefore become quite feasible.

Galerkin's procedure may be tedious when on-line designs are needed, in particular when lossy layers are present. For this reason, Kirschning and Jansen have proposed fitted formulas for the characteristics of single and coupled microstrips [3.32], obtained by intensive computations using Galerkin's procedure in the spectral domain. The resulting closed-form expressions are valid, however, for a specific range of values of the geometrical and physical parameters of the lines. For slot-lines, Janaswamy and Schaubert have proposed fitted formulas based on Galerkin's procedure [3.26], while Garg and Gupta [3.33] developed fitted formulas based on an intensive use of the transverse resonance method, described in a next section. The most simple use of Galerkin's procedure for planar lines uses only one basis function for each source component:

$$\bar{j}_t(x,0) = a_{lx}j_{lx}(x)\bar{a}_x + a_{lz}j_{lz}(x)\bar{a}_z \tag{3.61a}$$

where the basis functions are chosen such that $\bar{j}_t$ satisfies (3.39b):

$$j_{lu}(x) = 0 \quad \text{for} \quad |x| > \frac{W}{2} \quad \text{and} \quad u = x,z \tag{3.61b}$$

For slot-like problems, a similar formulation is applied to the source $\bar{e}_t$:

$$\bar{e}_t(x,0) = a_{lx}e_{lx}(x)\bar{a}_x + a_{lz}e_{lz}(x)\bar{a}_z \tag{3.62a}$$

with

$$e_{lu}(x) = 0 \quad \text{for} \quad |x| > \frac{W}{2} \quad \text{and} \quad u = x, z \tag{3.62b}$$

In this case, notations (3.60a,b) become

$$\det(\bar{\bar{\tilde{\mathcal{L}}}}_{st}(\gamma)) = 0 \qquad \text{for strip-like problems} \tag{3.63a}$$

$$\det(\bar{\bar{\tilde{\mathcal{L}}}}_{sl}(\gamma)) = 0 \qquad \text{for slot-like problems} \tag{3.63b}$$

with

$$\bar{\bar{\tilde{\mathcal{L}}}}_{st,sl}(\gamma) = \left[\begin{array}{cc} \tilde{\mathcal{L}}_{st,sl}(\gamma)_{xx} & \tilde{\mathcal{L}}_{st,sl}(\gamma)_{xz} \\ \tilde{\mathcal{L}}_{st,sl}(\gamma)_{zx} & \tilde{\mathcal{L}}_{st,sl}(\gamma)_{zz} \end{array} \right] \tag{3.63c}$$

where $\tilde{\mathcal{L}}_{st,sl}(\gamma)_{uv}$ are now scalar integral expressions involving the Green's operator:

$$\tilde{\mathcal{L}}_{st}(\gamma)_{uv} = \int_{-\infty}^{\infty} \tilde{j}_{lu}(k_x)\tilde{G}_{st,uv}(k_x,0;\gamma)\tilde{j}_{lv}(k_x)dk_x \tag{3.63d}$$

$$\text{for strip-like problems}$$

$$\tilde{\mathcal{L}}_{sl}(\gamma)_{uv} = \int_{-\infty}^{\infty} \tilde{e}_{lu}(k_x)\tilde{G}_{sl,uv}(k_x,0;\gamma)\tilde{e}_{lv}(k_x)dk_x \tag{3.63e}$$

$$\text{for slot-like problems}$$

with $k, n = 1, ..., N$ and $u, v = x, z$. This simplification is referred to, in the literature, as the first-order approximation. The zero-order approximation consists of neglecting the z-component of the source electric field in the case of slot-like lines, or the x-component of the source current density for strip-like lines. Under those assumptions, conditions (3.63d,e) simplify into

$$\int_{-\infty}^{\infty} \tilde{j}_{lz}(k_x)\tilde{G}_{st,zz}(k_x,0;\gamma)\tilde{j}_{lz}(k_x)dk_x = 0 \quad \text{for strip-like problems} \tag{3.64a}$$

$$\int_{-\infty}^{\infty} \tilde{e}_{lx}(k_x)\tilde{G}_{sl,xx}(k_x,0;\gamma)\tilde{e}_{lx}(k_x)dk_x = 0 \quad \text{for slot-like problems} \tag{3.64b}$$

since the matrix $\bar{\bar{\tilde{\mathcal{L}}}}_{st,sl}(\gamma)$ now reduces to a scalar. The zero-order approximation has been widely used for computing the propagation constant of planar transmission lines, because it is the less time consuming. However, it still requires the solution of an implicit equation for the propagation constant γ.

3.3.5   Approximate modeling for slot-lines: equivalent magnetic current

In his famous paper Cohn [3.14] presented the first attempt to model the transmission line behavior of a slot-line. His quasi-static approach is based on the definition of an equivalent magnetic current deduced from the transverse tangential electric field in the slot aperture. If the slot width W is much smaller than the free-space wavelength, this approximation is valid. This equivalent line current $\overline{M}$ is embedded in a equivalent homogeneous medium with a permittivity equal to $(\varepsilon_r + 1)/2$. Then $\overline{M}$ is considered as a source in Maxwell's equations, from which the magnetic field $\overline{H}$ is deduced. This model is useful for evaluating the far-field behavior of the electric and magnetic field. The line approximation made about the aperture current of course limits the efficiency of this model, which is relevant for narrow slots only.

3.3.6   Modeling slot-lines: transverse resonance method

The transverse resonance method was developed for slot-lines by Cohn [3.14] and applied by Mariani *et al.* [3.34]. In this method the slot-line is modeled as a rectangular waveguide, as illustrated in Figure 3.8. In Figure 3.8a the slot-line is bounded by electric walls placed perpendicular to the z-propagation axis, spaced by a distance $a = \lambda_s/2$. This combination will of course not disturb the fields of the structure. Next, in Figure 3.8b, electric or magnetic walls are inserted in planes parallel to the slot and perpendicular to the substrate at planes $x = \pm b/2$, where b is chosen large enough to assume that the walls have no effect on the fields. The introduction of electric walls generates the configuration of a capacitive iris (Figure 3.8c)in a waveguide having the y-propagation axis and is usually preferred.

The transverse resonance method imposes the sum of the susceptances along the y-axis to be zero. This sum includes the susceptances of the TE mode looking in the $+y$ and $-y$ directions and the capacitive iris susceptance due to higher order modes on both sides of the iris. ¿From this equation, the value of a is extracted, which provides the slot-line wavelength. The main drawbacks of the method are that the calculation of the susceptance involves series expansions that converge slowly and are complicated functions of a, and it is not valid for ratios W/H greater than unity.

3.3.7   Transmission Line Matrix (TLM) method

In the TLM method, invented by Johns and developed by Hoefer [3.35][3.36], the field problem is converted into a three-dimensional network problem. In its simplest form, the space is discretized into a three-dimensional lattice with period Δl, to describe six field components by a hybrid TLM cell obtained as a combination of shunt and series nodes (Fig. 3.9a,c) and modeled as equivalent transmission lines (Fig. 3.9b,d). Some equivalencies can be found between the field components and the voltages, as well as currents and lumped elements of the cell. This method has been developed for modeling time-domain propagation. It has also been improved by a number of authors [3.15] and has

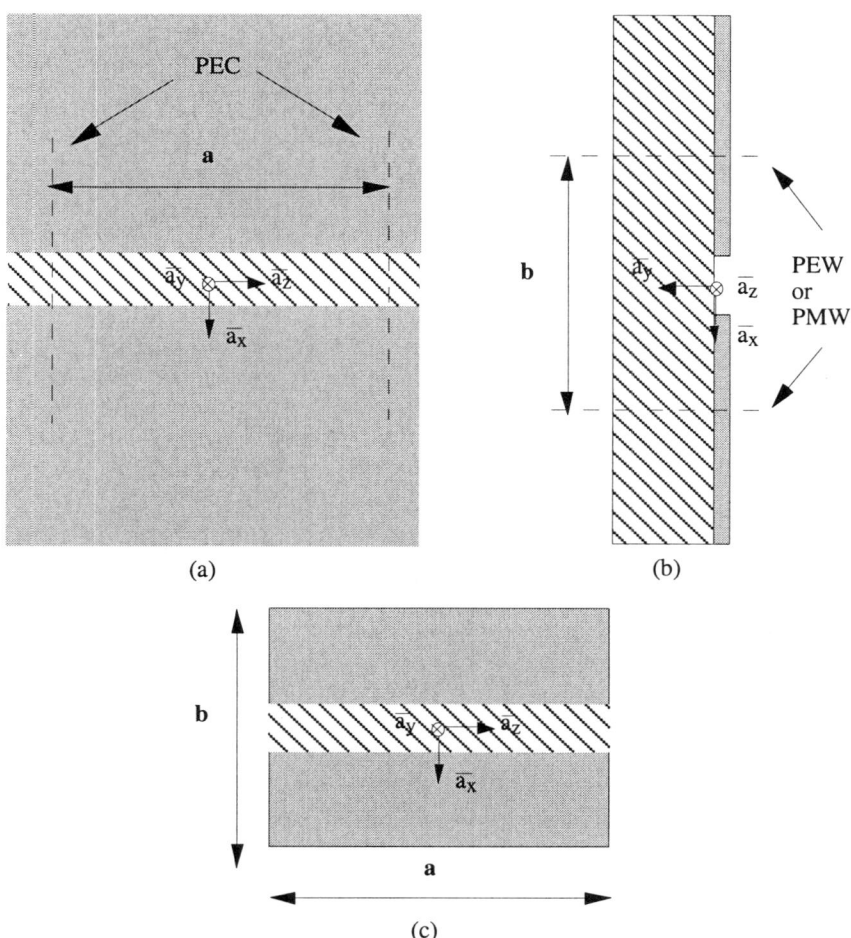

Fig. 3.8 *Equivalent waveguide for modeling slot-lines by transverse resonance technique (a) limitation of slot along z-axis by using perfect electric walls; (b) limitation of slot along x-axis by using perfect electric or magnetic walls; (c) equivalent capacitive iris*

become a subject in itself which we do not cover here in detail. Choi and Hoefer [3.37] have shown that the efficiency of the TLM method can be enhanced by using a finite-difference method in the time domain (FDTD): the combination of TLM and FDTD reduces by half the computation time and memory space needed for the TLM method only. The time-domain response

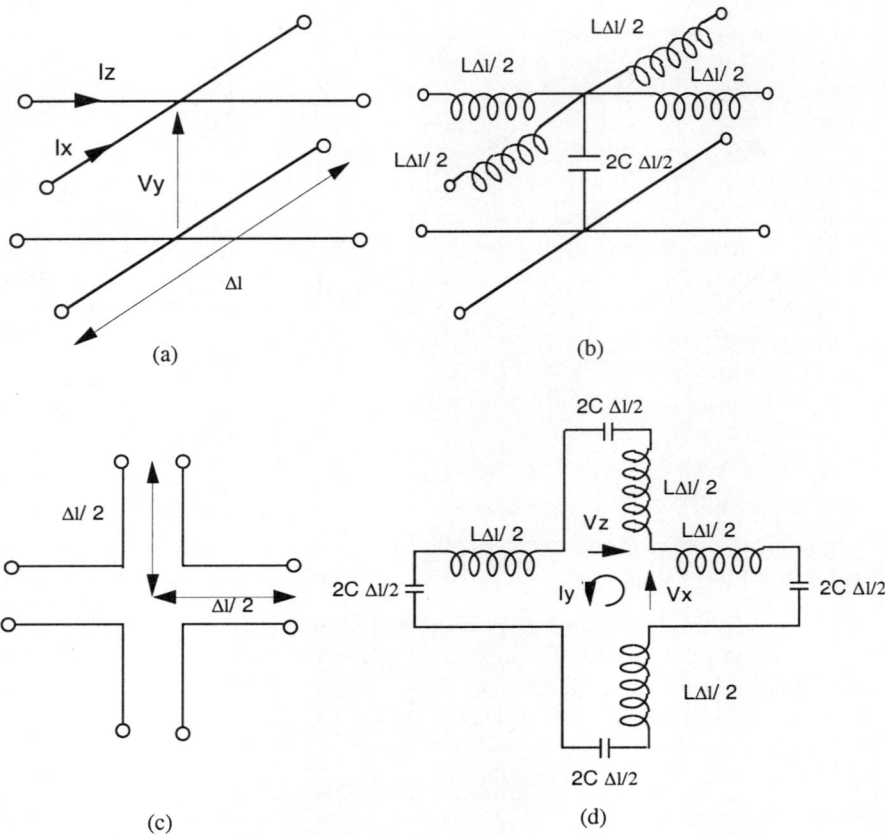

Fig. 3.9 *Equivalent nodes for Transmission Line Matrix method [3.38] (a) shunt node; (b) equivalent lumped element model of shunt node; (c) series node; (d) equivalent lumped element model of series node*

of the cell is computed, from which the frequency response is obtained by Fourier transform. From our point of view, it is important to note that the drawback of the TLM method is that a filtering effect is introduced in the frequency response, because of the spatial discretization. It is, however useful, for analyzing planar discontinuities.

3.4   Analytical variational formulations

The previous sections have presented a wide variety of methods for analyzing planar multilayered transmission lines. This section concentrates on variational methods. They are classified as explicit or implicit variational forms

(Chapter 2). As introduced in Chapter 1, a *variational principle* is usually formulated as a ratio of integral forms, yielding an explicit expression for a scalar parameter of an equivalent circuit or transmission line, while *variational methods* use the well-known variational character of a functional in order to compute nearly exact functions or field distributions. Those fields, however, may be a function of the propagation constant γ, so that rendering the functional extremum must also yield the actual value for the propagation constant.

Hence, the possible variational character of the propagation constant has to be investigated in connection with the possible variational character of the characteristic equation in the case of non-standard, implicit, eigenvalues problems described in the previous section. We first concentrate on variational principles providing an explicit expression for quasi-static equivalent transmission line parameters. We shall then clarify the variational character of Galerkin's procedure as implicit non-standard eigenvalue problem for the propagation constant of planar lines.

3.4.1 Explicit variational form for a quasi-static analysis

For planar transmission lines, the quasi-static analysis is based on the computation of a quasi-TEM lumped-circuit component, capacitance or inductance.

3.4.1.1 Quasi-TEM capacitance based on Green's formalism

Formulations (3.16) to (3.19) presented in Section 3.3 is a well-known variational principle for the capacitance. Hence, the charge density on the conductor is considered as the trial quantity, and the trial potential is deduced via (3.17) once the Green's function associated to the problem has been found. For the present purpose, the variational principle is rewritten using (3.16) to (3.19) and (3.22a) as

$$Y \triangleq \frac{1}{C} = \frac{1}{Q_l^2} \oint_\Gamma \rho(x,y)\Phi(x,y)d\Gamma \tag{3.65}$$

which yields

$$Y\{\int_{-\frac{W_c}{2}}^{\frac{W_c}{2}} \rho_s(x)dx\}^2 = \int_{-\frac{W_c}{2}}^{\frac{W_c}{2}} \rho_s(x)\{\int_{-\frac{W_c}{2}}^{\frac{W_c}{2}} G_P(x,0|x',0)\rho_s(x')dx'\}dx \tag{3.66}$$

where G_P is the scalar Green's function associated to the problem described by Poisson's equation. The proof of the variational behavior of Y with respect to distribution of surface charge $\rho_s(x)$ is given by Collin [3.39]. It is briefly reported here. We use a simplified notation for the integration on the strip conductor:

$$\int_{-\frac{W_c}{2}}^{\frac{W_c}{2}} \ldots \, dx = \int_{strip} \ldots \, dx$$

Varying Y and ρ_s in (3.66) and neglecting second-order variations, we obtain

$$2Y\{\int_{strip}\delta\rho_s(x)dx\}\{\int_{strip}\rho_s(x)dx\} + \delta Y\{\int_{strip}\rho_s(x)dx\}^2$$

$$= \int_{strip}\delta\rho_s(x)\{\int_{strip}G_P(x,0|x',0)\rho_s(x')dx'\}dx \qquad (3.67)$$

$$+ \int_{strip}\rho_s(x)\{\int_{strip}G_P(x,0|x',0)\delta\rho_s(x')dx'\}dx$$

Assuming that the Green's function is symmetrical in variables x and x', so that the reciprocity theorem holds, right-hand side of (3.67) is recombined into

$$\int_{strip}\delta\rho_s(x)\{\int_{strip}G_P(x,0|x',0)\rho_s(x')dx'\}dx = \int_{strip}V_0\delta\rho_s(x)dx \quad (3.68)$$

since the Green's integral in (3.68) is precisely the definition of (3.19), (3.20) and (3.22b,c) of the actual potential on the strip which is equal to V_0 :

$$V_0 = \int_{-\frac{W_c}{2}}^{\frac{W_c}{2}} G_P(x,0|x',0)\rho_s(x')dx' \qquad (3.69)$$

Equation (3.67) is finally written as

$$\delta Y\{\int_{strip}\rho_s(x)dx\}^2 = 2\{\int_{strip}\delta\rho_s(x)dx\}[V_0 - Y\{\int_{strip}\rho_s(x)dx\}] \quad (3.70a)$$

where Y and ρ_s are the exact quantities. By virtue of (3.18), the right-hand side of (3.70a) is transformed into

$$2\{\int_{strip}\delta\rho_s(x)dx\}(V_0 - YQ_l) \qquad (3.70b)$$

where (3.14) appears. Hence, expression (3.70a) vanishes. So, instead of solving Poisson's equation for a particular charge distribution, it is possible to obtain a value of the capacitance per unit length which is correct to the second-order, by taking a trial charge distribution and combining it with the exact Green's function associated to Poisson's equation. Since the Green's function is assumed to be symmetrical, Collin states that the integral in (3.66) is always positive and that the variational value obtained for the inverse of the capacitance is a minimum [3.40]. Hence, the value provided by (3.66) is always smaller than the actual one.

Section 3.3.4 briefly outlined that the spectral domain formulation offers an easy way to express Green's functions for planar circuits. Yamashita and Mittra have investigated this approach when using variational principle (3.16) applied to a multilayered microstrip line [3.41][3.42]. This is a good illustration

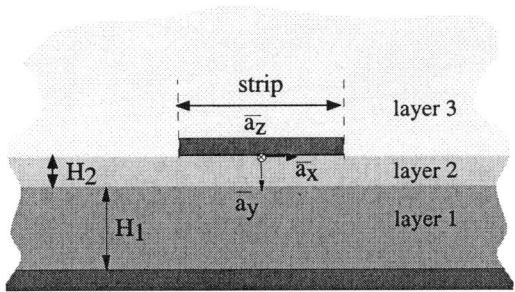

Fig. 3.10 *Microstrip line investigated by Yamashita and Mittra [3.41][3.42]*

of various concepts and methods presented in Sections 3.2 and 3.3. It is reported here as an example.

Yamashita and Mittra consider a multilayered microstrip (Fig. 3.10) and assume that the metallizations have a zero thickness. The density of surface charge on the strip satisfies (3.18) and (3.22a), is located at the interface between layer 1 and 2, and can be described as a function of the x-variable only. Hence, the static potential in the three layers satisfies the charge-free Poisson's equation, namely the two-dimensional Laplace's equation. An inhomogeneous boundary condition applies at interface between layers 1 and 2 to the normal component of the electric field. The x-Fourier transform of Laplace's equation (3.8) yields a second-order ordinary differential equation for the spectral form of the potential:

$$\frac{\partial^2 \tilde{\Phi}_i(k_x, y)}{\partial y^2} - k_x^2 \tilde{\Phi}_i(k_x, y) = 0 \tag{3.71}$$

where the subscript i refers to layer i, layer 1 is the conductor-backed layer, layer 3 the semi-infinite air layer, and $\tilde{\Phi}_i(k_x, y)$ the spectral potential. Equation (3.71) has the well-known general solution in each layer i:

$$\tilde{\Phi}_i(k_x, y) = A_i(k_x)e^{k_x y} + B_i(k_x)e^{-k_x y} \tag{3.72}$$

On the other hand, the transverse electrostatic field derives from the potential in each layer:

$$[E_{xi}(x, y), E_{yi}(x, y)] = \left(-\frac{\partial \Phi_i(x, y)}{\partial x}, -\frac{\partial \Phi_i(x, y)}{\partial y} \right) \tag{3.73a}$$

and transforms in the spectral k_x-domain into

$$[\tilde{E}_{xi}(k_x, y), \tilde{E}_{yi}(k_x, y)] = -\Big(jk_x \tilde{\Phi}_i(k_x, y), k_x\{A_i(k_x)e^{k_x y} $$
$$- B_i(k_x)e^{-k_x y}\} \Big) \tag{3.73b}$$

The knowledge of the six coefficients $\{A_i(k_x), B_i(k_x)\}$ is sufficient to describe the potential and the electric field. These fields have to satisfy the following Fourier-transformed boundary conditions at the various interfaces:

a. on the ground plane $y = -H_1 + H_2$

$$\tilde{E}_{x3}\left[k_x, -(H_1 + H_2)\right] = 0 \tag{3.74a}$$

b. on the dielectric interface $y = -H_2$

$$\tilde{E}_{x2}(k_x, -H_2) = \tilde{E}_{x3}(k_x, -H_2) \tag{3.74b}$$

$$\varepsilon_2 \tilde{E}_{y2}(k_x, -H_2) = \varepsilon_3 \tilde{E}_{y3}(k_x, -H_2) \tag{3.74c}$$

c. on the dielectric interface y = 0

$$\tilde{E}_{x2}(k_x, 0) = \tilde{E}_{x1}(k_x, 0) \tag{3.74d}$$

$$\varepsilon_1 \tilde{E}_{y1}(k_x, 0) - \varepsilon_2 \tilde{E}_{y2}(k_x, 0) = -\tilde{\rho}_s(k_x) \tag{3.74e}$$

where $\tilde{\rho}_s(k_x)$ is the x-Fourier transform of the charge distribution $\rho_s(x)$ modeling the strip

d. at $y = \infty$ the electric field and potential have to vanish, which imposes

$$A_1(k_x) = 0 \tag{3.74f}$$

The six equations (3.74a-f) form an inhomogeneous system of six linear equations to be solved for the six unknown coefficients $\{A_i(k_x), B_i(k_x)\}$. Since only the right-hand side of (3.74c) is different from zero and equal to $\tilde{\rho}_s(k_x)$, each coefficient $\{A_i(k_x), B_i(k_x)\}$ is proportional to $\tilde{\rho}_s(k_x)$:

$$A_i(k_x) = K_{A_i}(k_x)\tilde{\rho}_s(k_x) \tag{3.75a}$$

$$B_i(k_x) = K_{B_i}(k_x)\tilde{\rho}_s(k_x) \tag{3.75b}$$

with, of course, $K_{A_1}(k_x) = 0$. Hence, the potential in each layer is expressed as

$$\tilde{\Phi}_i(k_x, y) = \{K_{A_i}(k_x)e^{k_x y} + K_{B_i}(k_x)e^{-k_x y}\}\tilde{\rho}_s(k_x) \tag{3.76}$$

which yields the spectral form of the Green's function relating the potential to the charge in each layer:

$$\tilde{G}_P(k_x, y) = \{K_{A_i}(k_x)e^{k_x y} + K_{B_i}(k_x)e^{-k_x y}\} \tag{3.77}$$

As already mentioned, at this stage of the development the user has the choice either to invert the spectral Green's function to obtain its corresponding form in the space domain, or to complete the solution of the problem in the spectral domain. Yamashita *etal.* has chosen the second approach. He expresses the basic relation (3.16) in the spectral domain, using Parseval's theorem. Keeping in mind that the charge distribution $\rho_s(x)$ satisfies (3.65) and vanishes outside the strip yields

$$\frac{1}{C} = \frac{1}{Q_l^2} \int_{strip} \rho_s(x)\Phi(x,y)dx = \frac{1}{Q_l^2} \int_{-\infty}^{\infty} \rho_s(x)\Phi(x,y)dx \tag{3.78a}$$

$$= \frac{1}{2\pi Q_l^2} \int_{-\infty}^{\infty} \tilde{\rho}_s(k_x)\tilde{\Phi}_2(k_x,0)dk_x \tag{3.78b}$$

Introducing (3.66) and (3.76) in (3.78b), then yields

$$\frac{1}{C} = \frac{1}{2\pi Q_l^2} \int_{-\infty}^{\infty} \tilde{\rho}_s(k_x)\{K_{A_2}(k_x) + K_{B_2}(k_x)\}\tilde{\rho}_s(k_x)dk_x \tag{3.78c}$$

where Q_l is obtained from (3.18). Expression (3.78c) has the advantage that the tedious inversion of the Green's function (3.77) is avoided and that the integrands result in a product of simple algebraic expressions.

Yamashita has also investigated various shapes for the trial charge density and determined the most adequate by using the Rayleigh-Ritz method. His results agree with experiments up to 4 GHz. Obviously the model fails at higher frequencies because the static assumption is no longer valid.

3.4.1.2 Quasi-TEM inductance based on Green's formalism

Most of the quasi-static applications of variational principles for modeling quasi-TEM planar lines are based on the computation of the capacitance per unit length. This approach is not valid when magnetic materials are involved in the cross-section of the line, because the inductance of the line is no longer equal to the inductance in air. As shown in Section 3.2, instead of computing the pair (C, C_0) from which inductance L_0 is deduced, the pair (L, L_0) must be computed which yields C_0. Hence, a variational principle for the inductance is useful. It is expected that, by duality, the variational form for the inductance will be similar to (3.16).

In magnetostatics, the fields are derived from a magnetic vector potential $\overline{A}$ satisfying the vector equation

$$\nabla^2\overline{A}(x,y) = -\mu_0\overline{\overline{\mu}}_r \cdot \overline{J}(x,y) \tag{3.79}$$

The static electric and magnetic fields and the magnetic flux density satisfy respectively

$$\nabla \times \overline{A} = \overline{B}(x,y) \tag{3.80a}$$

$$\nabla \times \overline{H} = \overline{J}(x,y) \tag{3.80b}$$

$$\nabla \cdot \overline{B} = 0 \tag{3.80c}$$

$$\overline{B} = \mu_0\overline{\overline{\mu}}_r \cdot \overline{H} \tag{3.80d}$$

By virtue of (3.80b) the magnetic field can be derived from a scalar potential in the areas where the current density is zero. The first variational form presented for the inductance is mentioned by Collin [3.43]:

$$L = \frac{1}{I^2} \int_{strip} J_z(x,0)\Phi_0(x,0)dx \qquad (3.81)$$

where $J_z(x,0)$ is the distribution of the longitudinal current density flowing on the strip:

$$I = \int_{strip} J_z(x,0)dx \triangleq \int_{-\frac{W_c}{2}}^{\frac{W_c}{2}} J_z(x,0)dx \qquad (3.82)$$

and $\Phi_0(x)$ is the integrated induction flux per unit length between the ground plane and the strip (Fig 3.11). Indeed one has

$$\Phi_0(x)\Delta z = \int_z^{z+\Delta z} \int_{-H}^0 B_x(x,y)dydz \qquad (3.83a)$$

$$= \oint_\Gamma \overline{A} \cdot \overline{d\Gamma} \qquad (3.83b)$$

$$= [A_z(x,0) - A_z(x,-H)]\Delta z \qquad (3.83c)$$

$$= \Delta z \int_{strip} G_{W,zz}(x,0|x',0)J_z(x',0)dx' \qquad (3.83d)$$

using Stoke's theorem and noting that the integrals on the two vertical paths of Γ at z and $z+\Delta z$ cancel. The vector potential is defined to be zero on the ground plane.

The proof of the variational behavior of (3.81) is similar to that of equation (3.16) in the previous Section. Introducing (3.83d) in (3.81), varying L and J_z, and neglecting second-order variations, we have

$$2L\{\int_{strip} \delta J_z(x,0)dx\}\{\int_{strip} J_z(x,0)dx\} + \delta L\{\int_{strip} J_z(x,0)dx\}^2$$

$$= \int_{strip} \delta J_z(x,0)\{\int_{strip} G_{W,zz}(x,0|x',0)J_z(x',0)dx'\}dx \qquad (3.84)$$

$$+ \int_{strip} J_z(x,0)\{\int_{strip} G_{W,zz}(x,0|x',0)J_z(x',0)dx'\}dx$$

Assuming again that the Green's function satisfies reciprocity, the right-hand side of (3.84) is recombined into

$$\int_{strip}\{\int_{strip} G_{W,zz}(x,0|x',0)J_z(x',0)dx'\}\delta J_z(x,0)dx$$

$$= \int_{strip} \Phi_0(x)\delta J_z(x,0)dx \qquad (3.85)$$

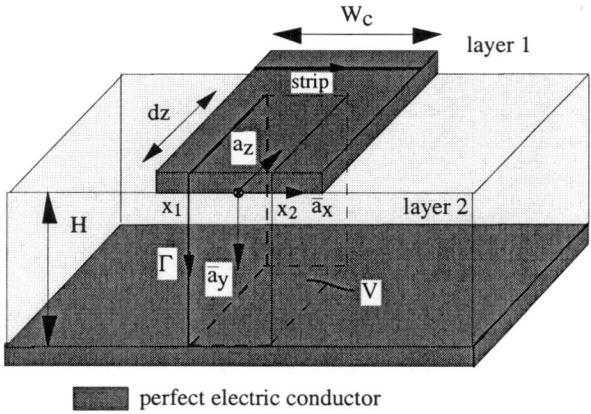

perfect electric conductor

Fig. 3.11 *Microstrip line with integration paths for calculating quasi-TEM inductance*

Equation (3.84) is finally written as

$$\delta L\{\int_{strip} J_z(x,0)dx\}^2 = 2\left[\int_{strip} \Phi_0(x)\delta J_z(x,0)dx\right.$$
$$\left. - L\{\int_{strip} \delta J_z(x,0)dx\}\{\int_{strip} J_z(x,0)dx\}\right] \tag{3.86}$$

As $\Phi_0(x)$ remains constant on the strip area, it may be extracted from the integral. Using (3.82), equation (3.86) finally reduces to

$$\delta L\{\int_{strip} J_z(x,0)dx\}^2 = 2\{\int_{strip} \delta J_z(x,0)dx\}(\Phi_0 - LI) \tag{3.87}$$

Since L and J_z in the right-hand side of equation (3.86) are exact quantities, they satisfy the commonly stated definition for the inductance per unit length:

$$L = \frac{\Phi_0}{I} \tag{3.88}$$

which causes the first-order error made on the inductance in (3.87) to vanish. That Φ_0 remains constant on the strip area is demonstrated by applying the integral formulation of divergence equation (3.80c) on volume V (Fig. 3.11). From the divergence theorem, the integral of the left-hand side of (3.80c) is equal to the flux integral of the induction field on the surface enclosing volume V. Since no variation of fields occurs along the z-axis, the contribution to the integral of the two planar faces perpendicular to this axis is canceled. On the other hand, the component of the actual magnetic flux density normal to the conductor vanishes on the conductor, and the contributions to the surface integral of the top and bottom faces of V are zero. The flux integrals between the ground plane and the strip on the two planes $x = x_1$ and $x = x_2$ are identical, because the integral of the right-hand side of (3.80c) remains equal to zero.

Another particular choice for the integration of the flux per unit length is depicted in Figure 3.12. Here $\Phi_0(x)$ is the integrated induction flux per unit length flowing through the surface at plane $y = 0$ limited by contour Γ (Fig. 3.12a,b):

$$\Phi_0(x)\Delta z = \int_z^{z+\Delta z} \int_{-H}^0 B_y(x',0)dx'dz \tag{3.89a}$$

$$= \oint_\Gamma \overline{A} \cdot \overline{d\Gamma} \tag{3.89b}$$

$$= [A_z(x,0) - A_z(-\infty,0)]\Delta z \tag{3.89c}$$

$$= \Delta z \int_\Lambda G_{W,zz}(x,0|x',0)J_z(x',0)dx' \tag{3.89d}$$

using Stoke's theorem, and noting that the integrals on two of the paths of Γ, respectively BA in $z + dz$ and DC in z, are zero, while the vector potential vanishes at $x = -\infty$. For coplanar waveguides (CPW) (Fig. 3.12b), $\Phi_0(x)$ corresponds to the actual flux. For strips, it is equivalent to the previous definition (3.83b), by applying the divergence theorem to volume V' (Fig. 3.12a) where surface integrals vanish at infinity. Hence, the contribution of surfaces $ABCD$ and $BCFE$ to the integrals compensate themselves, which demonstrates that the two contours of Figures 3.11 and 3.12a are equivalent for the strip problem.

Definition (3.89a) implies that no contribution to $\Phi_0(x)$ occurs on the strip, because the y-component of the magnetic flux density vanishes on a perfect conductor. As a consequence, $\Phi_0(x)$ remains constant during the integration on the strip and can be extracted from the integral.

As for the quasi-TEM capacitance, the knowledge of the Green's function on the strip area is required. Its derivation is very close to that discussed when presenting Yamashita's method. Now it is only briefly outlined. Since the current distribution is located on a strip of zero-thickness, the various layers are current source-free. Hence, a static magnetic scalar potential Φ_i^M is considered in each layer, from which the magnetic field:

$$\overline{H}_i = -\nabla\Phi_i^M \tag{3.90}$$

When the media are isotropic, combining (3.90) with (3.80c) yields Laplace's equation for Φ_i^M and the general solution (3.72) is applicable in each layer:

$$\tilde{\Phi}_i^M(k_x,y) = A_i^M(k_x)e^{k_x y} + B_i^M(k_x)e^{-k_x y} \tag{3.91}$$

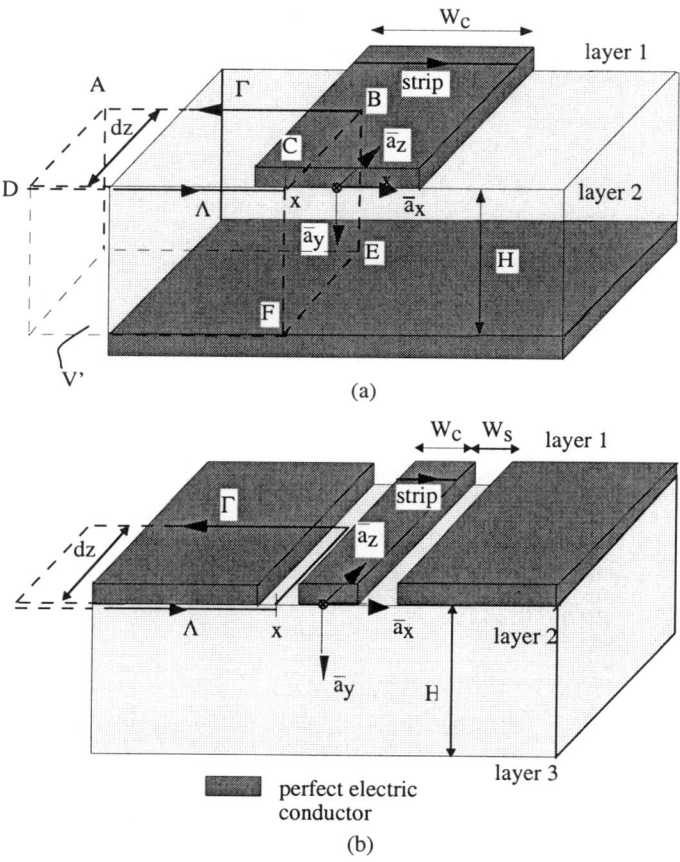

Fig. 3.12 *Integration paths for energetic variational principle (a) microstrip;*
(b) coplanar waveguide

The spatial and spectral magnetic fields in each layer are obtained by replacing $\overline{E}_i$ and Φ_i in (3.73a,b) by $\overline{H}_i$ and Φ_i^M, respectively:

$$[H_{xi}(x,y), H_{yi}(x,y)] = \left(-\frac{\partial \Phi_i^M(x,y)}{\partial x}, -\frac{\partial \Phi_i^M(x,y)}{\partial y} \right) \tag{3.92a}$$

which transform in the spectral domain into

$$[\tilde{H}_{xi}(k_x,y), \tilde{H}_{yi}(k_x,y)] = -\left(jk_x \tilde{\Phi}_i^M(k_x,y), k_x\{A_i^M(k_x)e^{k_x y}\right.$$
$$\left. - B_i^M(k_x)e^{-k_x y}\} \right) \tag{3.92b}$$

As previously for the electrostatic case, the superscript $\tilde{\ }$ holds for spectral quantities.

When anisotropy or gyrotropy occurs, Φ_i^M still remains the solution of a second-order differential equation. The y dependent coefficients in the general solution have a different expression, provided as an example in the next chapter. From the spectral magnetic field, the spatial and spectral magnetic flux densities are calculated and the following boundary conditions are applied to these spectral fields:

a. on the ground plane $y = -H$

$$\tilde{B}_{y2}(k_x, -H) = 0 \tag{3.93a}$$

b. on the interface $y = 0$

$$\tilde{B}_{y2}(k_x, 0) = \tilde{B}_{y1}(k_x, 0) \tag{3.93b}$$

$$\tilde{H}_{y1}(k_x, 0) - \tilde{H}_{y2}(k_x, 0) = \tilde{J}_z(k_x) \tag{3.93c}$$

where $\tilde{J}_z(k_x)$ is the x-Fourier transform of the current density distribution $J_z(x, 0)$ modeling the strip

c. at $y = \infty$, the magnetic field and potential have to vanish, which provides the same condition as (3.74f)

$$A_1^M(k_x) = 0 \tag{3.93d}$$

Four relations are obtained, relating the four coefficients $\{A_i^M(k_x), B_i^M(k_x)$ describing the magnetostatic potential in the two layers. Hence, the spectral magnetostatic potential in each layer is related to the spectral current distribution $\tilde{J}_z(k_x)$ when solving the resulting system. The spectral form of magnetic flux density is derived from (3.80d) and yields the spectral form of the induction flux per unit length

$$\Phi_0(k_x) = \frac{\tilde{B}_{y2}(k_x, y)}{jk_x} \tag{3.94}$$

The Fourier-transform of (3.89a) - equation (3.94) - yields the Green's function relating the spectral vector potential $\tilde{A}_z(k_x, 0)$ to $\tilde{J}_z(k_x)$. An alternate way would be to express all field components from a scalar expression for the $\tilde{A}_z$ component of the static vector potential.

The application of this variational principle mentioned by Collin has been carried out for a microstrip line configuration. Kitazawa [3.44] used expression (3.81) to calculate the quasi-TEM parameters of coplanar waveguides and microstrips on a magnetic substrate, possibly anisotropic. The derivation of the Green's function in the CPW case follows the method presented above, with the two integration contours Γ and Λ (Fig. 3.12). The results were compared to those obtained theoretically [3.45] and experimentally [3.46] by Pucel and Massé.

Kitazawa also proposed another variational principle, expressed as

$$
\frac{1}{L} = \frac{\int_{\frac{W_c}{2}}^{U} B_y(x, 0)\{\int_{-U}^{x} J_z(x', 0)dx'\}dx}{\{\int_{\frac{W_c}{2}}^{U} B_y(x, 0)dx\}^2}
$$

$$
= \frac{\int_{\frac{W_c}{2}}^{U} B_y(x, 0)\{\int_{-U}^{x} \int_{-\infty}^{\infty} G_{M,zy}(x', 0|x'', 0)B_y(x'', 0)dx''dx'\}dx}{\{\int_{\frac{W_c}{2}}^{U} B_y(x, 0)dx\}^2}
$$

$$(3.95)$$

where $U = W_c/2 + W_s$ for coplanar waveguides, and $U = \infty$ for microstrips. Expression (3.95) assumes that a suitable Green's function $\overline{\overline{G}}_M$ can be defined between the current density J_z flowing on the strip and the magnetic flux density $\overline{B}$. The spectral flux density $\tilde{\overline{B}}$ can be directly related to the spectral magnetic field via (3.80d). It is related to the spectral potential by (3.92b), and finally to the spectral coefficients $\{A_i^M(k_x), B_i^M(k_x)\}$ which have been shown to be proportional to the current density. So, the required $\overline{\overline{G}}_M$ is obtained by inverting the relation found between the spectral magnetic flux density $\tilde{\overline{B}}$ and $\tilde{J}_z$. Varying $(1/L)$ and B_y into (3.95) yields

$$
\delta(\frac{1}{L})\{\int_{\frac{W_c}{2}}^{U} B_y(x, 0)dx\}^2 + 2(\frac{1}{L})\{\int_{\frac{W_c}{2}}^{U} B_y(x, 0)dx\}\{\int_{\frac{W_c}{2}}^{U} \delta B_y(x, 0)dx\}
$$

$$
= \int_{\frac{W_c}{2}}^{U} \delta B_y(x, 0)\{\int_{-U}^{x} \int_{-\infty}^{\infty} G_{M,zy}(x', 0|x'', 0)B_y(x'', 0)dx''dx'\}dx
$$

$$
+ \int_{\frac{W_c}{2}}^{U} B_y(x, 0)\{\int_{-U}^{x} \int_{-\infty}^{\infty} G_{M,zy}(x', 0|x'', 0)\delta B_y(x'', 0)dx''dx'\}dx
$$

$$(3.96)$$

Under the same assumptions of the Green's function used before, the first-order error in $(1/L)$ is

$$\delta(\frac{1}{L})\{\int_{\frac{W_c}{2}}^{U} B_y(x,0)dx\}^2$$

$$= 2\int_{\frac{W_c}{2}}^{U} \delta B_y(x,0)\{\int_{-U}^{x}\int_{-\infty}^{\infty} G_{M,zy}(x',0|x'',0)B_y(x'',0)dx''dx'\}dx$$

$$-2(\frac{1}{L})\{\int_{\frac{W_c}{2}}^{U} B_y(x,0)dx\}\{\int_{\frac{W_c}{2}}^{U} \delta B_y(x,0)dx\} \qquad (3.97a)$$

$$= 2\int_{\frac{W_c}{2}}^{U} \delta B_y(x,0)\{\int_{-U}^{x} J_z(x',0)dx'\}dx$$

$$-2(\frac{1}{L})\{\int_{\frac{W_c}{2}}^{U} B_y(x,0)dx\}\{\int_{\frac{W_c}{2}}^{U} \delta B_y(x,0)dx\} \qquad (3.97b)$$

$$= 2\int_{\frac{W_c}{2}}^{U} \delta B_y(x,0)I dx$$

$$-2(\frac{1}{L})\{\int_{\frac{W_c}{2}}^{U} B_y(x,0)dx\}\{\int_{\frac{W_c}{2}}^{U} \delta B_y(x,0)dx\} \qquad (3.97c)$$

The first term of the right-hand side of (3.97b) contains an integral over x' which remains constant and equal to I during the integration over the specified domain for the x variable. Hence, (3.97c) is finally equivalent to:

$$[I - \frac{1}{L}\{\int_{\frac{W_c}{2}}^{U} B_y(x,0)dx\}]\{\int_{\frac{W_c}{2}}^{U} \delta B_y(x,0)dx\}$$

$$= [I - \frac{1}{L}\Phi_0]\{\int_{\frac{W_c}{2}}^{U} \delta B_y(x,0)dx\} = 0 \qquad (3.98)$$

and by virtue of equation (3.88), equation (3.95) is indeed variational.

3.4.1.3   Capacitance based on energy

The other variational principle for a capacitance proposed by Collin provides an upper bound for the capacitance. It is based on the assumption that the electrostatic energy W_e stored in a length Δz is stationary. It is

$$W_e = \frac{\Delta z}{2}\int_A \overline{E} \cdot (\varepsilon\overline{E})dA \qquad (3.99a)$$

$$= \frac{\Delta z}{2}\varepsilon\int_A (-\nabla\Phi) \cdot (-\nabla\Phi)dA \qquad (3.99b)$$

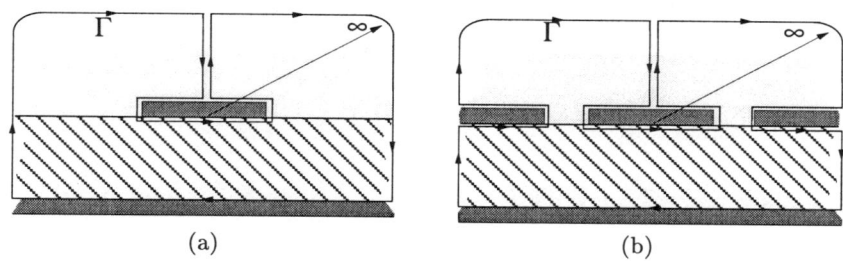

Fig. 3.13 *Integration paths over transverse section A for energetic variational principle (a) microstrip; (b) coplanar waveguide*

It is easily demonstrated that the first-order variation of W_e vanishes when Φ satisfies Laplace's equation. Varying indeed W_e and Φ in (3.99b) and applying Green's first identity (Appendix B) we obtain

$$\delta W_e = \Delta z \, \varepsilon \int_A (-\nabla\Phi) \cdot (-\nabla\delta\Phi) dA$$
$$= \Delta z \, \varepsilon \{ \int_A \delta\Phi(-\nabla^2\Phi) dA + \oint_\Gamma \delta\Phi(\nabla\Phi) \cdot \bar{n} d\Gamma \} \tag{3.100}$$

Contour Γ may be chosen such that it surrounds the various boundaries of transverse section A for a microstrip line (Fig. 3.13a) and for a coplanar waveguide (Fig. 3.13b). The line integrals at infinity vanish because of the decay of the fields and potentials at infinity. The contribution of the branch cuts is canceled, because the same values of potential and field are integrated in opposite directions and these branches are taken infinitely close to each other. On the conductor, only the tangential component of the exact electrostatic field $(-\nabla\Phi)$ vanishes. Hence, the only way to cancel the contribution of the total field is to impose $\delta\Phi$ being zero on the conductors. This is equivalent to imposing that the approximate solution satisfies the specified boundary conditions on the conductor. Under this assumption, the contribution of the line integral on contour Γ in (3.100) is zero. On the other hand, the actual electrostatic potential satisfies Laplace's equation in each layer, which demonstrates that the first-order error made on W_e vanishes, *i.e.* that W_e is stationary about the electrostatic potential. This proof is valid only when all the layers have the same permittivity, as assumed by Collin, who proposed the proof for a homogeneous medium. It is, however, possible to extend the proof to a non-homogeneous case, in the following manner. The electrostatic energy and its variation are still given by (3.99b) and (3.100) rewritten for each layer. Green's first identity is applied at the boundaries of each layer, for which a

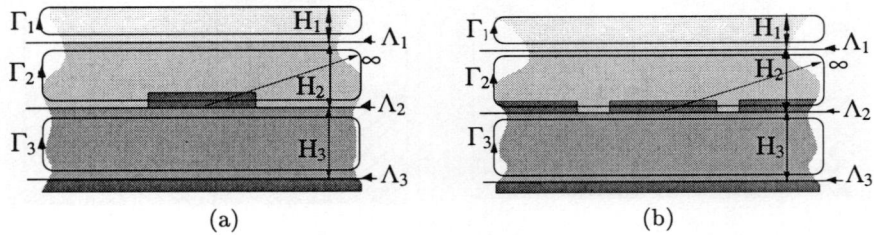

Fig. 3.14 *Integration paths for inhomogeneous energetic variational principle (a) microstrip; (b) coplanar waveguide*

line contour Γ_i is defined (Fig. 3.14):

$$W_{ei} = \frac{\Delta z}{2} \int_{A_i} \overline{E}_i \cdot (\varepsilon_i \overline{E}_i) dA_i \tag{3.101a}$$

$$= \frac{\Delta z}{2} \varepsilon_i \int_{A_i} (-\nabla \Phi_i) \cdot (-\nabla \Phi_i) dA_i \tag{3.101b}$$

$$\delta W_{ei} = \Delta z \; \varepsilon_i \{ \int_{A_i} \delta \Phi_i (-\nabla^2 \Phi_i) dA_i + \oint_{\Gamma_i} \delta \Phi_i (\nabla \Phi_i) \cdot \overline{n}_i d\Gamma_i \} \tag{3.101c}$$

The total variation of energy is the sum of the variations of energies contained in each layer, which yields

$$\delta W_e = \sum_{i=1}^{N} \delta W_{ei}$$

$$= \sum_{i=1}^{N} \Delta z \; \varepsilon_i \int_{A_i} \delta \Phi_i (-\nabla^2 \Phi_i) dA_i + \sum_{i=1}^{N} \Delta z \; \varepsilon_i \oint_{\Gamma_i} \delta \Phi_i (\nabla \Phi_i) \cdot \overline{n}_i d\Gamma_i$$

$$\tag{3.102}$$

The line integrals at infinity vanish because of the decay of the fields and potentials at infinity. The contribution of the line integrals at the interface between two adjacent layers i and $i+1$ may be rewritten as a line integral on this interface, with an associated line path denoted by Λ_i:

$$\delta W_e = \sum_{i=1}^{N} \delta W_{ei}$$

$$= \sum_{i=1}^{N} \Delta z \; \varepsilon_i \int_{A_i} \delta \Phi_i (-\nabla^2 \Phi_i) dA_i$$

$$+ \sum_{i=1}^{N} \Delta z \int_{\Lambda_i} \{ \delta \Phi_i(x, H_i) \, \varepsilon_i \nabla \Phi_i(x, H_i) \cdot \overline{n}_i$$

$$+ \delta \Phi_{i+1}(x, H_i) \, \varepsilon_{i+1} \nabla \Phi_{i+1}(x, H_i) \cdot \overline{n}_{i+1} \} d\Lambda_i \qquad (3.103)$$

$$= \sum_{i=1}^{N} \Delta z \; \varepsilon_i \int_{A_i} \delta \Phi_i (-\nabla^2 \Phi_i) dA_i$$

$$+ \sum_{i=1}^{N} \Delta z \int_{\Lambda_i} \Big[\delta \Phi_i(x, H_i) \{ \varepsilon_i \nabla \Phi_i(x, H_i) \}$$

$$- \delta \Phi_{i+1}(x, H_i) \{ \varepsilon_{i+1} \nabla \Phi_{i+1}(x, H_i) \} \Big] \cdot \overline{n}_i d\Lambda_i$$

The behavior of the integrands of the line integral has to be considered over two different areas. If no conductors are present between layers i and j, the normal component of the exact displacement field and the exact electrostatic potential are continuous on the two sides of this conductor-free area:

$$\{ \varepsilon_i \nabla \Phi_i(x, H_i) \} \cdot \overline{n}_i = \{ \varepsilon_{i+1} \nabla \Phi_{i+1}(x, H_i) \} \cdot \overline{n}_i \qquad (3.104a)$$

and

$$\Phi_i(x, H_i) = \Phi_{i+1}(x, H_i) \qquad (3.104b)$$

In this case the line integral is then rewritten as

$$\sum_{i=1}^{N} \Delta z \int_{\Lambda_i} [\delta \Phi_i(x, H_i) - \delta \Phi_{i+1}(x, H_i)] \{ \varepsilon_i \nabla \Phi_i(x, H_i) \} \cdot \overline{n}_i d\Lambda_i \qquad (3.105)$$

If a conductor is present on a portion of the interface between layers i and j, only the tangential component of the electrostatic field vanishes on it. Hence, no simplification of (3.103) occurs on the conductor. The only way to cancel the error arising from the line integral is to impose

$$\delta \Phi_i(x, H_i) = \delta \Phi_{i+1}(x, H_i) \qquad (3.106)$$

which implies the continuity of the trial potential on the conductor-free part of the interface. The exact value is continuous (3.104b). Equations (3.105) and (3.106) also imply that the trial solution has to be equal to the value of the exact potential on the conductor. Under these assumptions the contribution of the line integral on contour Γ in (3.100) is equal to zero. On the other hand, the exact electrostatic potential satisfies Laplace's equation in each layer, which demonstrates that the first-order error made on W_e vanishes, *i.e.* that W_e is stationary about the electrostatic potential. It must be emphasized that the stationary character of W_e is related to a fundamental concept of electrostatics. Thomson's theorem [3.47] states, indeed, that the charges present on conducting structures distribute themselves in such a way that the electrostatic energy is minimal. Hence, in some cases the variational principle is related to physical laws of the real world. One simply has to find a trial distribution for the potential, imposing that it has the value V_0 on one conductor and zero on the other, and that it is continuous at dielectric interfaces. Using then (3.99) and (3.101), a variational principle for the capacitance per unit length is obtained:

$$\frac{1}{2}CV_0^2 = \frac{1}{2}\sum_{i=1}^{N}\varepsilon_i\int_{A_i}(-\nabla\Phi_i)\cdot(-\nabla\Phi_i)dA_i \qquad (3.107)$$

Since the integral of the electrostatic energy density is always positive, Collin similarly states that (3.107) will always provide too large a value, *i.e.* an upper bound for the capacitance [3.48]. When the medium is homogeneous, the trial potential has only to be imposed equal to its exact value on the conductors.

3.4.1.4 Inductance based on energy

There is a duality between the equations of electrostatics and magnetostatics. It is interesting to investigate under which assumptions an expression similar to (3.99a) can be derived for the magnetostatic energy and proven to be variational. It is well-known that the electromagnetic power flowing into a volume V is related to the difference between two fundamental quantities, the time-average electric and magnetic energies contained in the volume:

$$W_e = \frac{1}{2}\int_V \overline{E}\cdot(\overline{D})^*dV \qquad (3.108a)$$

$$W_m = \frac{1}{2}\int_V \overline{B}\cdot(\overline{H})^*dV \qquad (3.108b)$$

Collin [3.47] states that these two quantities are positive functions, and proves it for the case of the electrostatic energy in the case of an isotropic homogeneous medium. For isotropic media, the constitutive parameters are scalar:

$$\overline{D} = \varepsilon\overline{E} \qquad (3.109a)$$

$$\overline{B} = \mu\overline{H} \qquad (3.109b)$$

Hence, the magnetostatic energy can be expressed either from a source-free scalar potential (3.90) or using the z-directed magnetic vector potential A_z

$$\overline{A} = \overline{a}_z A_z(x,y) \qquad (3.110)$$

yielding the magnetic flux density $\overline{B}$ via (3.80a). The magnetic flux density $\overline{B}$ is rewritten as

$$\overline{B} = -\overline{a}_z \times \nabla A_z \tag{3.111}$$

and the magnetic energy W_m in a section of line having a length Δz is

$$W_m = \Delta z \frac{1}{2} \int_A (-\{\overline{a}_z \times \nabla A_z\}) \cdot \{-\mu^{-1}\{\overline{a}_z \times \nabla A_z\}\} dA \tag{3.112a}$$

$$= \Delta z \frac{1}{2} \int_A (-\nabla A_z) \cdot \{\mu^{-1}(-\nabla A_z)\} dA \tag{3.112b}$$

where A is the cross-section of the line. Expression (3.112b) is valid because the fields and potentials are assumed to have no variation along the z-axis. Hence, if the medium is isotropic, the proof of the variational behavior leads to the following equation, obtained by replacing Φ by A_z in (3.100):

$$\delta W_m = \Delta z \mu^{-1} \int_A (-\nabla A_z) \cdot (-\nabla \delta A_z) dA \tag{3.113a}$$

$$= \Delta z \mu^{-1} \int_A \delta A_z(-\nabla^2 A_z) dA + \oint_\Gamma \delta A_z (\nabla A_z) \cdot \overline{n} d\Gamma \tag{3.113b}$$

Equation (3.111) is equivalent to

$$\nabla A_z = \overline{a}_z \times \overline{B} \tag{3.114}$$

Hence, the scalar product in (3.113b) is rewritten as

$$(\nabla A_z) \cdot \overline{n} = (\overline{a}_z \times \overline{B}) \cdot \overline{n} = (\overline{n} \times \overline{B}) \cdot \overline{a}_z \tag{3.115}$$

The actual magnetostatic vector potential A_z satisfies the source-free form of (3.79), and the surface integral in (3.113b) vanishes. Contour Γ may again be chosen as in Figure 3.13a,b, for a microstrip line and a coplanar waveguide. The line integrals at infinity vanish because of the decay of fields and potential at infinity. The contribution of the branch cuts cancel, because the same values of potential and field are integrated in opposite directions on branches which are close to each other. On the conductor, the exact actual magnetic flux density satisfies

$$\overline{n} \cdot \overline{B} = 0 \tag{3.116}$$

which implies that the only way to cancel the first-order error in the magnetic energy is to impose that the magnetic vector potential is equal to its exact value on the conductors. This is not a problem when remembering definition (3.83c), which shows that the difference of magnetic vector potential between the conductors is equal to the magnetic flux per unit length. Hence, since the magnetic energy is contained in the inductance, a variational principle for the inverse of the inductance of the line is obtained by imposing a value for the flux inside of the line:

$$W_m = \frac{1}{2}L\Delta z I^2 = \Delta z \frac{\Phi_0^2}{2L} \tag{3.117}$$

which finally yields

$$\frac{1}{L} = \frac{\displaystyle\int_A \nabla A_z \cdot \mu^{-1} \nabla A_z dA}{\Phi_0^2} \tag{3.118}$$

by virtue of (3.112b). Since the integral of the magnetic energy W_m is always positive, (3.118) provides a lower bound for the inductance per unit length.

When the medium is not magnetically homogeneous, the proof of stationarity is obtained similarly to the electrostatic inhomogeneous case. It can easily be shown that the variation (3.113b) of magnetic energy in each layer becomes

$$
\begin{aligned}
\delta W_m &= \sum_{i=1}^N \delta W_{mi} \\
&= \sum_{i=1}^N \Delta z \mu_i^{-1} \int_{A_i} \delta A_{zi}(-\nabla^2 A_{zi}) dA_i \\
&\quad + \sum_{i=1}^N \Delta z \int_{\Lambda_i} \{\delta A_{zi}(x, H_i)\mu_i^{-1} \nabla A_{zi}(x, H_i) \cdot \overline{n}_i \\
&\qquad + \delta A_{zi+1}(x, H_i)\mu_{i+1}^{-1} \nabla A_{zi+1}(x, H_i) \cdot \overline{n}_{i+1}\} d\Lambda_i \\
&= \sum_{i=1}^N \Delta z \mu_i^{-1} \int_{A_i} \delta A_{zi}(-\nabla^2 A_{zi}) dA_i \\
&\quad + \sum_{i=1}^N \Delta z \int_{\Lambda_i} \Big[\delta A_{zi}(x, H_i)\mu_i^{-1}\{\overline{n}_i \times \overline{B}_i(x, H_i)\}_z \\
&\qquad - \delta A_{zi+1}(x, H_i)\mu_{i+1}^{-1}\{\overline{n}_i \times \overline{B}_{i+1}(x, H_i)\}_z\Big] d\Lambda_i
\end{aligned}
\tag{3.119}
$$

where (3.115) has been used. Where no conductor exists, the tangential magnetic field is continuous:

$$\mu_i^{-1}\{\overline{n}_i \times \overline{B}_i(x, H_i)\} = \mu_{i+1}^{-1}\{\overline{n}_i \times \overline{B}_{i+1}(x, H_i)\} \tag{3.120}$$

and the line integral in (3.119) becomes

$$\sum_{i=1}^N \Delta z \int_{\Lambda_i} [\delta A_{zi}(x, H_i) - \delta A_{zi+1}(x, H_i)]\,\mu_i^{-1}\{\overline{n}_i \times \overline{B}_i(x, H_i)\}_z d\Lambda_i \tag{3.121}$$

This integral vanishes when the A_z-component of the trial vector potential is continuous at the conductor-free part of the interface. On the conductive part of the interface, the tangential magnetic field is no longer continuous. The only way to cancel the error produced by the line integral in (3.119) is to impose that the vector potential is constant and equal to its exact value on the conductor. This is equivalent to imposing that the A_z-component of the trial vector is continuous at any interface between layers. Its value on a conductor is put equal to a constant, which fixes again (by virtue of equation 3.83c) the value of the magnetic flux per unit length inside the line. Under such constraints, and keeping in mind that the exact vector potential satisfies Laplace's equation in each layer, the following variational principle is obtained for the inhomogeneous case:

$$\frac{1}{L} = \frac{\displaystyle\sum_{i=1}^{N} \int_{A_i} \nabla A_{zi} \cdot \mu_i^{-1} \nabla A_{zi} dA_i}{\Phi_0^2} \tag{3.122}$$

3.4.1.5   Generalization to non-isotropic media

Up to now the various explicit quasi-static principles have been proven variational under the assumption that the materials are characterized by scalar constitutive relations (3.109a,b). As a matter of fact, the various variational principles developed in the literature for planar lines are usually restricted to isotropic layers. More general constitutive relations, however, are

$$\overline{B} = \mu_0 \overline{\overline{\mu}}_r \cdot \overline{H} \tag{3.123a}$$

$$\overline{D} = \varepsilon_0 \overline{\overline{\varepsilon}}_r \cdot \overline{E} \tag{3.123b}$$

and the two static energetic formulations become, for homogeneous media:

$$W_e = \varepsilon_0 \frac{\Delta z}{2} \int_A \overline{E} \cdot (\overline{\overline{\varepsilon}}_r \cdot \overline{E}) dA \tag{3.124a}$$

$$= \varepsilon_0 \frac{\Delta z}{2} \int_A (-\nabla\Phi) \cdot \{\overline{\overline{\varepsilon}}_r \cdot (-\nabla\Phi)\} dA \tag{3.124b}$$

$$W_m = \frac{\Delta z}{2\mu_0} \int_A \overline{B} \cdot (\overline{\overline{\mu}}_r^{-1} \cdot \overline{B}) dA \tag{3.125a}$$

$$= \mu_0 \frac{\Delta z}{2} \int_A (-\nabla\Phi^M) \cdot \{\overline{\overline{\mu}}_r \cdot (-\nabla\Phi^M)\} dA \tag{3.125b}$$

Expression (3.125b) involves the scalar magnetic potential, because this is more convenient for dealing with the dyadic formulation. It only holds in areas with no current source. As in the previous examples, the various conductors are assumed to be of zero thickness, so that any current flow reduces to a current sheet at the interface between two layers. The layers are considered to be source-free. This is exactly the same concept as developed earlier for the Green's formalism, for which the integral equation may be solved either for a given source distribution and zero boundary conditions, or for a source-free situation with inhomogeneous boundary conditions. In the present case, the existence of current on the conductor is imposed as a boundary condition on a source-free form of the magnetostatic potential Φ^M. For both cases, the first-order variation yields

$$
\begin{aligned}
\delta W_e = \varepsilon_0 \frac{\Delta z}{2} \int_A \Big[&(-\nabla\delta\Phi) \cdot \{\bar{\bar{\varepsilon}}_r \cdot (-\nabla\Phi)\} \\
&+ (-\nabla\Phi) \cdot \{\bar{\bar{\varepsilon}}_r \cdot (-\nabla\delta\Phi)\} \Big] dA
\end{aligned}
\tag{3.126}
$$

$$
\begin{aligned}
\delta W_m = \mu_0 \frac{\Delta z}{2} \int_A \Big[&(-\nabla\delta\Phi^M) \cdot \{\bar{\bar{\mu}}_r \cdot (-\nabla\Phi^M)\} \\
&+ (-\nabla\Phi^M) \cdot \{\bar{\bar{\mu}}_r \cdot (-\nabla\delta\Phi^M)\} \Big] dA
\end{aligned}
\tag{3.127}
$$

In each case, there is a scalar product involving a tensor dyadic, which satisfies the following identity:

$$
\bar{a} \cdot (\bar{\bar{u}} \cdot \bar{b}) = \bar{b} \cdot (\bar{\bar{u}}^T \cdot \bar{a})
\tag{3.128}
$$

where superscript T denotes the transpose of the dyadic. Using this identity, (3.126) and (3.127) are rewritten as

$$
\delta W_e = \varepsilon_0 \frac{\Delta z}{2} \int_A \Big[(-\nabla\delta\Phi) \cdot \{(\bar{\bar{\varepsilon}}_r + \bar{\bar{\varepsilon}}_r^T) \cdot (-\nabla\Phi)\} \Big] dA
\tag{3.129}
$$

$$
\delta W_m = \mu_0 \frac{\Delta z}{2} \int_A \Big[(-\nabla\delta\Phi^M) \cdot \{(\bar{\bar{\mu}}_r + \bar{\bar{\mu}}_r^T) \cdot (-\nabla\Phi^M)\} \Big] dA
\tag{3.130}
$$

Making use of identity

$$
\bar{a} \cdot \nabla f = \nabla \cdot (f\bar{a}) - f\nabla \cdot \bar{a}
\tag{3.131}
$$

(3.129) and (3.130) transform respectively into

$$
\begin{aligned}
\delta W_e = \varepsilon_0 \Delta z \Big[&\int_A (\delta\Phi)\nabla \cdot \{(\bar{\bar{\varepsilon}}_r + \bar{\bar{\varepsilon}}_r^T) \cdot (-\nabla\Phi)\} dA \\
&- \oint_\Gamma (\delta\Phi)\{(\bar{\bar{\varepsilon}}_r + \bar{\bar{\varepsilon}}_r^T) \cdot (-\nabla\Phi)\} \cdot \bar{n}d\Gamma \Big]
\end{aligned}
\tag{3.132}
$$

$$
\begin{aligned}
\delta W_m = \mu_0 \Delta z \Big[&\int_A (\delta\Phi^M)\nabla \cdot \{(\bar{\bar{\mu}}_r + \bar{\bar{\mu}}_r^T) \cdot (-\nabla\Phi^M)\} dA \\
&- \oint_\Gamma (\delta\Phi^M)\{(\bar{\bar{\mu}}_r + \bar{\bar{\mu}}_r^T) \cdot (-\nabla\Phi^M)\} \cdot \bar{n}d\Gamma \Big]
\end{aligned}
\tag{3.133}
$$

When the tensors are simply symmetrical, that is anisotropy without gyrotropy, one has

$$\overline{\overline{u}} = \overline{\overline{u}}^T \tag{3.134}$$

and (3.132) and (3.133) respectively simplify into

$$
\begin{aligned}
\delta W_e &= 2\varepsilon_0 \Delta z \left[\int_A (\delta\Phi)\nabla \cdot \{\overline{\overline{\varepsilon}}_r \cdot \nabla\Phi\} dA \right. \\
&\quad \left. - \oint_\Gamma \delta\Phi\{\overline{\overline{\varepsilon}}_r \cdot \nabla\Phi\} \cdot \overline{n}d\Gamma \right] \\
&= 2\varepsilon_0 \Delta z \left[\int_A (\delta\Phi)(\nabla \cdot \overline{D}) dA \right. \\
&\quad \left. - \oint_\Gamma \delta\Phi\overline{D} \cdot \overline{n}d\Gamma \right]
\end{aligned} \tag{3.135}
$$

$$
\begin{aligned}
\delta W_m &= 2\mu_0 \Delta z \left[\int_A (\delta\Phi^M)\nabla \cdot \{\overline{\overline{\mu}}_r(\nabla\Phi^M)\} dA \right. \\
&\quad \left. - \oint_\Gamma (\delta\Phi^M)\{\overline{\overline{\mu}}_r \cdot \nabla\Phi^M\} \cdot \overline{n}d\Gamma \right] \\
&= 2\mu_0 \Delta z \left[\int_A (\delta\Phi^M)\nabla \cdot \overline{B} dA \right. \\
&\quad \left. - \oint_\Gamma (\delta\Phi^M)\overline{B} \cdot \overline{n}d\Gamma \right]
\end{aligned} \tag{3.136}
$$

where relationships (3.123a,b) have been used. The exact displacement field $\overline{D}$ satisfies the charge-free divergence equation

$$\nabla \cdot \overline{D} = 0 \tag{3.137}$$

and its normal component does not vanish on the conductors. It is found again that the only way to cancel the first-order error made on W_e in the case of non-gyrotropic anisotropic dielectric media is to impose an exact, constant value for the potential on the conductor, as detailed previously. As a consequence, the following expression is a variational principle for the capacitance per unit length in the case of anisotropic media whose permittivity tensor satisfies (3.134):

$$\frac{1}{2}CV_0^2 = \frac{\varepsilon_0}{2} \int_A (-\nabla\Phi) \cdot \{\overline{\overline{\varepsilon}}_r \cdot (-\nabla\Phi)\} dA \tag{3.138}$$

For the magnetostatic energy, the magnetic flux density $\overline{B}$ satisfies the charge-free divergence equation

$$\nabla \cdot \overline{B} = 0 \tag{3.139}$$

and its normal component vanishes on the conductor. This is why the magnetostatic energy (3.125b) has been proven to be variational about the scalar magnetic potential Φ^M. Keeping in mind the previous development leading to expressions (3.90) and (3.93c), it is possible to relate the potentials to an equivalent current density J_z flowing on the strip. Assuming a given current I on the strip and using (3.82) makes W_m proportional to I. Hence, a variational expression is deduced for the inductance per unit length:

$$L = \mu_0 \frac{\int_A (-\nabla \Phi^M) \cdot \{\overline{\overline{\mu}}_r \cdot (-\nabla \Phi^M)\} dA}{I^2} \tag{3.140}$$

This variational principle provides a lower bound for the inductance. Obviously, the two derivations (3.138) and (3.140) may be extended to the case of multilayered anisotropic and inhomogeneous structures, following the reasoning of the previous subsection.

We conclude this discussion by underlining that variational principles can usually be easily derived when symmetrical tensors characterize the materials involved or when the Green's function has specific symmetry properties. When this is not the case, that is when the medium is gyrotropic or non-Hermitian, the derivation by assuming simple trial fields or potentials may be impossible. This has been observed by a number of authors. Formulas derived from Rumsey [3.49] for closed waveguides are variational, provided the permittivity and permeability are, at most, symmetric tensors. This means that isotropic materials and materials with crystalline anisotropy, lossy or lossless, are allowed while gyrotropic media such as magnetized devices and magneto-ionic media are excluded. On the other hand, Berk [3.50] developed variational formulas applicable to gyrotropic media or, in general, media whose dielectric constant and permeability are Hermitian tensors, restricted, however, to lossless materials. All those formulations were limited to closed waveguides and resonators. More recently, Jin and Chew [3.51] have studied the conditions required for a variational formulation of electromagnetic boundary conditions involving anisotropic media. They found that the formulation is variational and can be expressed as a function of electric field and its complex conjugate. This is possible only in the presence of lossy anisotropic media having symmetric permittivity and permeability tensors, or in the presence of lossless anisotropic media having Hermitian permittivities and permeabilities. The functional is interesting for the application of the finite-element method to structures containing anisotropic materials. When the medium is non-Hermitian, however, the problem must be solved both for the electric field and for an adjoint field satisfying Maxwell's equations involving the transpose conjugate of the constitutive tensors. This is because the functional is a function of the field and of its adjoint. A similar conclusion will be obtained in the following section when investigating the variational behavior of implicit variational methods.

3.4.1.6 Link with eigenvalue formalism

Expressions (3.16) and (3.66), and (3.81) and (3.95), are variational principles for the capacitance and the inductance of the line, respectively. This can be shown quite easily by observing that the problem they solve is in fact closely related to a linear expression involving an integral operator. The definition of capacitance is

$$Q = CV_0 \tag{3.141}$$

If charge density and potential are related by the Green's formalism (3.20), (3.141) is rewritten using (3.20), (3.18) and (3.22b) as

$$Q_l = \int_{-\frac{W_c}{2}}^{\frac{W_c}{2}} \rho_s(x)dx = C \int_{-\frac{W_c}{2}}^{\frac{W_c}{2}} G_P(x,0|x',0)\rho_s(x')dx' \tag{3.142}$$

which is equivalent to the following eigenvalue formalism with two linear operators $\mathcal{L}$ and $\mathcal{M}$:

$$\mathcal{L}(\mathbf{\Psi}) = \lambda \mathcal{M}(\mathbf{\Psi}) \tag{3.143}$$

As introduced in Chapters 1 and 2, there is a general variational principle for the eigenvalue λ in the case of self-adjoint operators $\mathcal{L}$ and $\mathcal{M}$ [3.52]:

$$\lambda = \frac{\int \mathbf{\Psi} \mathcal{L}(\mathbf{\Psi})dS}{\int \mathcal{M}(\mathbf{\Psi})\mathbf{\Psi}dS} \tag{3.144}$$

Replacing $\mathcal{M}$ in this expression by integral (3.18) and $\mathcal{L}$ by the Green's formalism (3.20) and (3.22b), the result will yield the previous expressions (3.16) and (3.66).

For the variational principles (3.81) and (3.95) for inductances, the corresponding eigenvalue problems are, respectively

$$\Phi_0 = \int_\Gamma G_{W,zz}(x,0|x',0)J_z(x',0)dx'$$
$$= \lambda I = \lambda \int_\Gamma J_z(x,0)dx \quad \text{with} \quad \lambda = L \tag{3.145}$$

$$I = \int_{-A}^x \int_{-\infty}^\infty G_{M,zy}(x',0|x'',0)B_y(x'',0)dx''dx'$$
$$= \lambda \Phi_0 = \lambda \int_{\frac{W_c}{2}}^A B_y(x,0)dx \tag{3.146a}$$

with

$$\lambda = \frac{1}{L} \quad \text{and} \quad x \geq \frac{W}{2} \tag{3.146b}$$

3.4.2 Implicit variational methods for a full-wave analysis

The possible variational character of full-wave dynamic methods is closely related to their implicit nature. Most of them are implicit non-standard eigenvalue problems, related to the reaction concept and to the general theory of eigenvalue problems. The discussion concentrates first on the moment method, whose variational behavior may be studied using these three formalisms: implicit non-standard eigenvalue problems, reaction concept, and general eigenvalue problems. Since the implicit method yields a transcendental equation to be solved for the unknown propagation constant, it can first be studied following the non-standard eigenvalue formalism presented by Lindell [3.21].

3.4.2.1 Galerkin's determinantal equation

The general presentation of the moment method in Chapter 2 established that (3.54a,b) minimizes the error ε made on $\mathcal{L}_{st}(\bar{j}_t(x,0))$ (3.54a) or $\mathcal{L}_{sl}(\bar{e}_t(x,0))$ (3.54b). It is interesting to determine what error is made on γ when the determinantal equation (3.57a,b) is solved for a given set of basis functions satisfying (3.50b) or (3.51b). The discussion is adapted to the case of the slot-line (3.57b), a similar reasoning holding for other lines. Harrington [3.53] states that using Galerkin's procedure is equivalent to applying the Rayleigh-Ritz procedure to the error ε caused by applying the linear integral operator to the trial quantity instead of applying it to the exact quantity. Harrington expresses this error as

$$\varepsilon = \langle \bar{f}, \mathcal{L}(\bar{f}) \rangle - \langle \bar{f}, \bar{g} \rangle \qquad (3.147)$$

where the inner product is defined according to (3.52a). Hence, applying Galerkin's procedure is equivalent to minimizing the error, by virtue of the definition of the Rayleigh-Ritz procedure. This means that when the exact value $\bar{f}_0$ is replaced by a trial one, such as

$$\bar{f} = \bar{f}_0 + \delta\bar{f} \qquad (3.148)$$

the first-order error made on ε is zero. For the exact solution $(\bar{f}_0, \gamma_0)$ corresponding to the exact fields, these fields satisfy the boundary conditions expressed as integral equation (3.52a,b):

$$\langle \bar{f}_0, g \rangle = \langle \bar{f}_0, \mathcal{L}_0(\bar{f}_0) \rangle = \langle \bar{e}_{t0}(x,0), \mathcal{L}_{sl_0}(\bar{e}_{t0}(x,0)) \rangle = 0 \qquad (3.149a)$$

so that the corresponding exact value for ε is zero:

$$\varepsilon_0 = 0 \qquad (3.149b)$$

The subscript 0 for $\mathcal{L}_{sl}$ or $\mathcal{L}$ indicates that values of the Green's function are taken for the exact value γ_0. This is equivalent to the notation $\mathcal{L}_{sl}(\bar{e}_t(x,0); \gamma_0)$ in expressions (3.58a,b). Since the error ε is made stationary about $\bar{f}$ by

Galerkin's procedure, it also vanishes (to the second-order) in the presence of the trial $\overline{f}$, which imposes

$$\begin{aligned}
\varepsilon &= \langle \overline{f}, \mathcal{L}_0(\overline{f}) \rangle - \langle \overline{f}, g \rangle = \langle \overline{f}, \mathcal{L}_0(\overline{f}) \rangle - \langle \overline{f}, \mathcal{L}_0(\overline{f_0}) \rangle \\
&= \langle \overline{e}_t(x,0), \mathcal{L}_{sl_0}(\overline{e}_t(x,0)) \rangle - \langle \overline{e}_t(x,0), \mathcal{L}_{sl_0}(\overline{e}_{t0}(x,0)) \rangle = 0
\end{aligned} \tag{3.150a}$$

Expanding the trial field and neglecting second-order variations yields

$$\varepsilon = \langle \overline{e}_{t0}(x,0), \mathcal{L}_{sl_0}\{\delta \overline{e}_t(x,0)\} \rangle \tag{3.150b}$$

So that (3.150a) finally reduces to

$$\langle \overline{e}_{t0}(x,0), \mathcal{L}_{sl_0}\{\delta \overline{e}_t(x,0)\} \rangle = 0 \tag{3.151}$$

Considering the basis of Galerkin's procedure (3.52) applied to the integral equation for the boundary conditions, and developing its left-hand side around the exact solution, neglecting second-order variations, yields:

$$\begin{aligned}
\langle \overline{e}_t(x,0), \mathcal{L}_{sl}(\overline{e}_t(x,0)) \rangle &= \langle \overline{e}_{t0}(x,0), \mathcal{L}_{sl_0}(\overline{e}_{t0}(x,0)) \rangle \\
&+ \langle \delta \overline{e}_t(x,0), \mathcal{L}_{sl_0}(\overline{e}_{t0}(x,0)) \rangle + \langle \overline{e}_{t0}(x,0), \mathcal{L}_{sl_0}(\delta \overline{e}_t(x,0)) \rangle \\
&+ (\delta\gamma)\langle \overline{e}_{t0}(x,0), \frac{\partial \mathcal{L}_{sl_0}}{\partial \gamma}|\gamma_0(\overline{e}_{t0}(x,0)) \rangle = 0
\end{aligned} \tag{3.152}$$

If operator $\mathcal{L}$ is self-adjoint, one has

$$\langle \overline{e}_{t0}(x,0), \mathcal{L}_{sl_0}(\delta \overline{e}_t(x,0)) \rangle = \langle \delta \overline{e}_t(x,0), \mathcal{L}_{sl_0}(\overline{e}_{t0}(x,0)) \rangle \tag{3.153}$$

which, combined with (3.151), implies that only the term containing the first variation $(\delta\gamma)$ is *a priori* non-zero in the right-hand side of (3.152). Since the left-hand side of this equation is imposed to be zero by the formulation, it is concluded that the *first-order error made on γ using Galerkin's procedure with a self-adjoint operator is zero*. This also means that, to the second-order, the functional ρ that Harrington associates to the moment method and defines in the case of a self-adjoint operator as [3.54]

$$\rho = \frac{\langle \overline{f}, \overline{g} \rangle \langle \overline{f}, \overline{g} \rangle}{\langle \mathcal{L}(\overline{f}), \overline{f} \rangle} \tag{3.154a}$$

vanishes, which means

$$\langle \overline{e}_t(x,0), \mathcal{L}_{sl_0}(\overline{e}_{t0}(x,0)) \rangle = 0 \tag{3.154b}$$

This is equivalent to say that ρ is stationary about $\overline{e}_t(x,0)$. Equation (3.154b) also implies, by virtue of (3.150a),

$$\langle \overline{e}_t(x,0), \mathcal{L}_{sl_0}(\overline{e}_t(x,0)) \rangle = 0 \tag{3.154c}$$

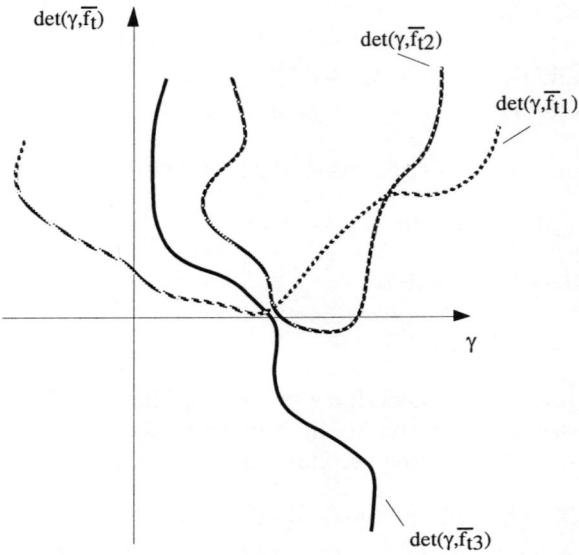

Fig. 3.15 *Possible behavior of determinantal equation as a function of propagation constant γ for various choices of trial quantity*

The self-adjoint character of the integral operator $\mathcal{L}$ - see equations (3.38a,b) - is closely related to the reciprocity and symmetry properties of the Green's dyadic function used for formulating the integral equation for boundary conditions. This concept will be clarified in Section 3.4.2.3.

It should be underlined that the previous development is an attempt to show that the solution found with Galerkin's procedure is variational about a variation of the basis functions. The method, however, still needs to solve the implicit equations (3.57a,b), which requires evaluation of the determinantal equation several times. This is because the method provides no information about the behavior of the implicit equation with respect to γ. At this stage of the discussion, there is no explicit variational principle available which would express γ as a ratio of integrals. The only feature highlighted by the previous discussion is that, whatever trial fields are present in the line, the product $\langle \overline{f}, \overline{g} \rangle$ crosses the γ axis at about the value γ_0 (which is the correct one, to the second-order) (Fig. 3.15). This, however, does provide neither information about the slope of the product $\langle \overline{f}, \overline{g} \rangle$ with respect to γ nor a guideline about the most efficient way to solve equation (3.58b):

$$\langle \overline{f}, \overline{g} \rangle (\gamma) = \langle \mathcal{L}_{sl}(\overline{e}_t(x,0); \gamma), \overline{e}_t(x,0) \rangle = 0 \tag{3.155}$$

Solving this kind of equation may be very difficult, as mentioned by Davies

[3.55], particularly when spurious solutions may occur [3.56]. It requires specific numerical techniques, especially when losses are present and evanescent modes are considered [3.57][3.58]. For these reasons, the method has usually been limited to lossless applications on dielectric multilayered media, as mentioned by Jansen [3.59].

3.4.2.2 Equivalent approach based on reaction concept

In Chapter 2, the reaction concept introduced by Rumsey has been demonstrated to be efficient for obtaining variational principles on a wide variety of electromagnetic parameters. Using the notations of Chapter 2, two sources located in different areas of space and noted a and b generate their associated fields denoted by A and B, respectively. Hence, the reaction of field A on source b has the form

$$\langle A, b \rangle = \int \{ \overline{E}^a \cdot \overline{dJ}^b - \overline{H}^a \cdot \overline{dM}^b \} \tag{2.31a}$$

Lorentz reciprocity principle is then rewritten using the reaction concept for isotropic media:

$$\langle A, b \rangle = \langle B, a \rangle \tag{2.32}$$

Equation (2.32) states that the reaction of field A on source b is equivalent to the reaction of field B on source a. When the medium is anisotropic, Rumsey proposes a correction to (2.32), attributed to Cohen [3.49]:

$$\langle A, b \rangle = \langle \hat{B}, a \rangle \tag{2.33}$$

where $\langle \hat{B}, a \rangle$ is written for the case corresponding to the same sources as $\langle A, b \rangle$ but where the fields generated by source b are solution of Maxwell's equations for a medium with transposed permittivity and permeability tensors.

It is demonstrated in Chapter 2 that the reaction $\langle A, b \rangle$ is variational about A and b provided that

$$\langle A, b \rangle = \langle A_0, b \rangle = \langle A, b_0 \rangle \tag{2.38}$$

In the case of self-reaction, assuming that the trial source and associated field generated by the source are functions of a parameter p and knowing a *priori* that the value of the actual reaction $\langle A_0, a_0 \rangle$ is zero, it is demonstrated that by removing the trial reaction $\langle A, a \rangle$ parameter p becomes stationary about the trial source and associated field:

$$\langle A, a \rangle = \langle A_0, a_0 \rangle = 0 \tag{2.43}$$

Referring to the characteristic equations of the moment method using Galerkin's procedure, it is now obvious that the left-hand side of equations (3.52a,b) are indeed self-reactions of the trial source of current and associated electric field, which are bounded by the Green's function. Since the exact value of the reaction is known *a priori* to be zero, (2.43) is applicable: imposing (2.43) on the trial reaction ensures that the first-order error made on the parameter $p = \gamma$ is zero, provided that reactions formed using both trial and exact quantities satisfy (2.38). In the case of Galerkin's procedure, cancelling (3.58b) ensures first that (2.43) is satisfied by the trials, while the exact fields are known to exist only on complementary areas of the domain x of the structure:

$$\langle A, a \rangle = \langle \mathcal{L}_{sl}(\overline{e}_t(x,0); \gamma), \overline{e}_t(x,0) \rangle = 0 \qquad (3.156a)$$

$$\langle A_0, a_0 \rangle = \langle \mathcal{L}_{sl_0}(\overline{e}_{t0}(x,0); \gamma_0), \overline{e}_{t0}(x,0) \rangle = 0 \qquad (3.156b)$$

Condition (2.38), rewritten for self-reaction, is partially ensured by the variational character of functional ε, which induces (3.151) by virtue of (3.156b):

$$\langle A, a \rangle = \langle A_0, a \rangle = \langle \overline{e}_{t0}(x,0), \mathcal{L}_{sl_0}(\overline{e}_t(x,0)) \rangle = 0 \qquad (3.157)$$

To complete the relationship between the reaction formalism and Galerkin's implicit procedure, Galerkin's formalism has to satisfy the right-hand part of equation (2.38):

$$\langle A_0, a \rangle = \langle \overline{e}_{t0}(x,0), \mathcal{L}_{sl_0}(\overline{e}_t(x,0)) \rangle \qquad (3.158a)$$

$$= \langle A, a_0 \rangle \qquad (3.158b)$$

$$= \langle \overline{e}_t(x,0), \mathcal{L}_{sl_0}(\overline{e}_{t0}(x,0)) \rangle \qquad (3.158c)$$

The right-hand sides of (3.158a) and (3.158c) are equal if the linear integral operator is self-adjoint, as obtained in (3.153):

$$\langle \overline{e}_t(x,0), \mathcal{L}_{sl_0}(\overline{e}_{t0}(x,0)) \rangle = \langle \overline{e}_{t0}(x,0), \mathcal{L}_{sl_0}(\overline{e}_t(x,0)) \rangle \qquad (3.159)$$

If conditions (3.40a) and (3.51b) for the trial field and associated source in the case of slot-like lines are satisfied by the exact field, the right-hand side of (3.158c) vanishes. Hence, since the cancellation of the right-hand side of (3.157) has been demonstrated to be equivalent to stating that error ε is stationary, it follows that ε is stationary if the operator is self-adjoint. This was not explicitly stated by Harrington [3.53]. Equation (3.159) also means that a reciprocity relationship must exist between the set $\{\mathcal{L}_{sl_0}(\overline{e}_{t0}(x,0), \overline{e}_t(x,0))\}$ and the set $\{\mathcal{L}_{sl_0}(\overline{e}_t(x,0), \overline{e}_{t0}(x,0))\}$. Hence, the symmetry and reciprocity properties of the Green's function acting as an integral linear operator finally determines the possible variational character of the spectral domain Galerkin's procedure. This depends, of course, on the (non-)isotropic character of the media. The properties of the Green's function are investigated in the next section.

3.4.2.3 Conditions for the Green's function

As we have seen, the variational character of both the propagation constant obtained using Galerkin's procedure and of the quasi-static parameters depends on the reciprocity properties of the Green's function relating the tangential current to the electric field at an interface. An interesting review of these properties is presented by Barkeshli and Pathak [3.60]. They report that the free-space dyadic Green's function has been investigated by Collin [3.61], Collin and Zucher [3.62], Tai [3.63] and others. It has been shown that the free-space dyadic Green's function satisfies reciprocity as well as symmetry relationships (Appendix A):

$$\overline{\overline{G}}(\overline{r}|\overline{r}_0) = \overline{\overline{G}}(\overline{r}_0|\overline{r}) \tag{A.37a}$$

with

$$\overline{\overline{G}} = \overline{\overline{G}}^T \tag{A.37b}$$

The additional symmetry property in free-space (*i.e.*, free-space dyadic equal to its transpose) derives from the fact that for an unbounded medium interchanging the locations of source and field points does not modify any existing boundary condition problem, because of the symmetry of the vacuum or free-space unbounded medium. In this simplified case, relation (A.37a) is valid. For other configurations, however, new boundary conditions problems are introduced, generally when source and field locations are interchanged. Overall, the Green's dyadic does not satisfy the symmetry relation (A.37a), even if it satisfies certain reciprocity relations. One of them is derived by Collin [3.64] from the reciprocity theorem (2.32) - in the case of a bounded homogeneous medium (assuming that no magnetic sources are present). Introducing the Green's dyadic operator as

$$\overline{P}(\overline{r},t) = \overline{\overline{\mathcal{G}}}\{\overline{X}(\overline{r}',t')\} \tag{A.17b}$$

with

$$P_u = \sum_v [\overline{\overline{\mathcal{G}}}\{\overline{X}(\overline{r}',t')\}]_{uv} \tag{A.17c}$$

$$P_u = \sum_v \int_{V_x} \int_{T_x} G_{uv}(\overline{r},t|\overline{r}',t')X_v(\overline{r}',t')dr'dt' \tag{A.17d}$$

into (2.32) yields

$$\int_V d\bar{J}^b(\bar{r}') \cdot [\overline{\overline{G}}\{\bar{J}^a(\bar{r})\}](\bar{r}')$$

$$= \int_V d\bar{J}^a(\bar{r}) \cdot [\overline{\overline{G}}\{\bar{J}^b(\bar{r}')\}](\bar{r}) \tag{3.160}$$

in which both sides are general forms for the right-hand sides of equations (3.52a,b), for instance. This equation has to be satisfied whatever sources a and b are, but one particular case is for unit point sources a and b located respectively at $\bar{r}_0$ and $\bar{r}_0'$ and having an arbitrary orientation. Assuming the presence of source points

$$\bar{J}^a(\bar{r}) = \bar{j}^a \delta(\bar{r} - \bar{r}_0)$$

$$\bar{J}^b(\bar{r}) = \bar{j}^b \delta(\bar{r} - \bar{r}_0')$$

(3.160) is rewritten as

$$\bar{j}^a \cdot \{\overline{\overline{G}}(\bar{r}_0|\bar{r}_0') \cdot \bar{j}^b\} = \bar{j}^b \cdot \{\overline{\overline{G}}(\bar{r}_0'|\bar{r}_0) \cdot \bar{j}^a\} \tag{3.161}$$

which implies, by application of the dyadic identity (3.128),

$$\overline{\overline{G}}(\bar{r}_0|\bar{r}_0') = \overline{\overline{G}}^T(\bar{r}_0'|\bar{r}_0) \tag{3.162}$$

Similarly Tai [3.63] presented reciprocity relations for the isotropic homogeneous half-space dyadic Green's function. On the other hand, Felsen and Marcuvitz [3.65] have investigated the reciprocity relationships for time-harmonic electromagnetic fields in spatially varying anisotropic media: they established a set of general reciprocity relationships for anisotropic and its associated transposed medium. Moreover Papayannakis *et al.* [3.66] have shown a more general form of the Love equivalence theorem applied to a finite-size space. It can be expected that, for media with anisotropic properties, a similar development is applicable, starting from the modified reciprocity theorem (2.33):

$$\bar{j}^a \cdot \{\overline{\overline{G}}(\bar{r}_0|\bar{r}_0') \cdot \bar{j}^b\} = \bar{j}^b \cdot \{\overline{\overline{G'}}(\bar{r}_0'|\bar{r}_0) \cdot \bar{j}^a\} \tag{3.163}$$

which implies

$$\overline{\overline{G}}(\bar{r}_0|\bar{r}_0') = \overline{\overline{G'}}^T(\bar{r}_0'|\bar{r}_0) \tag{3.164}$$

where the prime notation for the dyadic on the right-hand side of (3.163) and (3.164) refers to the Green's function established between the current source and the electric field in the transposed medium.

Up to now, and to the best of our knowledge, there is no satisfactory condition proving that the 2x2 dyadic Green's function involved in Galerkin's method used for planar line analysis satisfies the usual reciprocity theorem for arbitrary vector fields $\overline{A}$ and $\overline{B}$:

$$\langle \overline{A}, \mathcal{L}_{sl_0}(\overline{B}) \rangle = \langle \overline{B}, \mathcal{L}_{sl_0}(\overline{A}) \rangle \tag{3.165}$$

and for any medium involved in the planar layers. It is suspected that when lossy gyrotropic media are involved, equation (3.162) may not be valid anymore, because the Green's function acting as linear integral operator no longer satisfies (A.37a) or (3.160), and the variational character of the solution γ of (3.52) with respect to trials in the partially metallized interface of the planar lines disappears.

3.4.2.4 Link with explicit variational eigenvalue problems

The explicit quasi-static variational principles (3.16), (3.81) and (3.95) are related to an explicit eigenvalue formalism. They have a corresponding determinantal form, as explained by Schwinger [3.67]. Starting again from the general form (3.144), the Rayleigh-Ritz procedure is applied to the function Ψ expanded into

$$\Psi = \sum_{n=1}^{N} \alpha_n \Psi_n$$

Then (3.144) is rewritten for the three cases:

$$\lambda = \frac{\displaystyle\sum_{n=1}^{N} \sum_{m=1}^{M} \alpha_n \alpha_m \int \Psi_n \mathcal{L}(\Psi_m) dS}{\displaystyle(\sum_{n=1}^{N} \alpha_n \int \Psi_n dS)^2} \tag{3.166}$$

Since λ calculated by (3.144) is variational, taking the partial derivative with respect to each α_n and cancelling it yields a homogeneous system of N equations for the α_m:

$$\frac{\partial \lambda}{\partial \alpha_k} = 2 \frac{\displaystyle\sum_{m=1}^{M} \alpha_m \int \Psi_k \mathcal{L}(\Psi_m) dS - \lambda(\int \Psi_k dS)(\sum_{n=1}^{N} \alpha_n \int \Psi_n dS)}{\displaystyle(\sum_{n=1}^{N} \alpha_n \int \Psi_n dS)^2} \tag{3.167}$$

$$= 0$$

for $k = 1, \ldots, N$. This system has a non-trivial solution if, and only if, the determinant of its characteristic matrix vanishes, which yields the characteristic determinantal equation associated to those explicit variational principles:

$$\det \left[\int \Psi_n \mathcal{L}(\Psi_m) dS - \lambda(\int \Psi_n dS)(\int \Psi_m dS) \right] = 0$$

which is equivalent to

$$\det \left[\frac{\int \Psi_n \mathcal{L}(\Psi_m) dS}{(\int \Psi_n dS)(\int \Psi_m dS)} - \lambda \right] = 0 \tag{3.168}$$

Hence, it is possible to obtain a determinantal expression for the parameter under consideration. As underlined by Schwinger [3.67], however, the equation that is required is neither a polynomial expression of degree N in λ, nor an eigenvalue problem associated with the matrix. Indeed subtracting the first row from all the rest yields a matrix with λ present only in the first row. Hence the result of the determinant computation is a linear expression in λ. "It is also evident", says Schwinger, "that one cannot solve (3.168) rigorously. However, in view of the stationary nature of λ we do know that the error in the value of λ produced by an incorrect field or current function will be proportional not to the first power of the deviation of the field for the correct value, but to the square of the deviation. Thus, if a field or current function is chosen judiciously, the variational principle can yield remarkably accurate results with relatively little labor. Unfortunately, the ability to choose good trial functions generally comes only with experience" [3.67].

It is also to be pointed out that applying the Rayleigh-Ritz procedure to the general eigenvalue variational principle for calculating quasi-static parameters is equivalent to applying Galerkin's procedure of the moment method to the definitions (3.142), (3.145), and (3.146a) corresponding to these quasi-static parameters. Multiplying both sides of these equations by the appropriate form of the function Ψ and integrating the results is indeed applying Galerkin's procedure, but the result of this operation precisely yields the general variational principle (3.144). Hence, in the case of the standard explicit eigenvalue problem, applying Galerkin's procedure to the definition of the parameter under consideration is a variational method, because it is equivalent to applying the Rayleigh-Ritz procedure to a variational principle established for this parameter. However, the determinantal equation to be solved will provide a linear equation to be solved for the parameter, hence an explicit form for this parameter. This point has been mentioned by Collin for the quasi-static capacitance [3.43].

To conclude this section about quasi-static formulations, the comment by Schwinger also implies that, when choosing a judicious infinite series for the trial quantity, the Rayleigh-Ritz procedure finds an exact distribution for the quantity, provided that the trial fields derived from this trial quantity fulfill all the other boundary conditions satisfied by the actual exact fields and potentials. Since the proof of the variational behavior of the quasi-static formulations derived in the previous section usually does not imply specific boundary conditions on the trials, the trial fields used in the quasi-static variational principles usually do not satisfy all the boundary conditions of the actual field, and even an infinite summation for the trial quantity does not guarantee that exact fields are obtained.

3.4.3 Variational methods for obtaining accurate fields

Galerkin's procedure minimizes the error functional ε on the trial $\overline{f}$ with respect to the actual distribution solution of the problem

$$\mathcal{L}(\overline{f}_0) = \overline{g} \tag{3.169}$$

As seen previously, this is equivalent to saying that Gakerkin's procedure minimizes the functional ρ (3.154) and the self-reaction (2.43). This has been confirmed by Richmond [3.68] and Harrington [3.53]. Peterson *et al.* [3.69], however, have shown that non-Galerkin's moment methods are also able to minimize this error, even when the linear integral operator is not self-adjoint. This means that Galerkin's method may exhibit no significant advantages when compared to non-Galerkin's moment methods. Our attention, however, is focused on the commonly used analysis techniques for planar lines, so we start from Galerkin's procedure and the assumption of Richmond [3.68].

For planar lines, equation (3.169) has to be solved for a very reduced portion of the space, namely the area of the strip or of the slot $|x| \leq \frac{W}{2}$, where $\mathcal{L}(\overline{f})$ has to vanish. The proper choice of the basis functions ensures that integrating the product $\overline{f} \cdot \mathcal{L}(\overline{f})$ on the area ($|x| \leq \frac{W}{2}$) and making it vanish is equivalent to integrating it over the whole x-domain. In doing this, however, the user is only certain, after solving for the approximate field or current distribution $\overline{f}$, that the integral of the product $\overline{f} \cdot \mathcal{L}(\overline{f})$ on the area $|x| \leq \frac{W}{2}$ vanishes. Nothing guarantees that the quantity provided by the Green's relationship vanishes, which means that (3.40) is satisfied for every value of x. It will in fact be the case if the Green's relationships (3.34a,b) and (3.35a,b), yielding the electric field (for strip) or the current density (for slot), preserve the symmetry about the x-variable of the source of current (for strip) or of electric field (for slot). Since the Green's function contains the unknown propagation constant γ, it is *a priori* possible that some particular values of this propagation constant provide a Green's function such that the integral of the resulting field (3.40a,b) vanishes on the aperture, while the local value of the field varies in the aperture. This behavior is schematically depicted in Figure 3.16 for the strip and slot cases. If a symmetric (even) slot field or strip current distribution, noted $\overline{f}(x)$, is assumed over the area of interest, and if for particular values or ranges of values for γ the Green's function generates an odd behavior for the resulting quantity, the product of the two will of course vanish when integrated over the area. The local value of the product, however, may not be zero. Hence, in this instance, the value of γ provided by the determinantal equation does not correspond to an actual or physical field distribution.

On the other hand, it is quite obvious that actual physical solutions are obtained if the basis functions chosen for expanding the unknown trial field

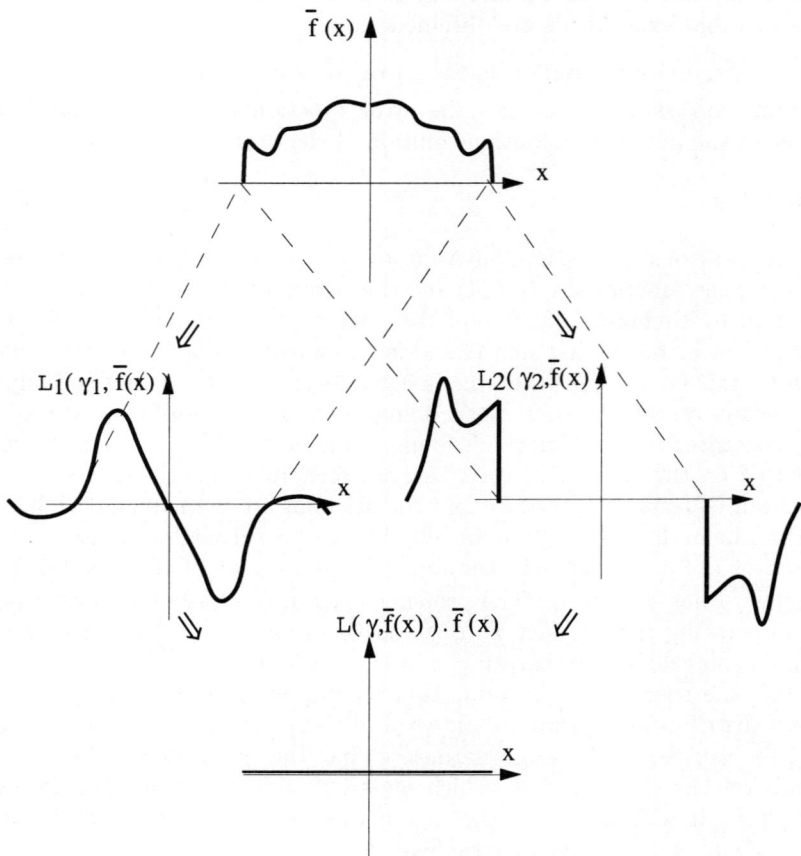

Fig. 3.16 *Possible behavior of product* $\mathcal{L}(\overline{f}) \cdot \overline{f}$ *as a function of the value of the propagation constant (even slot field or strip current distribution, noted* $\overline{f}$*)*

represent rigorously the physical solution [3.68]. In this case, and in this case only, it can be expected that considering an infinite number of terms in the expansion will provide a solution which corresponds to physical fields, provided that all the boundary conditions for the fields are satisfied by a proper expression of the Green's function. This also means that, if the trial and Green's functions for the problem are such that all the conventional boundary conditions are satisfied - except the conditions in the slot or on the strip (3.40) - then the problem is fully resolved once these conditions are satisfied.

As concluded before, some doubt may remain, since only an integral product containing (3.40) is forced to vanish. Moreover, it is impossible to take an infinite number of terms in the expansion of the trial - a truncation error will always be present. But as Schwinger mentioned for the Rayleigh-Ritz procedure, because of the stationary character of Galerkin's procedure with respect to γ under the assumption (3.165), the truncation error will have no effect to the first order on the value obtained for the propagation constant.

To conclude the discussion about the choice of the trial and test fields, it has to be underlined, following the theory presented in Chapter 2, that the stationarity of ε and ρ, and of the reaction calculated by the moment method, is proven only when Galerkin's procedure is used - that is when the basis and test functions belong to the same set. This has been confirmed by Richmond [3.68]. In the case of a mutual reaction, however, Mautz [3.70] has shown that another choice of basis and test functions yields a stationary expression for the mutual reaction $\langle A, b \rangle$. It consists of: taking as test functions for the integral dealing with the b source the basis functions used for expanding the a source, while the test functions for the integral containing the a source are the ones used for expanding the b source. This, however, is not directly part of the present discussion, since the spectral domain Galerkin's method works with a self-reaction only. As a matter of fact, Denlinger's method presented in Section 3.3 is a moment method using a Dirac function as a test function for a self-reaction. Taking indeed the inner product between the aperture field (3.44) in the slot or the current density (3.45) on the strip and the Dirac function in x-domain, is equivalent to system (3.46) and (3.47) or to equation (3.48). Hence, this method will not, to the best of our knowledge, provide a variational result for the propagation constant.

3.5 Discretized formulations

3.5.1 Finite-element method (FEM)

At the end of the previous section, the possibility of obtaining exact field distributions from variational principles has been discussed. The finite-element method (FEM) takes advantage of a variational quantity to solve differential equations for a specified scalar or vector quantity. The purpose of this subsection is not to describe extensively the FEM, but to point out where a variational behavior is invoked and what exactly are the possible links with our previous developments. An excellent review of the finite-element method is provided by Davies [3.55]. Basically, the FEM is used to find a scalar or vector field distribution satisfying a given differential equation. It discretizes a trial distribution and takes advantage of a variational form for adjusting the coefficients of the distribution in order to satisfy the equation. The most simple example is the static Laplace's equation (3.8) that we previously used for quasi-static applications. Instead of directly solving Laplace's equation, the finite-element method uses a functional of the unknown potential, denoted by $J(\Phi)$, which has the interesting property of being stationary about the unknown quantity, provided that the exact unknown satisfies the equation to be solved. Hence, the Rayleigh-Ritz procedure is applied: it imposes that the first-order derivative of $J(\Phi)$ is zero. The resulting equation provides the coefficients yielding an exact description of the unknown Φ.

As a first step, the domain of interest is discretized into cells (surfaces for 2-D or volumes for 3-D problems). For 2-D problems, rectangles or triangles are used. The quantity Φ is approximated in each of the N cells as a polynomial expression of the coordinates. The simplest expression is a linear function:

$$\Phi(x, y) = p_0 + p_x x + p_y y \tag{3.170}$$

which means that vector p may be expressed at each vertex of the triangle as a function of the value of the potential at this vertex k (or node), denoted by U^k:

$$\Phi(x_k, y_k) = p_0 + p_x x_k + p_y y_k = U^k \quad \text{with} \quad k = 1, 2, 3 \tag{3.171}$$

Each cell n has its own vector

$$\overline{p}^n = \begin{bmatrix} p_0^n \\ p_x^n \\ p_y^n \end{bmatrix} \tag{3.172}$$

From the set of three equations written at each vertex of each cell

$$p_0^n + p_x^n x_k + p_y^n y_k = U^k \tag{3.173}$$

the potential at the three nodes of the cell n is rewritten in matrix form. Inverting the resulting system yields the set of coefficients $\overline{p}^n$ associated with each cell. Introducing them in (3.170) yields

$$\Phi^n(x, y) = \begin{bmatrix} 1 \\ x \\ y \end{bmatrix}^T \overline{p}^n = \begin{bmatrix} 1 \\ x \\ y \end{bmatrix}^T \overline{\overline{A}}^{-1} \overline{U}^k \tag{3.174}$$

where

$$\overline{\overline{A}} \triangleq \begin{bmatrix} 1 & x_1 & y_1 \\ 1 & x_2 & y_2 \\ 1 & x_3 & y_3 \end{bmatrix}$$

Hence, the potential in each cell is considered as an interpolation polynomial between the values at the three vertices of the cell. Finally the functional to be minimized has to be computed. For Laplace's solution, it can be shown [3.71] that minimizing the following functional x-domain $J(\Phi)$ yields the solution Φ:

$$J(\Phi) = \int_S \Phi \nabla^2 \Phi \, dS = \int_S \Phi \left(\frac{\partial^2 \Phi}{\partial x^2} + \frac{\partial^2 \Phi}{\partial y^2} \right) dS$$
$$= -\int_S \left[\left(\frac{\partial \Phi}{\partial x} \right)^2 + \left(\frac{\partial \Phi}{\partial y} \right)^2 \right] dS \tag{3.175}$$

The right-hand side is related to the electrostatic energy (3.99a), which has been demonstrated to be variational with respect to trial fields satisfying Laplace's equation. As explained in [3.71], expression (3.100) also means that when $(\Phi + \delta\Phi)$ and Φ are two distributions satisfying the same boundary conditions on the conductor, that is $\delta\Phi = 0$ on the conductor, then the function which cancels δW_e is forced to satisfy Laplace's equation. Using (3.174) the two partial derivatives are rewritten as

$$\frac{\partial \Phi}{\partial x} = \begin{bmatrix} 0 \\ 1 \\ y \end{bmatrix}^T \overline{\overline{A}}^{-1} \overline{U}^k \tag{3.176a}$$

$$\frac{\partial \Phi}{\partial y} = \begin{bmatrix} 0 \\ x \\ 1 \end{bmatrix}^T \overline{\overline{A}}^{-1} \overline{U}^k \tag{3.176b}$$

and their squared power has the matrix form

$$\left(\frac{\partial \Phi}{\partial x}\right)^2 = \left(\overline{U}^k\right)^T \{\overline{\overline{A}}^{-1}\}^T \begin{bmatrix} 0 \\ 1 \\ y \end{bmatrix} \begin{bmatrix} 0 \\ 1 \\ y \end{bmatrix}^T \overline{\overline{A}}^{-1} \overline{U}^k \tag{3.177a}$$

$$\left(\frac{\partial \Phi}{\partial y}\right)^2 = \left(\overline{U}^k\right)^T \{\overline{\overline{A}}^{-1}\}^T \begin{bmatrix} 0 \\ x \\ 1 \end{bmatrix} \begin{bmatrix} 0 \\ x \\ 1 \end{bmatrix}^T \overline{\overline{A}}^{-1} \overline{U}^k \tag{3.177b}$$

Introducing (3.177a,b) into (3.175) yields the final expression for J:

$$J(\Phi) = \left(\overline{U}^k\right)^T \overline{\overline{G}} \, \overline{U}^k \tag{3.178}$$

where $\overline{\overline{G}}$ is a global matrix resulting from the integration over each cell of the product of the coordinate matrices by their transpose, their summation over all the cells, and a rearrangement as a function of the various products $U^k U^j$. Equation (3.178) is then rewritten using as boundary conditions, those nodes where the potential or its derivative is *a priori* known. These nodes are defined by vector U^{k_0}, while nodes where the potential is unknown are in vector U^{k_u}. The matrices are then rearranged and partitioned as

$$J(\Phi) = \begin{bmatrix} \overline{U}^{k_u} \\ \overline{U}^{k_0} \end{bmatrix}^T \begin{bmatrix} \overline{\overline{G}}_{uu} & \overline{\overline{G}}_{u0} \\ \overline{\overline{G}}_{0u} & \overline{\overline{G}}_{00} \end{bmatrix} \begin{bmatrix} \overline{U}^{k_u} \\ \overline{U}^{k_0} \end{bmatrix} \tag{3.179}$$

Applying the Rayleigh-Ritz method to (3.179) is equivalent to taking its first-derivative with respect to each component of $\overline{U}^{k_u}$:

$$\frac{\partial J(\Phi)}{\partial U^{k_u}} = \overline{\overline{G}}_{uu} \overline{U}^{k_u} + \overline{\overline{G}}_{u0} \overline{U}^{k_0} = 0 \tag{3.180a}$$

which yields the final equation for FEM

$$\overline{\overline{G}}_{uu}\overline{U}^{k_u} = -\overline{\overline{G}}_{u0}\overline{U}^{k_0} \tag{3.180b}$$

The above developments need some further comments.

First, the FEM is clearly most applicable to enclosed structures only, since it requires description of the space or the transverse section by defining a mesh and associated nodes. Open structures have to be approximated by closed ones, with end boundaries located far away so that fields and potentials have a negligible value. Such boundaries are usually difficult to predict. At these boundaries an artificial boundary condition has to be specified. Initially they were perfect magnetic or electric walls. Since then, however, various local absorbing boundary conditions have been implemented in the TLM method. As an example, for using the method in open problems, like antenna and scattering problems, it is necessary to terminate the mesh in such a way that it simulates free space. The perfect absorbing boundary condition should perfectly absorb incident waves with arbitrary angles and at all frequencies. Since such walls do not exist, a number of attempts have been made with a reasonable success for defining very good absorbing walls.

Secondly, the potential or field is still influenced by the choice of the polynomial expansion used to describe it in each cell. As we have previously seen, however, an exact solution will be obtained by the Rayleigh-Ritz method if the basis functions form a complete set. In the case of FEM, this is equivalent to saying that the polynomial expansion (3.170) describes the exact field when the order of the basis polynomial used for each cell goes to infinity. It should be noted that mesh refinement techniques do presently exist in the FEM and have already been successfully applied.

Thirdly, there is a crucial question after having solved equation (3.180b): what can we do with the result? We hope to have obtained the exact potential or field in the structure, but we have no idea about the way to use it for transmission lines. The result has to be introduced in another formulation yielding a parameter of interest (radar cross-section, characteristic or input impedance, propagation constant, cut-off frequency) as a function of the potential or field that we have just calculated. A judicious solution is to combine the FEM with a variational principle for a parameter of interest, such as a quasi-static capacitance or inductance, or a propagation constant. For example, we have seen in Chapter 2 that there is a variational principle for the cut-off frequency of a waveguide involved in the Helmholtz equation:

$$\omega_c^2 \varepsilon_0 \mu_0 = -\frac{\displaystyle\int_S \Psi \nabla^2 \Psi \, dS}{\displaystyle\int_S \varepsilon_r \Psi^2 \, dS} \tag{2.115}$$

Letting $p^2 = \omega_c^2 \varepsilon_0 \mu_0$, it can be shown [3.72] that equation (3.180b) has to be rewritten for this case as

$$\overline{\overline{X}}\,\overline{U}^{k_u} = p^2 \overline{\overline{Y}}\,\overline{U}^{k_u} \tag{3.181}$$

which is a matrix formulation for an eigenvalue problem, discussed in Section 3.4.2.2 and characterized by the determinantal equation (3.168). In both equations (3.168) and (3.181), the minimizing procedure makes the unknown appear as an eigenvalue for the final matrix or as a determinantal equation to be solved. This is a consequence of the formulation involving a ratio, and not a specific feature of the variational theory. In fact λ in (3.168) or p^2 in (3.181) may be replaced by their expressions given by (2.115) or (3.166) respectively. This yields a possibly more complicated equation. After solving it, (2.115) or (3.166) should be used for obtaining the required parameter. Hence, equations (3.168) and (3.181) are a shorter way for finding the solution.

The example above shows that FEM just performs a discretization of the trial fields. The trial fields are then adjusted using the Rayleigh-Ritz procedure in order to possibly yield at the same time exact fields associated with a parameter of interest. To summarize, the interest of FEM combined with a variational form is strongly related to the physical meaning of the functional J to be minimized. In the first case, it is just used to find exact fields (their accuracy being subject to the interpolation approximation, hence to the number of cells), while in the second case FEM is used to obtain a parameter of interest.

To complete this brief outline of FEM, it has to be mentioned that the method is also used to discretize equations for boundary conditions, such as condition (3.40). It provides, for example, a matrix formulation of those equations:

$$\overline{\overline{Z}}\,\overline{U}^{k_u} = 0 \tag{3.182a}$$

which yields a determinantal equation:

$$\det(\overline{\overline{Z}}) = 0 \tag{3.182b}$$

where each component of $\overline{\overline{Z}}$ is a function of the unknown propagation constant.

3.5.2 Finite-difference method (FDM)

The comments made about the FEM are also valid for the finite-difference method (FDM). At the start of this chapter, we have seen that finite differences may be useful for solving Laplace's equation for quasi-static problems. Discretizing the potential by using a Taylor's expansion results in a system (3.12) which yields a solution of Laplace's equation which is correct to the h^4 order. The same discretization technique may be used for solving Helmholtz equations (3.27) for TE and TM potentials. Rearranging (3.10) and neglecting

$O(h^4)$ yields:

$$(\frac{\partial^2 \Phi}{\partial x^2} + \frac{\partial^2 \Phi}{\partial x^2})\Big|_A = \frac{(\Phi_B + \Phi_C + \Phi_D + \Phi_E - 4\Phi_A)}{h^2} \tag{3.183}$$

Using this approximation in (3.27) yields

$$\Phi_i^{TE}\Big|_B + \Phi_i^{TE}\Big|_C + \Phi_i^{TE}\Big|_D + \Phi_i^{TE}\Big|_E + h^2(k_i^2 + \gamma^2)\,\Phi_i^{TE}\Big|_A = 0 \tag{3.184}$$

$$\Phi_i^{TM}\Big|_B + \Phi_i^{TM}\Big|_C + \Phi_i^{TM}\Big|_D + \Phi_i^{TM}\Big|_E + h^2(k_i^2 + \gamma^2)\,\Phi_i^{TM}\Big|_A = 0 \tag{3.185}$$

or in matrix form:

$$\overline{\overline{M}}_i \overline{\Phi}_i^{TE,TM} = -h^2(k_i^2 + \gamma^2)\overline{\Phi}_i^{TE,TM} \tag{3.186}$$

Up to now, no variational assumption has been made. We obtain using the finite differences an eigenvalue problem in each layer. The quantity $-h^2(k_i^2 + \gamma^2)$, however, has a different value in each layer. Also, the resulting matrix $\overline{\overline{M}}$ is asymmetric. To overcome those drawbacks, Corr and Davies [3.73] propose to use a variational expression suitable for the problem, which contains only the longitudinal components of the electric and magnetic fields:

$$J = \int_A \frac{1}{\varepsilon_r} \frac{1}{k_0^2}(\omega\varepsilon_0\varepsilon_r E_z \nabla_t^2 E_z + \omega\mu_0 H_z \nabla_t^2 H_z) + \omega\varepsilon_0\varepsilon_r E_z^2 + \omega\mu_0 H_z^2 dA$$

$$\tag{3.187}$$

Corr and Davies have rewritten this expression as a function of the Helmholtz potentials, since the longitudinal components of the fields are proportional to these potentials:

$$\begin{aligned} J(\Phi^{TM}, \Phi^{TE}) = \int_A &\tau A\varepsilon_r |\nabla_t \Phi^{TM}|^2 + \tau |\nabla_t \Phi^{TE}|^2 \\ &+ \tau 2A(\frac{\partial \Phi^{TM}}{\partial x}\frac{\partial \Phi^{TE}}{\partial y} - \frac{\partial \Phi^{TM}}{\partial y}\frac{\partial \Phi^{TE}}{\partial x}) \\ &- k_0^2\{(\Phi^{TE})^2 + A\varepsilon_r(\Phi^{TM})^2\}dA \end{aligned} \tag{3.188}$$

where

$$A = (\frac{\beta c_0}{\omega})^2$$

$$\tau = \frac{\omega^2 \varepsilon_0 \mu_0 - \beta^2}{\omega^2 \varepsilon_0 \varepsilon_r \mu_0 - \beta^2}$$

Using then, for example, definitions (3.9a-d) yields an approximation for the first derivative of the potential, correct to the h^3 order:

$$\frac{\partial \Phi}{\partial x}\Big|_A = \frac{\Phi_B - \Phi_D}{h} \tag{3.189a}$$

$$\left.\frac{\partial \Phi}{\partial y}\right|_B = \frac{\Phi_C - \Phi_E}{h} \tag{3.189b}$$

Introducing those approximations into functional (3.188) yields a matrix formulation similar to (3.182)

$$\overline{\overline{Z}} \left[\begin{array}{c} \overline{\Phi}^{TE} \\ \overline{\Phi}^{TM} \end{array} \right] = 0 \tag{3.190}$$

where the eigenvalues of $\overline{\overline{Z}}$ are functions of A and τ, hence of the unknown propagation constant. In the present case, we obtain an implicit eigenvalue problem, by imposing the first-derivative of the variational form to be zero. This approach is quite different from the implicit determinantal equation associated to Galerkin's procedure, presented in the previous sections. In Galerkin's case, we have shown that the eigenvalue obtained is variational about the trial potentials or fields, but this is obtained because we know *a priori* the exact value of the variational form, which was zero for our case. In the present case, we do not know the exact value of the functional J and we only impose its first-order derivative to be zero. Hence, we cannot conclude about any variational behavior of the obtained eigenvalue with respect to the TE and TM discretized potentials.

For homogeneously-filled waveguides, it is clear that simpler forms like

$$J(\Phi) = k^2 + \gamma^2 = \frac{\displaystyle\int_A \Phi \nabla^2 \Phi \, dA}{\displaystyle\int_A \Phi^2 \, dA} \tag{3.191}$$

may be used, in combination with (3.10) where the $O(h^4)$ term is neglected. The obtained matrix equation is then similar to (3.181).

As illustrated by the abundant literature, FDM and FEM are very powerful general purpose methods in their own right. In this context, however, we consider them mainly useful for obtaining trial fields. We do not wish to underrate them, we are simply limiting their application. With this in mind, it can be concluded that FDM and FEM are indeed efficient variational methods when combined with a variational principle yielding a circuit parameter of interest for the analysis of distributed circuits. In this case, applying one of these two methods is equivalent to discretizing in the space domain the trial field to put in the variational principle. They are of great interest for finding trial field distributions for structures which have an intricate geometry.

3.6 Summary

In the first two sections in this chapter we have reviewed quasi-static methods and full-wave dynamic methods. The quasi-static methods usually provide a straightforward expression for a line parameter. They are valid, however, in the low frequency range only, typically below 5 GHz. On the other hand, full-wave methods have as their main drawback that they provide no simple explicit behavior of a line parameter, despite their wider band application. The search for the root of the implicit equation is usually tedious, so that the use of those methods is preferably limited to lossless lines. The third section was devoted to detailed analytical formulations based on variational principles, applicable to a variety of configurations.

A general variational principle will be presented in detail in Chapter 4, and compared with the other methods. Chapter 5 will be devoted to various applications of the variational principle. They will include theoretical and experimental characterizations of multilayered lossy planar lines parameters, gyrotropic planar devices, and planar junctions and transitions involved in the design of MICs and monolithic microwave integrated circuits (MMICs) subsystems.

3.7 References

[3.1] R.E. Collin, *Field Theory of Guided Waves*. New York: McGraw-Hill, 1960, Ch. 2, p. 25.

[3.2] Z. Zhu, I. Huynen, A. Vander Vorst, "An efficient microwave characterization of PIN photodiodes", *Proc. 26th European Microwave Conference*, Prague, vol. 2, pp. 1010-1014, Sept. 1996.

[3.3] A. Vander Vorst, *Electromagnétisme*. Brussels: De Boeck-Wesmael, 1994, Ch. 1.

[3.4] A. Vander Vorst, D. Vanhoenacker-Janvier, *Bases de l'ingénierie micro-onde*. Brussels: De Boeck-Wesmael, 1996, Ch. 2.

[3.5] H.A. Wheeler, "Transmission line properties of parallel wide strips by conformal mapping approximation", *IEEE Trans. Microwave Theory Tech.*, vol. MTT-12, no. 3, pp. 280-289, Mar. 1964.

[3.6] H.A. Wheeler, "Transmission line properties of parallel wide strips separated by a dielectric sheet", *IEEE Trans. Microwave Theory Tech.*, vol. MTT-13, no. 2, pp. 172-185, Feb. 1965.

[3.7] C.P. Wen, "Coplanar waveguide: A surface strip transmission line suitable for nonreciprocal gyromagnetic device application", *IEEE Trans. Microwave Theory Tech.*, vol. MTT-16, no. 12, pp. 1087-1090, Dec. 1969.

[3.8] M.E. Davis et al., "Finite-boundary corrections to the coplanar waveguide analysis", *IEEE Trans. Microwave Theory Tech.*, vol. MTT-21, no. 6, pp. 594-596, June 1973.

[3.9] G. Kompa, "Dispersion measurements of the first two higher-order modes in open microstrip", *Arch. Elek. Übertragung.*, vol. AEÜ-27, no. 4, pp.182-184, Apr. 1973.

[3.10] R. Mehran, "The frequency-dependent scattering matrix of microstrip right angle bends", *Arch. Elek. Übertragung.*, vol. AEÜ-29, no. 11, pp. 454-460, Nov. 1975.

[3.11] K.C. Gupta, R. Garg, I. Bahl, P. Bhartia, *Microstrip Lines and Slot-lines*, 2nd ed. Boston: Artech House, 1996.

[3.12] A. Vander Vorst, *Electromagnétisme*. Brussels: De Boeck-Wesmael, 1994, Ch. 2.

[3.13] R.E. Collin, *Field Theory of Guided Waves*. New York: McGraw-Hill, 1960, Ch. 4.

[3.14] S.B. Cohn, "Slot-line on a dielectric substrate", *IEEE Trans. Microwave Theory Tech.*, vol. MTT-30, no. 10, pp. 259-269, Oct. 1969.

[3.15] T. Itoh, *Numerical Techniques for Microwave and Millimeter-Wave Passive Structures*. New York: John Wiley&Sons, 1989, Ch. 3, p. 139.

[3.16] R.E. Collin, *Field Theory of Guided Waves*, 2nd ed. New York: IEEE Press, 1991, Ch. 2, pp. 91-96.

[3.17] R.E. Collin, *Field Theory of Guided Waves*, 2nd ed. New York: IEEE Press, 1991, Ch. 2, pp. 95.

[3.18] R.E. Collin, *Field Theory of Guided Waves*, 2nd ed. New York: IEEE Press, 1991, Ch. 3, p. 230.

[3.19] J.R. Wait, *Electromagnetic Waves in Stratified Media*, 2nd ed. Oxford: Pergamon Press, 1970.

[3.20] E.J. Denlinger, "A frequency dependent solution for microstrip transmission lines", *IEEE Trans. Microwave Theory Tech.*, vol. MTT-19, no. 1, pp. 30-39, Jan. 1971.

[3.21] I.V. Lindell, "Variational methods for nonstandard eigenvalue problems in waveguide and resonator analysis", *IEEE Trans. Microwave Theory Tech.*, vol. MTT-30, no. 8, pp. 1194-1204, Aug. 1982.

[3.22] T. Itoh, R. Mittra, "Spectral-domain approach for calculating the dispersion characteristics of microstrip lines", *IEEE Trans. Microwave Theory Tech.*, vol. MTT-21, no. 7, pp. 496-498, July 1973.

[3.23] J.B. Knorr, K. D. Kuchler, "Analysis of coupled slots and coplanar strips on dielectric substrate", *IEEE Trans. Microwave Theory Tech.*, vol. MTT-23, no. 7, pp. 541-548, July 1975.

[3.24] T. Itoh, R. Mittra, "Dispersion characteristics of slot-line", *Electron. Lett.*, vol. 7, pp. 364-365, Jan. 1971.

[3.25] R. Janaswamy, D. Schaubert, "Dispersion characteristics for wide slot-lines on low permittivity substrate", *IEEE Trans. Microwave Theory Tech.*, vol. MTT-33, pp. 723-726, July 1985.

[3.26] R. Janaswamy, D. Schaubert, "Characteristic impedance of a wide slot on low permittivity substrates", *IEEE Trans. Microwave Theory Tech.*, vol. MTT-34, pp. 900-902, July 1986.

[3.27] T. Itoh, R. Mittra, "A technique for computing dispersion characteristics of shielded microstrip lines", *IEEE Trans. Microwave Theory Tech.*, vol. MTT-22, no. 1, pp. 896-898, Jan. 1974.

[3.28] L.P. Schmidt, T. Itoh, "Spectral domain analysis of dominant and higher order modes in finlines", *IEEE Trans. Microwave Theory Tech.*, vol. MTT-28, no. 9, pp. 981-985, Sept. 1980.

[3.29] J.B. Knorr, P. Shayda, "Millimeter wave finline characteristics", *IEEE Trans. Microwave Theory Tech.*, vol. MTT-28, no. 7, pp. 737-743, July 1980.

[3.30] T. Itoh, "Spectral-domain immitance approach for dispersion characteristics of generalized printed transmission lines", *IEEE Trans. Microwave Theory Tech.*, vol. MTT-28, no. 7, pp. 733-736, July 1980.

[3.31] T. Uwano, T. Itoh, "Spectral domain analysis", in *Numerical Techniques for Microwave and Millimeter Wave Passive Structures* (Ed. T. Itoh). New York: John Wiley&Sons, 1989, Ch. 5.

[3.32] M. Kirschning, R. Jansen, "Accurate wide-range design of the frequency-dependent characteristics of parallel coupled microstrips", *IEEE Trans. Microwave Theory Tech.*, vol. MTT-32, no. 1, pp. 83-90, Jan. 1984.

[3.33] R. Garg, K.C. Gupta, "Expressions for wavelength and impedance of a slot-line", *IEEE Trans. Microwave Theory Tech.*, vol. MTT-24, no. 8, pp. 532, Aug. 1976.

[3.34] E.A. Mariani, C.P. Heinzman, J.P. Agrios, S.B. Cohn, "Slot-line characteristics", *IEEE Trans. Microwave Theory Tech.*, vol. MTT-17, no. 12, pp. 1091-1096, Dec. 1969.

[3.35] W.J. Hoefer, A. Ros, "Fin line parameters calculated with the TLM-method", *IEEE MTT-S Int. Microwave Symp. Dig.*, pp. 341-343, Orlando, Apr. 1979.

[3.36] S. Akhtarzad, P.B. Johns, "Three-dimensional transmission-line matrix computer analysis of microstrip resonators", *IEEE Trans. Microwave Theory Tech.*, vol. MTT-23, no. 12, pp. 990-997, Dec. 1975.

[3.37] D.H. Choi, J.R. Hoefer, "The finite-difference-time-domain method and its application to eigenvalue problems", *IEEE Trans. Microwave Theory Tech.*, vol. MTT-34, no.12, pp. 1464-1470, Dec. 1986.

[3.38] J.R. Hoefer, "The transmission line matrix method - Theory and applications", *IEEE Trans. Microwave Theory Tech.*, vol. MTT-33, no.10, pp. 882-893, Oct. 1985.

[3.39] R.E. Collin, *Field Theory of Guided Waves*. New York: McGraw-Hill, 1960, Ch. 4, pp. 152-155.

[3.40] R.E. Collin, *Field Theory of Guided Waves*, 2nd ed. New York: IEEE Press, 1991, Ch. 4, p. 279.

[3.41] E. Yamashita, R. Mittra, "Variational method for the analysis of microstrip lines", *IEEE Trans. Microwave Theory Tech.*, vol. MTT-16, no. 8, pp. 251-256, Apr. 1968.

[3.42] E. Yamashita, "Variational method for the analysis of microstrip-like transmission lines", *IEEE Trans. Microwave Theory Tech.*, vol. MTT-16, no. 8, pp. 529-535, Aug. 1968.

[3.43] R.E. Collin, *Field Theory of Guided Waves*, 2nd ed. New York: IEEE Press, 1991, Ch. 4, p. 328, ex. 4.17.

[3.44] T. Kitazawa, "Variational method for planar transmission lines with anisotropic magnetic materials", *IEEE Trans. Microwave Theory Tech.*, vol. 37, no. 11, pp. 1749-1754, Nov. 1989.

[3.45] R.A. Pucel, D.J. Massé, "Microstrip propagation on magnetic substrates - Part I: Design theory", *IEEE Trans. Microwave Theory Tech.*, vol. MTT-20, no. 5, pp. 304-308, May 1972.

[3.46] R.A. Pucel, D.J. Massé, "Microstrip propagation on magnetic substrates - Part II: Experiment", *IEEE Trans. Microwave Theory Tech.*, vol. MTT-20, no. 5, pp. 309-313, May 1972.

[3.47] R.E. Collin, *Field Theory of Guided Waves*. New York: McGraw-Hill, 1960, Ch. 1, pp. 11-13.

[3.48] R.E. Collin, *Field Theory of Guided Waves*, 2nd ed. New York: IEEE Press, 1991, Ch. 4, p. 274-275.

[3.49] V.H. Rumsey, "Reaction concept in electromagnetic theory", *Phys. Rev.*, vol. 94, pp. 1483-1491, June 1954, and vol. 95, p. 1705, Sept. 1954.

[3.50] A.D. Berk, "Variational principles for electromagnetic resonators and waveguides", *IEEE Trans. Antennas Propagat.*, vol. AP-4, no. 2, pp. 104-111, Apr. 1956.

[3.51] J.Jin, W.C. Chew, "Variational formulation of electromagnetic boundary-value problems involving anisotropic media", *Microwave and Opt. Technol. Lett.*, vol. 7, no. 8, pp. 348-351, June 1994.

[3.52] P.M. Morse, H. Feshbach, *Methods of Theoretical Physics*. New York: McGraw-Hill, 1953, vol. 2 , Ch. 9.

[3.53] R.F. Harrington, *Field Computation by Moment Methods*. New York: Macmillan, 1968, Ch. 1.

[3.54] R.F. Harrington, *Field Computation by Moment Methods*. New York: Macmillan, 1968, Ch. 1, p. 19.

[3.55] J.B. Davies in T. Itoh, *Numerical Techniques for Microwave and Millimeter Wave Passive Structures*. New York: John Wiley&Sons, 1989, Ch. 2.

[3.56] M. Essaaidi, M. Boussouis, "Comparative study of the Galerkin and the least squares boundary residual method for the analysis of microwave integrated circuits", *Microwave and Opt. Technol. Lett.*, vol. 7, no. 3, pp. 141-146, Mar. 1992.

[3.57] F.L. Mesa, M. Horno, "Quasi-TEM and full wave approach for coplanar multistrip lines including gyromagnetic media longitudinally magnetized", *Microwave and Opt. Technol. Lett.*, vol. 4, no. 12, pp. 531-534, Nov. 1991.

[3.58] N.K. Das, D.M. Pozar, "Full-wave spectral-domain computation of material, radiation, and guided wave losses in infinite multilayered printed transmission lines", *IEEE Trans. Microwave Theory Tech.*, vol. MTT-39, no. 1, pp. 54-63, Jan. 1991.

[3.59] R.H. Jansen, "The spectral domain approach for microwave integrated circuits", *IEEE Trans. Microwave Theory Tech.*, vol. MTT-33, no. 10, pp. 1043-1056, Oct. 1985.

[3.60] S. Barkehli, P.H. Pathak, "Reciprocal properties of the dyadic Green's function for planar multilayered dielectric/magnetic media", *Microwave and Opt. Technol. Lett.*, vol. 4, no. 9, pp. 333-335, Aug. 1991.

[3.61] R.E. Collin, *Field Theory of Guided Waves*. New York: McGraw-Hill, 1960, Ch. 2.

[3.62] R.E. Collin, F.J. Zucher, *Antenna Theory*. New York: McGraw-Hill, 1969, Ch. 2.

[3.63] C.T. Tai, *Dyadic Green Functions in Electromagnetic Theory*, 2nd ed. New York: IEEE Press, 1994.

[3.64] R.E. Collin, *Field Theory of Guided Waves*, 2nd ed. New York: IEEE Press, 1991, Ch. 2, p. 102.

[3.65] L.B. Felsen, N. Marcuvitz, *Radiation and Scattering of Waves*. New York: Prentice-Hall, 1973, Ch. 1.

[3.66] A. Papayannakis, T.D. Tsiboukis, E.E. Kriezis, "An investigation of the theory and the properties of the dyadic Green's function for the vector Helmholtz equation", *Electromagnetics*, vol. 7, pp. 91-100, 1987.

[3.67] J. Schwinger, D.S. Saxon, *Discontinuities in Waveguides - Notes on Lectures by J. Schwinger*. London: Gordon and Breach, p. 36-37, 1968.

[3.68] J.H. Richmond, "On the variational aspects of the moment method", *IEEE Trans. Antennas Propagat.*, vol. 39, no. 4, pp. 473-479, Apr. 1991.

[3.69] A.F. Peterson, D.R. Wilton, R.E. Jorgenson, "Variational nature of Galerkin and non-Galerkin moment method solutions", *IEEE Trans. Antennas Propagat.*, vol. 44, no. 4, pp. 500-503, Apr. 1996.

[3.70] J.R. Mautz, "Variational aspects of the reaction in the method of moments", *IEEE Trans. Antennas Propagat.*, vol. 42, no. 12, pp. 1631-1638, Dec. 1994.

[3.71] R.A. Waldron, *Theory of Guided Electromagnetic Waves*. London: Van Nostrand Reinhold Company, 1969.

[3.72] J.B. Davies in T. Itoh, *Numerical Techniques for Microwave and Millimeter Wave Passive structures*. New York: John Wiley&Sons, 1989, Ch. 2, pp. 79-80.

[3.73] D.J. Corr, J.B. Davies, "Computer analysis of the fundamental and higher order modes in single and coupled microstrip lines - Finite difference methods", *IEEE Trans. Microwave Theory Tech.*, vol. MTT-20, no. 10, pp. 669-678, Oct. 1972.

CHAPTER 4

General variational principle

4.1 Generalized topology

Variational principles were used by Schwinger to solve discontinuity problems in waveguides [4.1]. Formulas from Rumsey [4.2] were shown to be variational for closed waveguides provided that the permittitivity and permeability are, at most, symmetric tensors. As a result, isotropic materials and materials with crystalline anisotropy, lossy or lossless, are allowed while gyrotropic media such as magnetized devices and magneto-ionic media are excluded. On the other hand, Berk [4.3] developed variational formulas applicable to gyrotropic media or, in general, media whose dielectric constant and permeability are Hermitian tensors, restricted to lossless substances. The practical use, however, of a formula developed for lossless gyrotropic media is quite limited, because gyrotropy implies losses, except in very narrow-band circumstances. All these formulations were limited to closed waveguides and resonators.

The general variational formulation developed here is applicable to closed as well as open planar and coplanar line structures, which can be multilayered, with gyrotropic lossy inhomogeneous media. The conductors must be lossless. The validity of the formulation is demonstrated by original measurements on transmission lines, as well as on more complicated structures. Figure 4.1 shows a schematic representation of a general transmission line analyzed by using this variational principle [4.4]. It consists of plane conductive layers inserted between planar layers, each of them characterized by permittivity and permeability tensors. There are no limitations on the tensors.

Furthermore, each conductive layer may have several conductors. It should be noted that a distinction is usually made between slot-like conductors and strip-like conductors, based on the amount of metallization extending in the x-direction. For strip-like conductors, the metal is confined to finite areas along the x-axis, while slot-like conductors contain semi-infinite metal sheets and can be viewed as conductive layers with slots of finite extent. This distinction is of prime interest for the choice of trial fields, because their formulation is based on a trial quantity which is non-zero only on an area of finite extent along the x-axis, namely a current for strip-like problems and an electric field for slot-like problems. This will be detailed in Section 4.5. The whole multi-layered structure may be fully open, or enclosed in a waveguide, or partially bounded by electric or magnetic shielding. When using the moment method,

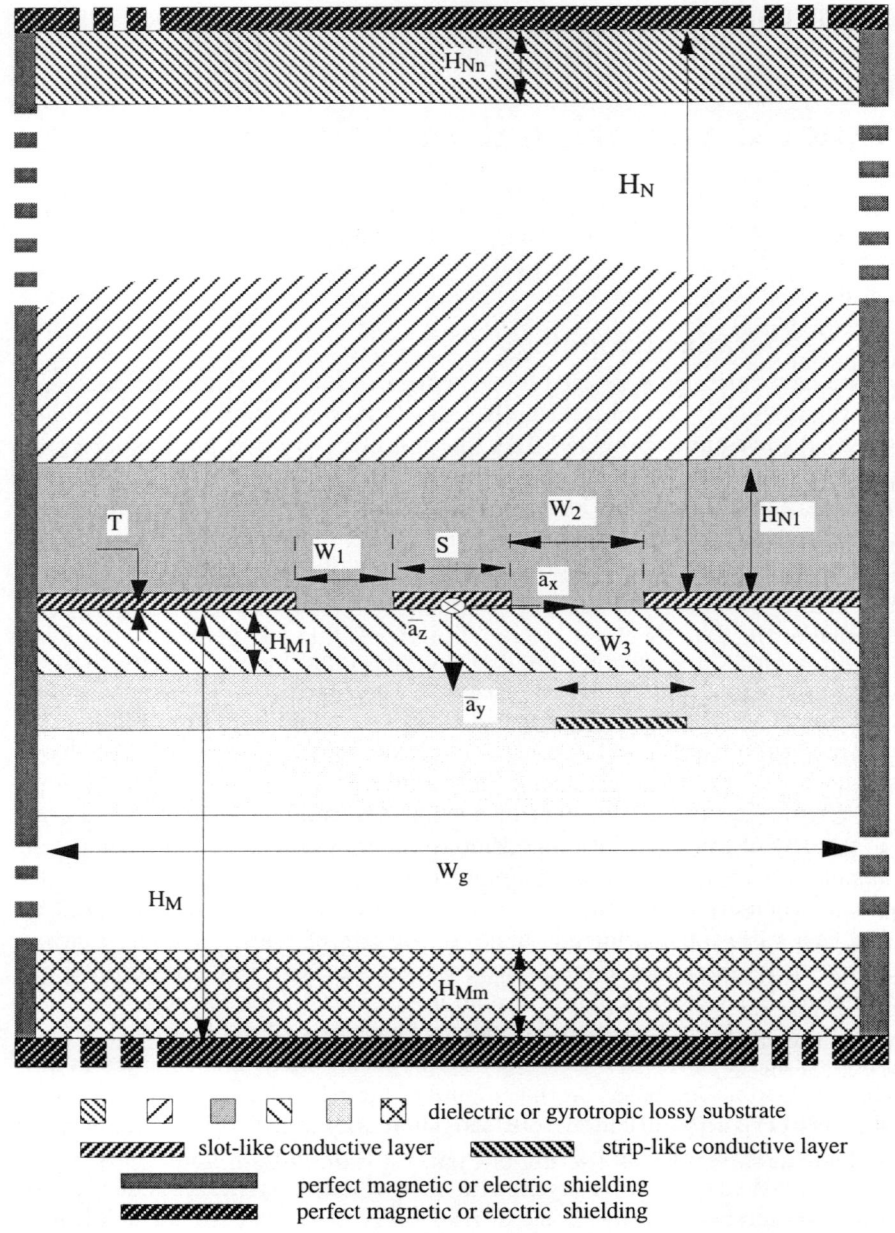

Fig. 4.1 *Geometry of transmission line analyzed by using general variational principle* $(0 \leq H, H_N, H_M, T, S, W_1, W_2 \leq \infty$ *with* $H + H_M + H_N + T \neq 0, W_1 + W_2 + S \leq W_g \leq \infty$ *and* $W_1 + W_2 \neq 0)$

fully open structures are usually considered as shielded lines whose shielding are far away from the active part of the line [4.5][4.6], but not at infinity. By doing so, advantage is taken of the simplification of integrations into summations resulting from the particular formulation of the problem for shielded lines. This artefact is not required with the formulation developed here, because of its convenient convergence properties even for fully open lines, as has been shown in [4.7], and will be explained in Section 4.6. The particular case of dielectric [4.8] or gyrotropic-loaded waveguides may also be calculated. It consists of a multilayered structure, without planar conductive layers, and enclosed into a waveguide.

In this chapter, the general form of a variational principle is derived in the spatial domain in Section 4.2 and in the spectral domain in Section 4.3. In Section 4.4, the methodology for formulating suitable expressions for trial fields is developed. Constraints, degrees of freedom, and formulations in terms of Hertzian potentials and of electric field components are presented and discussed. The choice of adequate trial components at discontinuous conductive interfaces is developed in Section 4.5, like conformal mapping for slot- and strip-like structures. The Rayleigh-Ritz procedure is shown to be compatible with the general variational principle, while Mathieu functions are demonstrated to be extremely efficient as trial components. The importance of the influence of the ratio between the longitudinal and transverse components of the trials is discussed. The advantages of the variational behavior are demonstrated and illustrated in Section 4.6. The influence of the value chosen for the propagation constant in the trial fields, the shape of trial components, and the number of Mathieu functions are calculated, as well as the truncation error of integrations in the spectral domain. Some illustrative cases are calculated in Section 4.7 where the efficiency of the variational calculation is demonstrated. On-line results are obtained with a regular PC, in a few seconds for the whole frequency range, in the case of lossy multilayered planar transmission lines. The spectral domain approach using Galerkin's procedure is compared with the variational principle. It will be shown that the numerical complexity of the first procedure is much larger than that induced by the variational principle. Finite-elements solvers are also compared, with the same advantage in favor of the variational procedure, with an extra-advantage in the case of very lossy structures.

4.2 Derivation in the spatial domain

The fields in the general structure of Figure 4.1 are defined as

$$\overline{E}(x, y, z) \triangleq \overline{e}(x, y)e^{-\gamma z} \tag{4.1a}$$

$$\overline{H}(x, y, z) \triangleq \overline{h}(x, y)e^{-\gamma z} \tag{4.1b}$$

with $\gamma = \alpha + j\beta$ and where a complex $e^{-\gamma z}$ dependence along the propagation axis $\overline{a}_z$ is assumed. It will now be shown that the solution of the following complex equation is a variational formulation of the complex propagation constant γ [4.9]:

$$\gamma^2 \int_A (\bar{a}_z \times \bar{e}^*) \cdot [\overline{\overline{\mu}}^{-1} \cdot (\bar{a}_z \times \bar{e})] \, dA$$

$$+ \gamma \int_A \{(\nabla \times \bar{e}^*) \cdot [\overline{\overline{\mu}}^{-1} \cdot (\bar{a}_z \times \bar{e})] - (\bar{a}_z \times \bar{e}^*) \cdot (\overline{\overline{\mu}}^{-1} \cdot \nabla \times \bar{e})\} \, dA \qquad (4.2)$$

$$- \int_A (\nabla \times \bar{e}^*) \cdot (\overline{\overline{\mu}}^{-1} \cdot \nabla \times \bar{e}) \, dA + \omega^2 \int_A \bar{e}^* \cdot (\overline{\overline{\varepsilon}} \cdot \bar{e}) \, dA = 0$$

In this notation $\bar{e}(x,y)$ and $\bar{h}(x,y)$ have both transverse and longitudinal components, while $\bar{e}^*$ denotes the complex conjugate of $\bar{e}$, A is the area of the cross-section, and ∇ is the classical 3-dimensional del-operator.

4.2.1 Derivation

Using the notation of (4.1), Maxwell's equations are written as

$$\nabla \times \bar{e} + j\omega\overline{\overline{\mu}} \cdot \bar{h} = \gamma \bar{a}_z \times \bar{e} \qquad (4.3)$$

$$\nabla \times \bar{h} = j\omega\overline{\overline{\varepsilon}} \cdot \bar{e} + \gamma \bar{a}_z \times \bar{h} \qquad (4.4)$$

with no assumptions about the permittivity and permeability tensors. Entering (4.3) into (4.4) yields the second-order equation for the fields

$$\nabla \times [\overline{\overline{\mu}}^{-1} \cdot (\gamma \bar{a}_z \times \bar{e} - \nabla \times \bar{e})]$$
$$= \gamma^2 \bar{a}_z \times [\overline{\overline{\mu}}^{-1} \cdot (\bar{a}_z \times \bar{e})] - \gamma \bar{a}_z \times (\overline{\overline{\mu}}^{-1} \cdot \nabla \times \bar{e}) - \omega^2 \overline{\overline{\varepsilon}} \cdot \bar{e} \qquad (4.5)$$

Multiplying by the complex conjugate $\bar{e}^*$ and rearranging yields

$$\gamma \nabla \times [\overline{\overline{\mu}}^{-1} \cdot (\bar{a}_z \times \bar{e})] \cdot \bar{e}^* - \nabla \times (\overline{\overline{\mu}}^{-1} \cdot \nabla \times \bar{e}) \cdot \bar{e}^*$$
$$= \gamma^2 \left\{ \bar{a}_z \times [\overline{\overline{\mu}}^{-1} \cdot (\bar{a}_z \times \bar{e})] \right\} \cdot \bar{e}^* \qquad (4.6)$$
$$- \gamma \bar{a}_z \times (\overline{\overline{\mu}}^{-1} \cdot \nabla \times \bar{e}) \cdot \bar{e}^* - \omega^2 (\overline{\overline{\varepsilon}} \cdot \bar{e}) \cdot \bar{e}^*$$

written for convenience as

$$\gamma A - B = \gamma^2 C - \gamma D - E \qquad (4.7)$$

where coefficients A, B, C, D, and E are obtained by identification with (4.6). Using vector identity

$$\overline{X} \cdot \nabla \times \overline{Y} = \overline{Y} \cdot \nabla \times \overline{X} - \nabla \cdot (\overline{X} \times \overline{Y}) \qquad (4.8)$$

the coefficients can be written as

$$A \overset{\Delta}{=} [\overline{\overline{\mu}}^{-1} \cdot (\bar{a}_z \times \bar{e})] \cdot \nabla \times \bar{e}^* - \nabla \cdot \{\bar{e}^* \times [\overline{\overline{\mu}}^{-1} \cdot (\bar{a}_z \times \bar{e})]\}$$

$$B \stackrel{\Delta}{=} (\nabla \times \overline{e}^*) \cdot [\overline{\overline{\mu}}^{-1} \cdot (\nabla \times \overline{e})] - \nabla \cdot [\overline{e}^* \times (\overline{\overline{\mu}}^{-1} \cdot \nabla \times \overline{e})]$$

$$C \stackrel{\Delta}{=} -(\overline{a}_z \times \overline{e}^*) \cdot [\overline{\overline{\mu}}^{-1} \cdot (\overline{a}_z \times \overline{e})]$$

$$D \stackrel{\Delta}{=} -(\overline{a}_z \times \overline{e}^*) \cdot (\overline{\overline{\mu}}^{-1} \cdot \nabla \times \overline{e})$$

After introducing into (4.7) and rearranging, the equation becomes

$$\begin{aligned}
&\gamma^2 (\overline{a}_z \times \overline{e}^*) \cdot [\overline{\overline{\mu}}^{-1} \cdot (\overline{a}_z \times \overline{e})] \\
&\quad - \gamma \{ (\overline{a}_z \times \overline{e}^*) \cdot (\overline{\overline{\mu}}^{-1} \cdot \nabla \times \overline{e}) - [\overline{\overline{\mu}}^{-1} \cdot (\overline{a}_z \times \overline{e})] \cdot \nabla \times \overline{e}^* \} \\
&\quad - (\nabla \times \overline{e}^*) \cdot (\overline{\overline{\mu}}^{-1} \cdot \nabla \times \overline{e}) + \omega^2 (\overline{\overline{\epsilon}} \cdot \overline{e}) \cdot \overline{e}^* \\
&\quad + \nabla \cdot \{ \overline{e}^* \times [\overline{\overline{\mu}}^{-1} \cdot \nabla \times \overline{e} - \gamma \, \overline{\overline{\mu}}^{-1} \cdot (\overline{a}_z \times \overline{e}^*)] \} = 0
\end{aligned} \tag{4.9}$$

Integrating in the transverse plane yields surface integrals for each medium and line integrals at the boundaries:

$$\begin{aligned}
&\gamma^2 \int_A (\overline{a}_z \times \overline{e}^*) \cdot [\overline{\overline{\mu}}^{-1} \cdot (\overline{a}_z \times \overline{e})] \, dA \\
&\quad + \gamma \int_A \{ (\nabla \times \overline{e}^*) \cdot [\overline{\overline{\mu}}^{-1} \cdot (\overline{a}_z \times \overline{e})] - (\overline{a}_z \times \overline{e}^*) \cdot (\overline{\overline{\mu}}^{-1} \cdot \nabla \times \overline{e}) \} \, dA \\
&\quad - \int_A (\nabla \times \overline{e}^*) \cdot (\overline{\overline{\mu}}^{-1} \cdot \nabla \times \overline{e}) \, dA + \omega^2 \int_A \overline{e}^* \cdot (\overline{\overline{\epsilon}} \cdot \overline{e}) \, dA \\
&\quad + \oint_\Gamma \{ \overline{e}^* \times [\overline{\overline{\mu}}^{-1} \cdot \nabla \times \overline{e} - \gamma \, \overline{\overline{\mu}}^{-1} \cdot (\overline{a}_z \times \overline{e})] \} \cdot \overline{n} \, d\Gamma = 0
\end{aligned} \tag{4.10}$$

The line integral can be written as

$$\oint_\Gamma (\overline{n} \times \overline{e}^*) \cdot [\overline{\overline{\mu}}^{-1} \cdot \nabla \times \overline{e} - \gamma \, \overline{\overline{\mu}}^{-1} \cdot (\overline{a}_z \times \overline{e})] \, d\Gamma \tag{4.11}$$

It vanishes when $\overline{e}$ is the exact field and if there are no losses at the conductive boundaries, because:

- at conducting planes $\overline{n} \times \overline{e}$ vanishes provided no losses are present in the conductor

- at interfaces between media $\overline{e}$ and $\overline{h}$ have continuous tangential components, and the effect of integration on both sides of the interface vanishes.

So (4.10) reduces to (4.2) and is satisfied by the exact field.

4.2.2   Proof of variational behavior

Replacing the exact field into equations (4.2) and (4.10) by a trial field

$$\bar{e}_{trial} = \bar{e} + \delta\bar{e} \tag{4.12}$$

induces a variation on γ. Neglecting the second-order variations, the equation becomes

$$
\begin{aligned}
(\gamma^2 + 2\gamma\delta\gamma) &\bigg\{ \int_A (\bar{a}_z \times \bar{e}^*) \cdot [\overline{\overline{\mu}}^{-1} \cdot (\bar{a}_z \times \bar{e})]\, dA \\
&+ \int_A (\bar{a}_z \times \delta\bar{e}^*) \cdot [\overline{\overline{\mu}}^{-1} \cdot (\bar{a}_z \times \bar{e})]\, dA + \int_A (\bar{a}_z \times \bar{e}^*) \cdot [\overline{\overline{\mu}}^{-1} \cdot (\bar{a}_z \times \delta\bar{e})]\, dA \bigg\} \\
&+ (\gamma + \delta\gamma) \bigg\{ \int_A \{ (\nabla \times \bar{e}^*) \cdot [\overline{\overline{\mu}}^{-1} \cdot (\bar{a}_z \times \bar{e})] - (\bar{a}_z \times \bar{e}^*) \cdot (\overline{\overline{\mu}}^{-1} \cdot \nabla \times \bar{e}) \}\, dA \\
&+ \int_A \{ (\nabla \times \bar{e}^*) \cdot [\overline{\overline{\mu}}^{-1} \cdot (\bar{a}_z \times \delta\bar{e})] - (\bar{a}_z \times \delta\bar{e}^*) \cdot (\overline{\overline{\mu}}^{-1} \cdot \nabla \times \bar{e}) \}\, dA \\
&+ \int_A \{ (\nabla \times \delta\bar{e}^*) \cdot [\overline{\overline{\mu}}^{-1} \cdot (\bar{a}_z \times \bar{e})] - (\bar{a}_z \times \bar{e}^*) \cdot (\overline{\overline{\mu}}^{-1} \cdot \nabla \times \delta\bar{e}) \}\, dA \bigg\} \\
&+ \omega^2 \{ \int_A \bar{e}^* \cdot (\overline{\overline{\varepsilon}} \cdot \bar{e})\, dA + \int_A \delta\bar{e}^* \cdot (\overline{\overline{\varepsilon}} \cdot \bar{e})\, dA + \int_A \bar{e}^* \cdot (\overline{\overline{\varepsilon}} \cdot \delta\bar{e})\, dA \} \\
&- \int_A [(\nabla \times \bar{e}^*) \cdot (\overline{\overline{\mu}}^{-1} \cdot \nabla \times \bar{e}) \\
&+ (\nabla \times \delta\bar{e}^*) \cdot (\overline{\overline{\mu}}^{-1} \cdot \nabla \times \bar{e}) + (\nabla \times \bar{e}^*) \cdot (\overline{\overline{\mu}}^{-1} \cdot \nabla \times \delta\bar{e})]\, dA = 0
\end{aligned}
\tag{4.13}
$$

It can easily be seen that the terms without variation form the left-hand side of equation (4.10), hence the sum of those terms vanishes. The first-order variations also vanish, under specific conditions. Re-arranging equation (4.13) yields indeed the following results.

1. The term containing the first-order variation $\delta\bar{e}$ of the field is

$$
\begin{aligned}
\gamma^2 &\int_A (\bar{a}_z \times \bar{e}^*) \cdot [\overline{\overline{\mu}}^{-1} \cdot (\bar{a}_z \times \delta\bar{e})]\, dA \\
&+ \gamma \int_A \{ (\nabla \times \bar{e}^*) \cdot [\overline{\overline{\mu}}^{-1} \cdot (\bar{a}_z \times \delta\bar{e})] - (\bar{a}_z \times \bar{e}^*) \cdot (\overline{\overline{\mu}}^{-1} \cdot \nabla \times \delta\bar{e}) \}\, dA \\
&+ \omega^2 \int_A \bar{e}^* \cdot (\overline{\overline{\varepsilon}} \cdot \delta\bar{e})\, dA - \int_A (\nabla \times \bar{e}^*) \cdot (\overline{\overline{\mu}}^{-1} \cdot \nabla \times \delta\bar{e})\, dA
\end{aligned}
$$

which can be written as

$$
\begin{aligned}
&- \gamma^2 \int_A \overline{e}^* \cdot \left\{ \overline{a}_z \times [\overline{\overline{\mu}}^{-1} \cdot (\overline{a}_z \times \delta\overline{e})] \right\} dA \\
&+ \gamma \left\{ \int_A \overline{e}^* \cdot \left\{ \nabla \times [\overline{\overline{\mu}}^{-1} \cdot (\overline{a}_z \times \delta\overline{e})] \right\} dA \right. \\
&\quad + \oint_\Gamma \left\{ \overline{e}^* \times [\overline{\overline{\mu}}^{-1} \cdot (\overline{a}_z \times \delta\overline{e})] \right\} \cdot \overline{n} \, d\Gamma \\
&\quad + \left. \int_A \overline{e}^* \cdot [\overline{a}_z \times (\overline{\overline{\mu}}^{-1} \cdot \nabla \times \delta\overline{e})] \, dA \right\} \\
&+ \omega^2 \int_A \overline{e}^* \cdot (\overline{\overline{\varepsilon}} \cdot \delta\overline{e}) \, dA - \int_A \overline{e}^* \cdot [\nabla \times (\overline{\overline{\mu}}^{-1} \cdot \nabla \times \delta\overline{e})] \, dA \\
&- \oint \overline{e}^* \times [\overline{\overline{\mu}}^{-1} \cdot (\nabla \times \delta\overline{e})] \cdot \overline{n} \, d\Gamma
\end{aligned}
\tag{4.14}
$$

where vector identity (4.8) has been used. Recombining the integrands of the various surface integrals yields the surface integral of $\overline{e}^*$ dot-multiplied by equation (4.5), written in terms of $\delta\overline{e}$ instead of the exact field. It vanishes if $\delta\overline{e}$ satisfies Maxwell's equations (4.3) and (4.4). Hence, the trial fields have to be chosen such that they satisfy those two equations. On the other hand, the condition to cancel the sum of the line integrals in (4.14) is that the quantity

$$
\overline{\overline{\mu}}^{-1} \cdot [\gamma \overline{a}_z \times \delta\overline{e} - \nabla \times \delta\overline{e}]
$$

has continuous tangential components so that the effect of integration on both sides of an interface vanishes. This is equivalent to saying that the trial magnetic field has continuous tangential components at any interface. Finally, on conducting planes, $\overline{n} \times \overline{e}$ vanishes provided the conductor is lossless.

2. Similarly, it can be shown that the term containing the conjugate of the first-order field variation vanishes provided the tangential components of the trial electric field are made to vanish at the perfect conducting boundaries, and are continuous at any interface. The term containing $\delta\overline{e}^*$ is

$$
\begin{aligned}
&\gamma^2 \int_A (\overline{a}_z \times \delta\overline{e}^*) \cdot [\overline{\overline{\mu}}^{-1} \cdot (\overline{a}_z \times \overline{e})] \, dA \\
&+ \gamma \int_A \left\{ (\nabla \times \delta\overline{e}^*) \cdot [\overline{\overline{\mu}}^{-1} \cdot (\overline{a}_z \times \overline{e})] - (\overline{a}_z \times \delta\overline{e}^*) \cdot (\overline{\overline{\mu}}^{-1} \cdot \nabla \times \overline{e}) \right\} dA \\
&+ \omega^2 \int_A \delta\overline{e}^* \cdot (\overline{\overline{\varepsilon}} \cdot \overline{e}) \, dA - \int_A (\nabla \times \delta\overline{e}^*) \cdot (\overline{\overline{\mu}}^{-1} \cdot \nabla \times \overline{e}) \, dA
\end{aligned}
$$

which can be written as

$$
\begin{aligned}
&- \gamma^2 \int_A \delta \bar{e}^* \cdot \left\{ \bar{a}_z \times [\overline{\overline{\mu}}^{-1} \cdot (\bar{a}_z \times \bar{e})] \right\} dA \\
&+ \gamma \left\{ \int_A \delta \bar{e}^* \cdot [\bar{a}_z \times (\overline{\overline{\mu}}^{-1} \cdot \nabla \times \bar{e})] \, dA \right. \\
&\quad + \oint_\Gamma \left\{ \delta \bar{e}^* \times [\overline{\overline{\mu}}^{-1} \cdot (\bar{a}_z \times \bar{e})] \right\} \cdot \bar{n} \, d\Gamma \\
&\quad \left. + \int_A \delta \bar{e}^* \cdot \nabla \times [\overline{\overline{\mu}}^{-1} \cdot (\bar{a}_z \times \bar{e})] \, dA \right\} \\
&+ \omega^2 \int_A \delta \bar{e}^* \cdot (\overline{\overline{\varepsilon}} \cdot \bar{e}) \, dA - \int_A \delta \bar{e}^* \cdot [\nabla \times (\overline{\overline{\mu}}^{-1} \cdot \nabla \times \bar{e})] \, dA \\
&- \oint \delta \bar{e}^* \times [\overline{\overline{\mu}}^{-1} \cdot (\nabla \times \bar{e})] \cdot \bar{n} \, d\Gamma
\end{aligned}
\tag{4.15}
$$

where vector identity (4.8) has been used again. Recombining the integrands of the various surface integrals yields the surface integral of $\delta \bar{e}^*$ dot-multiplied by equation (4.5), written in terms of the exact field so that the sum of the surface integrals vanishes. Similarly, the condition sufficient to cancel the sum of the line integrals in the term in $\delta \bar{e}^*$ is that the tangential components of $\bar{e}_{trial}$ are continuous at any interface and vanish at conducting planes, so that $\delta \bar{e}$ also vanishes with $\bar{e}$.

Hence the error $\delta \gamma$ vanishes, because:

1. the term without variation has been shown to vanish

2. the terms containing variations of $\delta \bar{e}$ and $\delta \bar{e}^*$ have also been shown to vanish too, by equations (4.14) and (4.15),

so that the remaining term of the left hand of equation (4.13) containing $\delta \gamma$ is forced to vanish.

4.2.3   Specificities of the general variational principle

The solution of equation (4.2) has been proved to be variational with respect to the electric field $\bar{e}$ and to its complex conjugate $\bar{e}^*$. From the proof just given, it is concluded that, when the exact electric field in expression (4.2) is replaced by a trial field denoted $\bar{e}_{trial}$, no first-order error is made on the solution γ of equation (4.2), provided the trial fields meet the following conditions:

1. $\bar{e}_{trial}$ satisfies Maxwell's equations (4.3) and (4.4)

2. $\bar{e}_{trial}$ and $\bar{h}_{trial}$ have continuous tangential components at any surface of discontinuity of the media in the transverse plane

3. $\bar{e}_{trial}$ has vanishing tangential components at conducting planes,

and provided the conductors are lossless. It should be emphasized that, in the derivation, no assumption is made on the dielectric and magnetic properties of the materials of the line. The derivation does not require complex conjugates of the constitutive tensors of the layers. Hence the Hermitian nature of $\overline{\overline{\epsilon}}$ and $\overline{\overline{\mu}}$ is not required here. Furthermore, the expression is variational even when losses are present in the media.

Formulation (4.2) has a number of advantages.

1. It implies that the error on the value of γ calculated by the equation is small if the trial electric field is an adequate approximation. Its efficiency is due to the fact that the error made on any component of the field is compensated by an exact analytical spatial integration performed over the whole cross-section, and not only over a line boundary at an interface containing conductive layers, as was the case for the integral equation approach introduced in Chapters 2 and 3.

2. It is a dynamic formulation, whilst usually explicit variational formulations are quasi-static [4.10], providing values for quasi-TEM parameters such as capacitances or inductances.

3. It provides a straightforward evaluation of the complex propagation constant by solving a second-order equation, which is much simpler and faster than the classical extraction of the root of the determinantal equation in the Galerkin's procedure [4.11][4.12]. Davies mentions in [4.13] that the solution of the determinantal equation is usually difficult to obtain.

4. The solution is variational even when the medium is inhomogeneous, lossy, and gyrotropic. Only the conductors have to be lossless. This is a major advantage when compared to the moment method used in the spectral domain. Indeed in this case, the problem is usually simplified by assuming lossless dielectric layers with a view to searching for a real root, as mentioned by Jansen [4.14]. The variational formulation will also be useful for modeling structures supporting leaky modes or slow waves, characterized by non-negligible values of both real and imaginary parts of the propagation constant. These structures are usually only calculated by using perturbational methods, as discussed by Das and Pozar [4.15].

5. The constitutive tensors of the layers are involved in the surface integrals so that non-uniformities in a layer, if any, can be taken into account. Those advantages will be illustrated by examples in Sections 4.6 and 4.7.

4.3 Derivation in the spectral domain

The spatial form of equation (4.2) is difficult to use for open and shielded planar lines. These present discontinuities of the function describing the field

components along the x-axis at the interfaces including the slots. The major
consequence is that the x- and y-dependencies of the field cannot be easily
separated, and the integration over the cross-sections in (4.2) becomes difficult. The spectral domain technique overcomes this problem, as will now be
shown.

4.3.1   Definition of spectral domain and spectral fields

The spectral domain technique simply consists of a Fourier transform performed along the x-axis on each component of the electric and magnetic fields
supported by the multilayered structure (Appendix C):

$$\tilde{e}_v(k_x, y) = \int_{-\infty}^{\infty} e_v(x,y)e^{-jk_x x}\, dx \tag{4.16a}$$

$$\tilde{h}_v(k_x, y) = \int_{-\infty}^{\infty} h_v(x,y)e^{-jk_x x}\, dx \quad \text{where} \quad v = x, y, z \tag{4.16b}$$

In all future discussions, these Fourier transforms of the spatial fields will be
referred to as spectral fields. Trial quantities (electric field for slot-like lines
and difference of magnetic fields for strip-like lines) are then chosen such that
their Fourier transform is a continuous function of the spectral variable k_x.

The Fourier transform of Maxwell's equations (4.3) and (4.4) provides a
general solution for the electromagnetic fields. These are usually separable
into a hyperbolic y-dependence multiplied by spectral coefficients, functions
of the spectral variable k_x. Hence, the Fourier transform of the boundary
conditions specified in Section 2 provides boundary conditions at the y-plane
interfaces which result in a set of simple algebraic relations between the spectral coefficients affecting the y-dependence and the trial spectral quantities.
Because of the linearity properties of the Fourier transform, this is equivalent
to saying that the solutions of the Fourier-transformed Maxwell's equations
are the Fourier transform of the trial fields, namely the trial spectral fields.
Different ways to obtain the trial spectral fields will be illustrated in the next
section.

4.3.2   Formulation of variational principle in spectral domain

The basic variational equation (4.2) can be rewritten in the spectral domain
as a function of the spectral trial fields. It is rewritten for convenience as

$$-\gamma^2 \sum_i A_i - \gamma \sum_i (B_i - C_i) + \sum_i D_i - \omega^2 \sum_i E_i = 0 \tag{4.17a}$$

with

$$A_i \triangleq \int_{A_i} (\bar{a}_z \times \bar{e}^*) \cdot [\overline{\overline{\mu}}^{-1} \cdot (\bar{a}_z \times \bar{e})]\, dA \tag{4.17b}$$

$$B_i \triangleq \int_{A_i} (\nabla \times \bar{e}^*) \cdot [\overline{\overline{\mu}}^{-1} \cdot (\bar{a}_z \times \bar{e})]\, dA \tag{4.17c}$$

$$C_i \triangleq \int_{A_i} (\overline{a}_z \times \overline{e}^*) \cdot [\overline{\overline{\mu}}^{-1} \cdot (\nabla \times \overline{e})] \, dA \tag{4.17d}$$

$$D_i \triangleq \int_{A_i} (\nabla \times \overline{e}^*) \cdot [\overline{\overline{\mu}}^{-1} \cdot (\nabla \times \overline{e})] \, dA \tag{4.17e}$$

$$E_i \triangleq \int_{A_i} \overline{e}^* \cdot (\overline{\overline{\epsilon}}_i \cdot \overline{e}) \, dA \tag{4.17f}$$

where subscript i holds for layer i. Use is made of Parseval's theorem [4.16] (Appendix C), which establishes the equality

$$\int_{-\infty}^{\infty} f_1(x) f_2^*(x) dx = \frac{1}{2\pi} \int_{-\infty}^{\infty} \tilde{f}_1(k_x) \tilde{f}_2^*(k_x) dk_x \tag{4.18}$$

where the functions in the integrands are related by the Fourier transform (4.16). Hence, assuming that the media of each layer is uniform, a new set of coefficients is defined, each of them being equal to the corresponding coefficient in set (4.17) by virtue of (4.18):

$$\tilde{A}_i \triangleq \frac{1}{2\pi} \int_{y_i} \int_{-\infty}^{\infty} [\overline{a}_z \times \tilde{\overline{e}}_i(k_x, y)^*] \cdot \{\overline{\overline{\mu}}_i^{-1} \cdot [\overline{a}_z \times \tilde{\overline{e}}_i(k_x, y)]\} \, dk_x dy \tag{4.19a}$$

$$\tilde{B}_i \triangleq \frac{1}{2\pi} \int_{y_i} \int_{-\infty}^{\infty} [\widetilde{\overline{\mathrm{curl}}}_i(k_x, y)^*] \cdot \{\overline{\overline{\mu}}_i^{-1} \cdot [\overline{a}_z \times \tilde{\overline{e}}_i(k_x, y)]\} \, dk_x dy \tag{4.19b}$$

$$\tilde{C}_i \triangleq \frac{1}{2\pi} \int_{y_i} \int_{-\infty}^{\infty} [\overline{\overline{\mu}}_i^{-1} \cdot \widetilde{\overline{\mathrm{curl}}}_i(k_x, y)] \cdot [\overline{a}_z \times \tilde{\overline{e}}_i(k_x, y)^*] \, dk_x dy \tag{4.19c}$$

$$\tilde{D}_i \triangleq \frac{1}{2\pi} \int_{y_i} \int_{-\infty}^{\infty} [\widetilde{\overline{\mathrm{curl}}}_i(k_x, y)^*] \cdot [\overline{\overline{\mu}}_i^{-1} \cdot \widetilde{\overline{\mathrm{curl}}}_i(k_x, y)] \, dk_x dy \tag{4.19d}$$

$$\tilde{E}_i \triangleq \frac{1}{2\pi} \int_{y_i} \int_{-\infty}^{\infty} \tilde{\overline{e}}_i(k_x, y)^* \cdot [\overline{\overline{\epsilon}}_i \cdot \tilde{\overline{e}}_i(k_x, y)] \, dk_x dy \tag{4.19e}$$

with $\tilde{A}_i = A_i, \tilde{B}_i = B_i, \tilde{C}_i = C_i, \tilde{D}_i = D_i, \tilde{E}_i = E_i$, and where $\widetilde{\overline{\mathrm{curl}}}_i(k_x, y)$, defined as the Fourier transform of $\nabla \times \overline{e}$, is expressed as

$$\widetilde{\overline{\mathrm{curl}}}_i(k_x, y) = \frac{\partial \tilde{\overline{e}}_z(k_x, y)}{\partial y} \overline{a}_x - j k_x \tilde{\overline{e}}_z \overline{a}_y + (j k_x \tilde{\overline{e}}_y - \frac{\partial \tilde{\overline{e}}_x(k_x, y)}{\partial y}) \overline{a}_z \tag{4.20}$$

The integrals along the y-axis can be performed analytically, before the integration along the k_x variable, because the trial spectral fields are usually chosen to have a hyperbolic y-dependence, as will be shown later. Because of the equality between each of the coefficients in set (4.19) with the corresponding coefficients in set (4.17), the variational equation (4.17a) can be written as

$$-\gamma^2 \sum_i \tilde{A}_i - \gamma \sum_i (\tilde{B}_i - \tilde{C}_i) + \sum_i \tilde{D}_i - \omega^2 \sum_i \tilde{E}_i = 0 \tag{4.21}$$

which is the variational spectral equation, used as the basis of the method in the spectral domain.

4.4   Methodology for the choice of trial fields

4.4.1   Constraints and degrees of freedom

It has been shown in Section 4.2 that adequate spatial trial fields must satisfy
Maxwell's equations (4.3) and (4.4) and have continuous tangential components at interfaces between layers, except on conducting parts where only the
trial tangential electric field has to vanish. By virtue of the linearity properties
of the Fourier transform, the spectral trial fields are a solution of the Fourier
transform of Maxwell's equations, and satisfy the Fourier transforms of the
spatial boundary conditions at y-planes corresponding to interfaces between
two layers.

There are two degrees of freedom for determining the trial field. First,
trials may be expressed in the spatial or spectral domain. Secondly, adequate
trial fields may be composed either from a well-known solution of Maxwell's
equations in each layer of the structure under consideration, or from a reasonable guess at a solution to Maxwell's equations. In the first case, it will be
shown that field solutions of Helmoltz equations rewritten in each layer are
adequate trials satisfying (4.3) and (4.4), so that one only has to impose the
continuity constraint on those fields. In the second case, the reasonable guess
is deduced from physical considerations and a rigorous application of (4.3)
and (4.4), and of the boundary conditions. In the following subsection, the
two approaches are developed extensively in the spectral domain. The same
reasoning, however, may be applied to trial fields in the spatial domain, as
exemplified in the next subsection, where the fields in a YIG-film are obtained
from a scalar potential.

4.4.2   Choice of integration domain

In Sections 4.2 and 4.3 the spatial and spectral forms of the variational principle were introduced. They differ with regard to the integration domain related
to the x-axis, respectively over the x-domain or over its Fourier transform,
the spectral k_x-domain. The choice depends essentially on the topology of the
structure (presence or absence of discontinuous conductive layers in y-planes),
and on the x-dependence of the constitutive parameters of each layer. The
spectral form may indeed be difficult to use when the constitutive parameters
of one of the layers are functions of the x-variable. Parseval's identity (4.18)
must be applied with care for each term of the scalar product of fields appearing in coefficients (4.17b-f) when the constitutive permeability tensor is
a function of the position. For coefficient A_i for instance, holds:

$$f_1(x,y) = [\overline{a}_z \times \overline{e}(x,y)^*]_v \tag{4.22a}$$

$$\tilde{f}_1(k_x,y) = [\overline{a}_z \times \tilde{\overline{e}}_i(k_x,y)^*]_v \tag{4.22b}$$

$$f_2(x,y) = \{\overline{\overline{\mu}}_i^{-1}(x,y) \cdot [\overline{a}_z \times \overline{e}_i(x,y)]\}_v \tag{4.22c}$$

$$\tilde{f}_2(k_x, y) = \int_{-\infty}^{\infty} \{\overline{\overline{\mu}}_i^{-1}(x, y) \cdot [\overline{a}_z \times \overline{e}_i(x, y)]\}_v e^{-jk_x x} \, dx \tag{4.22d}$$

$$= \int_{-\infty}^{\infty} \{\tilde{\overline{\overline{\mu}}}_i^{-1}(k_x - u, y) \cdot [\overline{a}_z \times \tilde{e}_i(u, y)]\}_v \, du \tag{4.22e}$$

where subscript i is for layer i and where $v = x, y, z$. As is well known, the Fourier transform of the product of two functions is not equal to the product of the Fourier transforms of the two functions, but to its convolution product. Hence, triple integrals appear in the expression of

$$\tilde{A}_i = \frac{1}{2\pi} \int_{y_i} \int_{-\infty}^{\infty} [\overline{a}_z \times \tilde{e}_i(k_x, y)^*]$$
$$\cdot \left\{ \int_{-\infty}^{\infty} \overline{\overline{\mu}}_i^{-1}(k_x - u, y) \cdot [\overline{a}_z \times \tilde{e}_i(u, y)] \, du \right\} dk_x dy \tag{4.23}$$

The same remark is valid for the calculation of $\tilde{B}_i, \tilde{C}_i, \tilde{D}_i$ respectively, and for coefficient $\tilde{E}_i$ if the constitutive permittivity tensor is also a function of the position. As a conclusion, the spatial form (4.2) is preferred when layer i is described by a non-uniform permeability or permittivity constitutive tensor.

Various applications covering the spatial- and spectral-domain use of the variational formulation will be illustrated in Chapter 5. The next sections, devoted to the choice of spectral trial fields, will extensively evaluate two topologies in the spectral domain, illustrating the use of the spectral form of the variational formulation. Before using the spectral domain technique however, the choice of a spatial formulation will be illustrated for the trial fields by an example requiring the use of the spatial form of the variational formulation. The methodology used for the search of trial fields is similar in the spatial and spectral domains.

4.4.3 Spatial domain: propagation in YIG-film of finite width

Yttrium-Iron-Garnet (YIG) is considered in the film configuration illustrated in Figure 4.2. The magnet poles generate a DC-magnetic field along the y-axis. The film is assumed to be infinite along the z-axis and has a finite width W along the x-axis. Taking advantage of the ferrite nature of the film, the propagation constant of this guiding structure can be modified by varying the applied DC magnetic field. These films find applications in planar microwave devices operating in the magnetostatic wave range. The reader will find a description of these waves in Appendix D as well as a rigorous derivation of the approximations. A simplified analysis of the structure of Figure 4.2 is available [4.17][4.18], based on the following assumptions:

1. The DC-magnetic field inside of the film is uniform, which implies that the demagnetizing effect and the external DC-magnetic field are uniform.

2. The edges of the film can be approximated by Perfect Magnetic Walls (PMW).

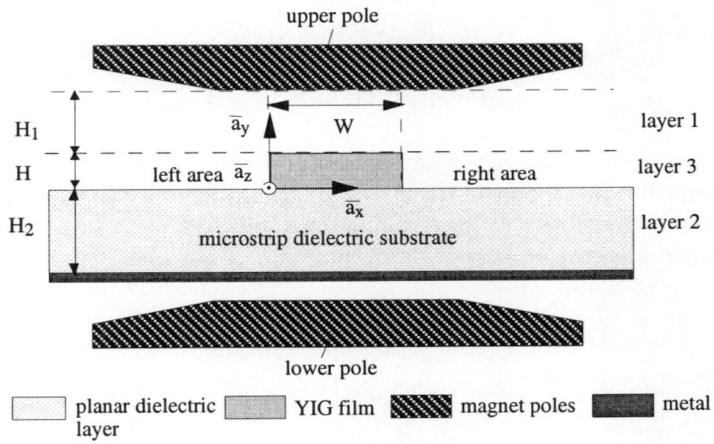

Fig. 4.2 *Configuration of YIG-film for YIG-tuned devices*

3. Low losses occur in the film, so that they can be obtained by a perturbational method.

These hypotheses are usually made for the practical design of Magneto-
Static Waves (MSW) devices, because of the lack of simple models taking into
account the non-uniformity of the internal field, the losses in the film, and the
finite width effect. For the present example only, the second assumption is
retained. The purpose is to show that the variational equation (4.2) is efficient
for the calculation of the propagation constant of MSW in YIG-films in the
presence of losses and of a non-uniform DC-magnetic field.

The following parameters are defined (Fig. 4.2):

x horizontal coordinate, centered at one edge of YIG-film
y vertical coordinate, centered at interface between YIG and
     layer 2
H thickness of substrate, called layer 3
H_i distance between the interface layer i - layer 3 and the metallic
     housing, if any
W width of YIG-film.

Maxwell's equations are rewritten under the magnetostatic assumption in
all the three media of Figure 4.2:

$$\nabla \times \overline{E}_i + j\omega\overline{\overline{\mu}}_i \cdot \overline{H}_i = 0 \tag{4.24}$$

$$\nabla \times \overline{H}_i \approx 0 \quad \text{(magnetostatic assumption)} \tag{4.25}$$

$$\nabla \cdot \overline{B}_i = 0 \tag{4.26}$$

where $\overline{\overline{\mu}}_i$ is the permeability tensor of layer i, defined as

$$\overline{\overline{\mu}}_i = \begin{bmatrix} \mu_{11,i} & 0 & \mu_{21,i} \\ 0 & \mu_0 & 0 \\ \mu_{12,i} & 0 & \mu_{22,i} \end{bmatrix} \tag{4.27}$$

The expressions for $\mu_{11,3}$, $\mu_{21,3}$, $\mu_{12,3}$ in layer 3 are given in Appendix D, which describes the permeability tensor of a YIG-film. They depend on the value of the internal DC magnetic field. As detailed in the appendix the internal DC magnetic field in a YIG-film having a finite width is spatially non-uniform, because of a non-uniform demagnetizing effect [4.19]. Hence, the permeability tensor $\overline{\overline{\mu}}_3$ is spatially non-uniform.

In layers 1 and 2, the medium is assumed to be isotropic :

$$\mu_{11,i} = \mu_{22,i} = \mu_0 \quad \text{and} \quad \mu_{12,i} = \mu_{21,i} = 0 \quad \text{for} \quad i = 1, 2 \tag{4.28}$$

Using the notations of (4.1), equations (4.24) and (4.25) may be rewritten as the corresponding magnetostatic form of equations (4.3) and (4.4):

$$\nabla \times \overline{e}_i + j\omega\overline{\overline{\mu}}_i \cdot \overline{h}_i = \gamma\overline{a}_z \times \overline{e}_i \tag{4.29a}$$

$$\nabla \times \overline{h}_i = \gamma\overline{a}_z \times \overline{h}_i \tag{4.29b}$$

From (4.25), it can be stated that the magnetic field is derived from a scalar potential Φ:

$$\overline{H}_i = -\nabla\Phi_i \tag{4.30}$$

The magnetic flux density $\overline{B}$ is easily deduced in each layer from the magnetic field as

$$\overline{B}_i = \overline{\overline{\mu}}_i\overline{H}_i$$

i.e.:

$$\overline{B}_{1,2} = \mu_0\overline{H}_{1,2} \tag{4.31a}$$

$$B_{y3} = \mu_0 H_{y3} \tag{4.31b}$$

$$B_{x3} = \mu_{11,3}H_{x3} + \mu_{21,3}H_{z3} \tag{4.31c}$$

$$B_{z3} = \mu_{12,3}H_{x3} + \mu_{11,3}H_{z3} \tag{4.31d}$$

Combining (4.30) and (4.31a-d) into (4.26) yields in each region after simplification

$$\mu_{11,i}\Big[\frac{\partial^2\Phi_i}{\partial x^2} + \frac{\partial^2\Phi_i}{\partial z^2}\Big] + \mu_0\frac{\partial^2\Phi_i}{\partial y^2} = 0 \tag{4.32}$$

which has the following general solution, obtained by separating the three variables x, y and z:

$$\Phi_i(x, y, z) = [A \sin k_x x + B \cos k_x x][C_i e^{s_{0i} y} + D_i e^{-s_{0i} y}] e^{-\gamma z} \qquad (4.33a)$$

with

$$s_{0i} = \sqrt{(\mu_{11,i}/\mu_0)(k_x^2 - \gamma^2)} \quad \text{wavenumber along } y\text{-axis} \qquad (4.33b)$$

k_x wavenumber along x-axis

MagnetoStatic Forward Volume Waves (MSFVW) occur in the ferrite film when $\mu_{11,3}$ is negative, *i.e.* when s_{03} is purely imaginary (no losses are considered) [4.20]. The next step is the determination of magnetic fields and magnetic fluxes in each layer. Equation (4.30) is used in each layer to obtain

$$\begin{aligned} H_{xi} &= -k_x[A \cos k_x x - B \sin k_x x][C_i e^{s_{0i} y} + D_i e^{-s_{0i} y}] e^{-\gamma z} \\ &= h_{xi} e^{-\gamma z} \end{aligned} \qquad (4.34a)$$

$$\begin{aligned} H_{yi} &= -s_{0i}[A \sin k_x x + B \cos k_x x][C_i e^{s_{0i} y} - D_i e^{-s_{0i} y}] e^{-\gamma z} \\ &= h_{yi} e^{-\gamma z} \end{aligned} \qquad (4.34b)$$

$$\begin{aligned} H_{zi} &= \gamma[A \sin k_x x + B \cos k_x x][C_i e^{s_{0i} y} + D_i e^{-s_{0i} y}] e^{-\gamma z} \\ &= h_{zi} e^{-\gamma z} \end{aligned} \qquad (4.34c)$$

The electric field in each layer is obtained by a straightforward application of equations (4.24) and (4.29a). Taking into account (4.25), MSFVW are modeled as TE-waves with respect to the z-axis, which implies

$$E_{zi} = e_{zi} e^{-\gamma z} = 0 \quad \text{for} \quad i = 1, 2, 3 \qquad (4.35)$$

This assumption is confirmed in the literature [4.20][4.21]. Hence, entering (4.35) and definition (4.1) into equation (4.29a) results in

$$e_{yi} = -(\frac{j\omega}{\gamma})[\mu_{11,i} h_{xi} + \mu_{21,i} h_{zi}] \qquad (4.36a)$$

$$e_{xi} = +(\frac{j\omega}{\gamma})\mu_0 h_{yi} \qquad (4.36b)$$

$$e_{zi} = 0 \qquad (4.36c)$$

Up to now, the electric field $\bar{e}$ obtained by (4.36a-c) and the magnetic field $\bar{h}$ defined by (4.34a-c) satisfy the magnetostatic form (4.29a,b) of Maxwell's equations (4.3) and (4.4), since (4.32) is derived from (4.25), which is equivalent to (4.29b).

The x-dependence of the fields is governed by the edge boundary condition at planes $x = 0$ and $x = W$. The perfect magnetic wall assumption

(PMW), introduced by O'Keefe and Paterson [4.22], is widely used for the analysis of MSW devices [4.23][4.24], and will be used here for simplicity. In the next chapter however, a significant theoretical improvement of this approximation will be developed which, combined with variational formulation (4.2), provides a better agreement with experiment. As a matter of fact, the structure of Figure 4.2 analyzed here under the PMW assumption consists of a gyrotropic-loaded waveguide with lateral PMW shielding, as was introduced in Section 4.1. The PMW condition is expressed as

$$H_{yi}(0, y, z) = H_{zi}(0, y, z) = 0 \quad \forall y, z \tag{4.37a}$$

$$H_{yi}(W, y, z) = H_{zi}(W, y, z) = 0 \quad \forall y, z \tag{4.37b}$$

which, combined with (4.34b-c), results in

$$B = 0 \tag{4.37c}$$

$$k_x = \frac{m\pi}{W} \quad \text{with } m \text{ arbitrary} \tag{4.37d}$$

Hence, the structure supports an infinite number of propagating modes, characterized by their width-harmonic number m.

Coefficients C_i and D_i of (4.34a-c) can now be calculated in each layer. They are obtained by applying the following boundary conditions at respective interfaces $y = -H_2$, $y = 0$, $y = H$ and $y = H_1 + H$.

1. In the plane $y = -H_2$.
The tangential electric field has to vanish:

$$e_{x2}(x, y) = 0 \quad \forall x \tag{4.38a}$$

$e_{z2} = 0$ is satisfied by virtue of (4.36c).
Equations (4.34b), (4.36b), and (4.38a) yield

$$D_2 = e^{-2s_{02}H_2} C_2 \tag{4.38b}$$

2. In the plane $y = H_1 + H$.
The tangential electric field has to vanish:

$$e_{x1}(x, y) = 0 \quad \forall x \tag{4.39a}$$

$e_{z1} = 0$ is satisfied by virtue of (4.36c).
Equations (4.34b), (4.36b), and (4.39a) yield

$$D_1 = e^{2s_{01}(H_1+H)} C_1 \tag{4.39b}$$

3. In the plane $y = 0$.
The continuity of h_x and h_z is imposed:

$$h_{x2} = h_{x3} \quad \forall x \tag{4.40a}$$

$$h_{z2} = h_{z3} \quad \forall\, x \tag{4.40b}$$

Equations (4.34a) and (4.40a) yield after simplification

$$C_3 + D_3 = C_2 + D_2 \tag{4.40c}$$

The continuity of e_x is imposed:

$$e_{x2} = e_{x3} \quad \forall\, x \tag{4.41a}$$

Equations (4.34b), (4.36b), and (4.41a) yield after simplification

$$s_{02}(C_2 - D_2) = s_{03}(C_3 - D_3) \tag{4.41b}$$

4. In the plane $y = H$.
The continuity of h_x and h_z is imposed:

$$h_{x1} = h_{x3} \quad \forall\, x \tag{4.42a}$$

$$h_{z1} = h_{z3} \quad \forall\, x \tag{4.42b}$$

Equations (4.34a) and (4.42a) yield after simplification

$$C_3 e^{s_{03}H} + D_3 e^{-s_{03}H} = C_1 e^{s_{01}H} + D_1 e^{-s_{01}H} \tag{4.42c}$$

The continuity of e_x is imposed:

$$e_{x1} = e_{x3} \quad \forall\, x \tag{4.43a}$$

Equations (4.34b), (4.36b), and (4.43a) yield after simplification

$$s_{01}(C_1 e^{s_{01}H} - D_1 e^{-s_{01}H}) = s_{03}(C_3 e^{s_{03}H} - D_3 e^{-s_{03}H}) \tag{4.43b}$$

Finally, the variational equation (4.2) is used. The solution is based on a ratio of field quantities, so that the knowledge of the absolute value of the electric field to put in the equation is not required. From the whole set of equations (4.38) to (4.43), relationships between C_1, C_2, C_3, D_1, D_2, and D_3 are obtained. They can be expressed as

$$D_3 = \frac{j\eta - \tanh s_{02}H_2}{j\eta + \tanh s_{02}H_2} C_3 \tag{4.44a}$$

$$C_1 = \frac{C_3 e^{s_{03}H} + D_3 e^{-s_{03}H}}{e^{s_{01}H} + e^{s_{01}(2H_1+H)}} \tag{4.44b}$$

$$C_2 = \frac{C_3 + D_3}{e^{-2s_{02}H_2} + 1} \tag{4.44c}$$

with

$$\eta = \sqrt{-\mu_{11,3}/\mu_0} \tag{4.44d}$$

while introducing (4.44) in (4.38) and (4.39) yields D_1 and D_2.

Using the same equations, the unknown coefficients C_i and D_i can be eliminated, which yields a transcendental equation containing the required propagation exponent

$$\tanh s_{01} H_1 = jn \frac{K e^{-s_{03} H} - e^{s_{03} H}}{K e^{-s_{03} H} - e^{s_{03} H}} \tag{4.45}$$

with

$$K = \frac{j\eta - \tanh s_{02} H_2}{j\eta + \tanh s_{02} H_2}$$

This is the usual simple equation widely used for the analysis of MSFVW in YIG-films [4.17]. Its solution, however, is only valid for a uniform permeability tensor, which is not the case here. However, inserting this solution in the trial field expressions (4.33a,b), (4.34a-c), and (4.36a-c), provides trial fields which satisfy all the boundary conditions and are therefore acceptable for the variational equation (4.2).

The trial electric field in each layer is finally expressed as the product of an x-dependence, obtained by combining relations (4.34) and (4.36), by a hyperbolic y-dependence whose coefficients are given by (4.44) and are proportional to the same undetermined factor expressed as the product

$$AC_3 \tag{4.46}$$

Once the x- and y-dependence of the trial fields have been determined, the various integrands of coefficients (4.17) are expressed using expressions (4.36) for the trial electric field. The resulting formulations are

$$(\overline{a}_z \times \overline{e}^*) \cdot \overline{\overline{\mu}}_i^{-1} (\overline{a}_z \times \overline{e})$$
$$= e^{2\alpha z} \left| \frac{\omega}{\gamma} \right|^2 \left(\mu_{11,i}^{-1} |\mu_{11,i} h_{xi} + \mu_{21,i} h_{zi}|^2 + \mu_0 |h_{yi}|^2 \right) \tag{4.47a}$$

$$\nabla \times \overline{e}^* \cdot (\overline{\overline{\mu}}_i^{-1} \overline{a}_z \times \overline{e})$$
$$= -e^{2\alpha z} \frac{\omega^2}{\gamma} \mu_{21,i}^{-1} [\mu_{11,i} h_{zi} + \mu_{12,i} h_{xi}]^* [\mu_{11,i} h_{xi} + \mu_{21,i} h_{zi}] \tag{4.47b}$$

$$(\overline{a}_z \times \overline{e}^*) \cdot (\overline{\overline{\mu}}_i^{-1} \nabla \times \overline{e})$$
$$= -e^{2\alpha z} \frac{\omega^2}{\gamma^*} \mu_{12,i}^{-1} [\mu_{11,i} h_{xi} + \mu_{21,i} h_{zi}]^* [\mu_{11,i} h_{zi} + \mu_{12,i} h_{xi}] \tag{4.47c}$$

$$\nabla \times \overline{e}^* \cdot (\overline{\overline{\mu}}_i^{-1} \nabla \times \overline{e}) = e^{2\alpha z} \omega^2 \mu_{11,i}^{-1} |\mu_{11,i} h_{zi} + \mu_{12,i} h_{xi}|^2 \tag{4.47d}$$

$$\overline{e}^* \overline{\overline{\varepsilon}}_i \cdot \overline{e} = \varepsilon_i e^{2\alpha z} \left| \frac{\omega}{\gamma} \right|^2 \left(|\mu_{11,i} h_{xi} + \mu_{21,i} h_{zi}|^2 + \mu_0^2 |h_{yi}|^2 \right) \tag{4.47e}$$

because the following conditions are satisfied:

$$\bar{a}_z \times \bar{e} = (-e_{yi}, e_{xi}, 0) \tag{4.48a}$$

$$\nabla \times \bar{e} = (0, 0, \frac{-j\omega B_{zi}}{e^{-\gamma z}}) \tag{4.48b}$$

$$\overline{\overline{\mu}}_i^{-1} \cdot (\bar{a}_z \times \bar{e}) = (-\mu_{11,i}^{-1} e_{yi}, e_{xi}, -\mu_{21,i}^{-1} e_{yi}) \tag{4.48c}$$

$$\overline{\overline{\mu}}_i^{-1} \cdot \nabla \times \bar{e} = (\frac{-j\omega\mu_{12,i}^{-1}B_{zi}}{e^{-\gamma z}}, 0, \frac{-j\omega\mu_{11,i}^{-1}B_{zi}}{e^{-\gamma z}}) \tag{4.48d}$$

The last step is to evaluate the integrals in (4.17b-f). The y-dependence of the fields consists of simple exponentials and the integration along the y-axis can be performed analytically. The x-dependence of the integrands is the product of the x-dependence of the inverted non-uniform permeability tensor by simple sinusoidal functions. This integral is evaluated numerically by a simple trapezoidal rule algorithm.

Figure 4.3 shows the results obtained when combining the spatial trial fields obtained in this paragraph with variational equation (4.2). The real and imaginary parts of the propagation constant solution of (4.2) are calculated for the first width-harmonic mode (solid line) - non-uniform case - and compared to the solution provided by equation (4.45) (dashed line), valid under the assumption of a uniform internal field. A significant shift of the dispersion relation is observed when the non-uniformity of the internal field is taken into account. This is of prime interest for the design of YIG-tuned devices, for which the center operating frequency is a function of the size of the YIG-film used as resonator. Since magnetostatic waves are slow waves, the size of the resonator to be used may be considerably reduced. For sizes of the order of 1 mm along the z-axis, the propagation constant at the resonance is $\pi 10^3$ rad/m. Figure 4.3 shows that neglecting the non-uniformity of the field in the sample yields an error of about 1 % in the resonant frequency. It can be seen that variational equation (4.2) in the spatial domain yields better results when modeling propagation effects in a non-uniformly biased gyrotropic YIG-film.

Two methods for deriving the trial fields will now be described in the spectral domain. The first deals with Hertzian potentials obtained from Maxwell's equations. The second deduces the fields from physical considerations and rigorous application of the boundary conditions required by the application of variational formulation (4.2). The two approaches are developed extensively in the spectral domain. The same reasoning, however, may be applied to trial fields in the spatial domain. This is exemplified for the last example, where the fields in a YIG-film are obtained from a scalar potential. Each approach will be illustrated by a practical example.

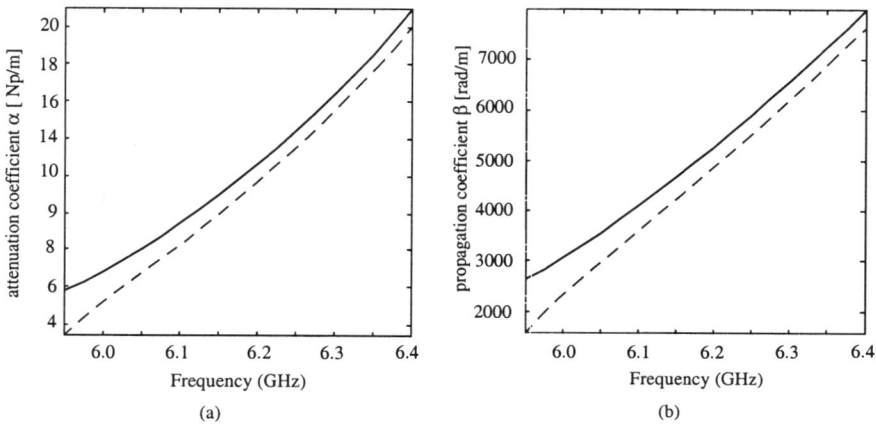

Fig. 4.3 *Propagation constant of YIG-film calculated for non-uniform internal DC-field with variational principle (solid line) and for uniform DC-field (dashed line) (a) attenuation coefficient; (b) propagation coefficient*

4.4.4 Trial fields in terms of Hertzian potentials

The classical electric and magnetic Hertzian potentials, solutions of Helmoltz equation in the various layers, are derived from the source-free formulation of Maxwell's equations [4.25]. For classical waveguides, they are expressed in each layer as a function of modes which are respectively TE (or H) and TM (or E) with respect to the z-axis of propagation:

$$\overline{\Pi}_i^H(x,y,z) = \Phi_i^H(x,y)e^{-\gamma z}\overline{a}_z \qquad (4.49a)$$

$$\overline{\Pi}_i^E(x,y,z) = \Phi_i^E(x,y)e^{-\gamma z}\overline{a}_z \qquad (4.49b)$$

where Φ_i^H and Φ_i^E are scalar potentials. Hence the electric and magnetic fields in (4.1) can be derived in each layer from expressions in [4.25] as

$$\overline{E}_i = e^{-\gamma z}[-\gamma\nabla\Phi_i^E + (k_i^2 + \gamma^2)\Phi_i^E\overline{a}_z + j\omega\mu_i\overline{a}_z \times \nabla\Phi_i^H] \qquad (4.50a)$$

$$\overline{H}_i = e^{-\gamma z}[-\gamma\nabla\Phi_i^H + (k_i^2 + \gamma^2)\Phi_i^H\overline{a}_z - j\omega\varepsilon_i\overline{a}_z \times \nabla\Phi_i^E] \qquad (4.50b)$$

with $k_i^2 = \omega^2\mu_i\varepsilon_i$. The potentials are solutions of the following Helmoltz equations, obtained from Maxwell's equations (4.3) and (4.4):

$$\nabla^2\Phi_i^H + (k_i^2 + \gamma^2)\Phi_i^H = 0 \qquad (4.51a)$$

$$\nabla^2\Phi_i^E + (k_i^2 + \gamma^2)\Phi_i^E = 0 \qquad (4.51b)$$

The electric and magnetic fields can be deduced in each layer from the scalar potentials $\Phi_i^H(x,y)$ and $\Phi_i^E(x,y)$:

$$e_{xi} = -\gamma\frac{\partial\Phi_i^E}{\partial x} - j\omega\mu_i\frac{\partial\Phi_i^H}{\partial y} \tag{4.52a}$$

$$e_{yi} = -\gamma\frac{\partial\Phi_i^E}{\partial y} + j\omega\mu_i\frac{\partial\Phi_i^H}{\partial x} \tag{4.52b}$$

$$e_{zi} = (k_i^2 + \gamma^2)\Phi_i^E \tag{4.52c}$$

$$h_{xi} = -\gamma\frac{\partial\Phi_i^H}{\partial x} + j\omega\varepsilon_i\frac{\partial\Phi_i^E}{\partial y} \tag{4.53a}$$

$$h_{yi} = -\gamma\frac{\partial\Phi_i^H}{\partial y} - j\omega\varepsilon_i\frac{\partial\Phi_i^E}{\partial x} \tag{4.53b}$$

$$h_{zi} = (k_i^2 + \gamma^2)\Phi_i^H \tag{4.53c}$$

The x-Fourier transform of equations (4.51a,b) results in the spectral equation

$$\frac{\partial^2\tilde{\Phi}_i^{H,E}(k_x,y)}{\partial^2 y} + (\gamma^2 - k_x^2 + k_i^2)\tilde{\Phi}_i^{H,E}(k_x,y) = 0 \tag{4.54a}$$

which has the general solution

$$\tilde{\Phi}_i^{H,E}(k_x,y) = \tilde{X}_i^{H,E}(k_x)\sinh(s_{0i}y) + \tilde{Y}_i^{H,E}(k_x)\cosh(s_{0i}y) \tag{4.54b}$$

with

$$s_{0i} = \sqrt{k_x^2 - \gamma^2 - k_i^2} = jk_{yi} \tag{4.54c}$$

k_{yi} wavenumber along y-axis

$\tilde{X}_i^{H,E}$ and $\tilde{Y}_i^{H,E}$ spectral coefficients $\hspace{2cm}$ (4.54d)

In the following discussion the coefficients $\tilde{X}_i^{H,E}$ and $\tilde{Y}_i^{H,E}$ affecting the hyperbolic dependencies of the spectral potentials will be referred to as spectral coefficients. The spectral fields associated with the spectral potentials are:

$$\tilde{e}_{xi}(k_x,y) = -\gamma jk_x\tilde{\Phi}_i^E(k_x,y) - j\omega\mu_i\frac{\partial\tilde{\Phi}_i^H(k_x,y)}{\partial y} \tag{4.55a}$$

$$\tilde{e}_{yi}(k_x,y) = -\gamma\frac{\partial\tilde{\Phi}_i^E(k_x,y)}{\partial y} + j\omega\mu_i jk_x\tilde{\Phi}_i^H(k_x,y) \tag{4.55b}$$

$$\tilde{e}_{zi} = (k_i^2 + \gamma^2)\tilde{\Phi}_i^E(k_x,y) \tag{4.55c}$$

$$\tilde{h}_{xi}(k_x,y) = -\gamma jk_x\tilde{\Phi}_i^H(k_x,y) + j\omega\varepsilon_i\frac{\partial\tilde{\Phi}_i^E(k_x,y)}{\partial y} \tag{4.56a}$$

$$\tilde{h}_{yi}(k_x, y) = -\gamma\frac{\partial\tilde{\Phi}_i^H(k_x, y)}{\partial y} - j\omega\varepsilon_{ij}k_x\tilde{\Phi}_i^E(k_x, y) \tag{4.56b}$$

$$\tilde{h}_{zi} = (k_i^2 + \gamma^2)\tilde{\Phi}_i^H(k_x, y) \tag{4.56c}$$

Considering the $(m + n)$ layer structure of Figure 4.1, and looking for simplifications of the general solution of equation (4.54) in the case of shielded and open lines, one defines (Fig. 4.1):

x horizontal coordinate, centered in middle of the waveguide in case of shielded lines

y vertical coordinate, centered in the plane containing the conducting layer between two layers, called respectively layer N_1 and layer M_1

H_{M_i, N_i} thickness of layer M_i, N_i

W_g width of waveguide in the case of shielded lines

H_M $\overset{\Delta}{=} \sum_{j=1}^m H_{M_j}$

H_N $\overset{\Delta}{=} \sum_{j=1}^n H_{N_j}$

The general form of solution (4.54) in layer M_i (y positive) is

$$\tilde{\Phi}_{M_i}^{H,E}(k_x, y) = \tilde{X}_{M_i}^{H,E}(k_x)\sinh[s_{0Mi}(y - a_i)] + \tilde{Y}_{M_i}^{H,E}(k_x)\cosh[s_{M0i}(y - a_i)] \tag{4.57a}$$

with

$$a_i = \sum_{j=1}^{i-1} H_{M_j} \tag{4.57b}$$

The general form of solution (4.54) in layer N_i (y negative) is

$$\tilde{\Phi}_{N_i}^{H,E}(k_x, y) = \tilde{X}_{N_i}^{H,E}(k_x)\sinh[s_{0Ni}(y + b_i)] + \tilde{Y}_{N_i}^{H,E}(k_x)\cosh[s_{N0i}(y + b_i)] \tag{4.58a}$$

with

$$b_i = \sum_{j=1}^{i-1} H_{N_j} \tag{4.58b}$$

This general solution may be customized to various topologies in the following manner.

1. Covered lines are defined by $0 < H_{N_n} < \infty$ and $0 < H_{M_m} < \infty$
The tangential electric field has to vanish in planes $y = H_M$ and $y = -H_N$, and one deduces from (4.55a) and (4.55c) the equivalent conditions

$$\tilde{\Phi}_{M_m}^E(k_{xn}, H_M) = 0 \quad \text{and} \quad \tilde{\Phi}_{N_n}^E(k_{xn}, -H_N) = 0 \tag{4.59a}$$

$$\left.\frac{\partial \tilde{\Phi}_{M_m}^H(k_{xn}, y)}{\partial y}\right|_{y=H_M} = 0 \quad \text{and} \quad \left.\frac{\partial \tilde{\Phi}_{N_n}^H(k_{xn}, y)}{\partial y}\right|_{y=-H_N} = 0 \qquad (4.59b)$$

which yield the particular solution for the spectral scalar potentials in layers M_m and N_n:

$$\tilde{\Phi}_{M_m}^E(k_x, y) = \tilde{X}_{M_m}^E(k_x) \sinh[s_{0Mm}(y - H_M)] \qquad (4.60a)$$

$$\tilde{\Phi}_{M_m}^H(k_x, y) = \tilde{Y}_{M_m}^H(k_x) \cosh[s_{0Mm}(y - H_M)] \qquad (4.60b)$$

$$\tilde{\Phi}_{N_n}^E(k_x, y) = \tilde{X}_{N_n}^E(k_x) \sinh[s_{0Nn}(y + H_N)] \qquad (4.61a)$$

$$\tilde{\Phi}_{N_n}^H(k_x, y) = \tilde{Y}_{N_n}^H(k_x) \cosh[s_{0Nn}(y + H_N)] \qquad (4.61b)$$

2. Open lines are defined by $H_{M_m} \to \infty$ and $H_{N_n} \to \infty$.
All the fields have to vanish at infinity, so that the particular solution for the spectral scalar potentials in layers M_m and N_n is

$$\tilde{\Phi}_{M_m}^E(k_x, y) = \tilde{X}_{M_m}^E(k_x) e^{-[s_{0Mm}(y - a_m)]} \qquad (4.62a)$$

$$\tilde{\Phi}_{M_m}^H(k_x, y) = \tilde{Y}_{M_m}^H(k_x) e^{-[s_{0Mm}(y - a_m)]} \qquad (4.62b)$$

$$\tilde{\Phi}_{N_n}^E(k_x, y) = \tilde{X}_{N_n}^E(k_x) e^{[s_{0Nn}(y + b_n)]} \qquad (4.63a)$$

$$\tilde{\Phi}_{N_n}^H(k_x, y) = \tilde{Y}_{N_n}^H(k_x) e^{[s_{0Nn}(y + b_n)]} \qquad (4.63b)$$

3. Shielded lines are characterized by electric or magnetic lateral shielding at planes $x = \pm W_g/2$. Some particularities in the formulations of these lines are used in the literature, without any satisfactory mention of their origin. In the following discussion, a rigorous derivation of the simplifications is presented. In the case of perfect electric shielding, the tangential electric field has to vanish on lateral planes, while for perfect magnetic shielding, the tangential magnetic field has to vanish at those planes. Hence, the spatial field may be described as a summation of periodic functions of period $n\pi/W_g$. As a consequence, the spectral fields and potentials exist only for discrete values k_{xn} of the k_x variable

$$k_{xn} = \frac{n\pi}{W_g}$$

The choice of n is related to the symmetry of the scalar potentials describing the particular mode under investigation, the nature of shielding and the relations existing between fields and potentials by virtue of equations (4.52a-c) and (4.53a-c). The odd part of $\Phi_i^H(x, y)$ is described by sine functions of x while the even part of $\Phi_i^E(x, y)$ is described by cosine functions of x. Both parts contribute to the sine component of the tangential magnetic field and the cosine component of the tangential electric field. On the other hand, the even part of $\Phi_i^H(x, y)$ is described by cosine functions of x while the odd part of $\Phi_i^E(x, y)$ is described by sine functions of x. Both parts contribute to the cosine component of the tangential magnetic field and the sine component of the tangential electric field. Hence, when looking for a particular mode having an even symmetry of e_{yi} and h_{xi}, we impose $\Phi_i^H(x, y)$ even and $\Phi_i^E(x, y)$ odd with

$$k_{xn} = (2n + 1)\pi/W_g \qquad \text{for perfect electric lateral shieldings}$$
$$k_{xn} = 2n\pi/W_g \qquad \text{for perfect magnetic lateral shielding}$$

while, when looking for a particular mode having an odd symmetry of e_{yi} and h_{xi}, we impose $\Phi_i^H(x, y)$ odd and $\Phi_i^E(x, y)$ even with

$$k_{xn} = (2n + 1)\pi/W_g \qquad \text{for perfect magnetic lateral shieldings}$$
$$k_{xn} = 2n\pi/W_g \qquad \text{for perfect electric lateral shielding}$$

The fundamentals of the spectral domain technique have been described. The method has been widely used many years. It originated in a paper by Yamashita [4.10]. Itoh and Mittra combined it with Galerkin's procedure to solve an integral equation by the moment method [4.5],[4.26], as introduced in Chapter 2 and developed in Chapter 3 for planar lines. As seen in Chapter 3, the main advantage of the method is that the Green's function relating the currents and the electric fields in the structure is an algebraic expression in the spectral domain. The next example will illustrate the point in the case of the variational principle. An important feature of this new formulation is that the spectral potentials, hence the spectral fields, are expressed as the product of a simple hyperbolic y-dependence with spectral coefficients as described in Section 4.3. Hence the integrations (4.19a-e) along the y-axis are performed analytically before the integration along the spectral k_x-axis.

A comment is necessary about the variational spectral equation (4.21) when shielded lines are considered. The analysis of shielded lines is characterized by the use of values of spectral potentials associated with discrete values of the spectral variable k_x, as explained above. Moreover, these potentials have a physical meaning only over the area of the waveguide. As a consequence, the integrals based on Parseval's equality (4.18) are replaced by an infinite summation over the integrands evaluated in $k_x = k_{xn}$. From now on, the distinction between shielded and open lines will be omitted in the text. In the case of open or covered lines, integrals over the whole k_x spectrum are considered in expressions (4.19), while the integrals are replaced by discrete summations in the case of shielded lines. Two structures differing only by the absence or presence of lateral shielding will have exactly the same expressions for the potentials, fields and integrands of (4.19). Those quantities will simply be evaluated respectively over the whole spectrum k_x or at discrete values k_{xn}.

As discussed at the beginning of this section, the development will now be centered on an example, to ensure a better illustration of the method. A microstrip line is considered, as illustrated in Figure 4.4 [4.4]. Since the line has discontinuities at its conductive layer in plane $y = 0$, the spectral domain

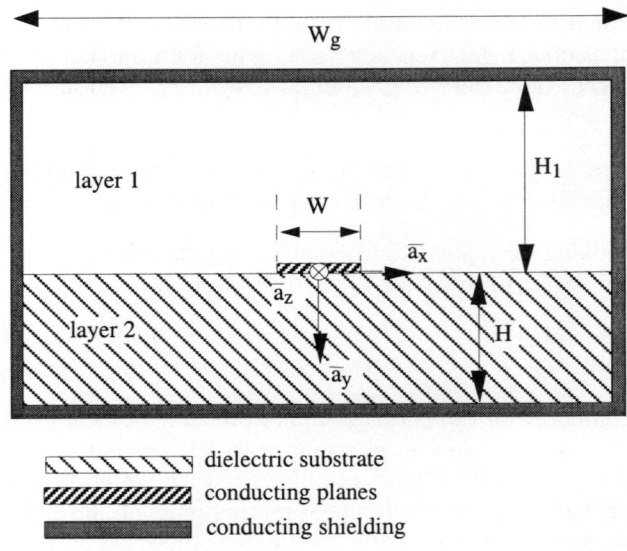

Fig. 4.4 *Geometry of shielded microstrip line analyzed by general variational principle (Reprinted by kind permission of GET/Hermes-Science [4.4])*

formulation of the variational principle will be used here. According to the above notations, the following parameters are defined:

H thickness of substrate, called layer 2
H_1 distance between interface layer 1 - layer 2 and metallic housing
W_g width of waveguide
W width of strip

The structure of Figure 4.4 has only two layers and is fully shielded. The dominant mode of this shielded stripline obviously has an even x-symmetry of the e_y and h_x field components. So, the following particular solution is considered for the potentials, derived from (4.60a,b) and (4.61a,b):

$$\tilde{\Phi}_2^E(k_x, y) = \tilde{X}_2^E(k_x) \sinh[s_{02}(y - H)] \tag{4.64a}$$

$$\tilde{\Phi}_2^H(k_x, y) = \tilde{Y}_2^H(k_x) \cosh[s_{02}(y - H)] \tag{4.64b}$$

$$\tilde{\Phi}_1^E(k_x, y) = \tilde{X}_1^E(k_x) \sinh[s_{01}(y + H_1)] \tag{4.64c}$$

$$\tilde{\Phi}_1^H(k_x, y) = \tilde{Y}_1^H(k_x) \cosh[s_{01}(y + H_1)] \tag{4.64d}$$

with

$$k_x = k_{xn} = \frac{(2n + 1)\pi}{W_g} \tag{4.64e}$$

The spectral fields satisfy Maxwell's equations (4.3) and (4.4), rewritten in the spectral domain in terms of Helmoltz equations applied to the spectral scalar potentials (4.54b).

The next step is the application of the boundary conditions at the various y-interfaces, as for the preceding example. Because of the linearity of the Fourier transform and of the required boundary conditions on the trial fields, the boundary conditions on the spatial fields are equivalent to the same boundary conditions expressed on the spectral fields.

1. In the plane $y = -H_1$.
The tangential electric field has to vanish. Combining the spectral definition (4.55a-c) for the spectral electric field with expressions (4.64a-e) yields

$$\tilde{e}_{x1}(k_x, y) = [-\gamma j k_x \tilde{X}_1^E(k_x) - j\omega\mu_1 \tilde{Y}_1^H(k_x)s_{01}] \sinh[s_{01}(y + H_1)] \quad (4.65a)$$

$$\tilde{e}_{z1}(k_x, y) = (k_1^2 + \gamma^2)\tilde{X}_1^E(k_x) \sinh[s_{01}(y + H_1)] \quad (4.65b)$$

The two expressions vanish for $y = -H_1$.
2. In the plane $y = H$.
The tangential electric field has to vanish. Combining again the spectral definition (4.55a-e) for the spectral electric field with expressions (4.64a,b) yields

$$\tilde{e}_{x2}(k_x, y) = [-\gamma j k_x \tilde{X}_2^E(k_x) - j\omega\mu_2 \tilde{Y}_2^H(k_x)s_{02}] \sinh[s_{02}(y - H)] \quad (4.66a)$$

$$\tilde{e}_{z2}(k_x, y) = (k_2^2 + \gamma^2)\tilde{X}_2^E(k_x) \sinh[s_{02}(y - H)] \quad (4.66b)$$

These two expressions vanish for $y = H$.
3. In the plane $y = 0$.
The continuity of e_x and e_z is imposed. Using expressions (4.65a,b) and (4.66a,b) for the trial electric field in each layer, the boundary conditions are rewritten as

$$[-\gamma j k_x \tilde{X}_1^E(k_x) - j\omega\mu_1 \tilde{Y}_1^H(k_x)s_{01}] \sinh(s_{01}H_1)$$
$$= [\gamma j k_x \tilde{X}_2^E(k_x) + j\omega\mu_2 \tilde{Y}_2^H(k_x)s_{02}] \sinh(s_{02}H) \quad (4.67)$$

$$(k_1^2 + \gamma^2)\tilde{X}_1^E(k_x) \sinh(s_{01}H_1) = -(k_2^2 + \gamma^2)\tilde{X}_2^E(k_x) \sinh(s_{02}H) \quad (4.68)$$

The h_x and h_z components of the magnetic field are related to the current densities flowing on the strip in the x- and z-directions:

$$h_{x1}(x, 0) - h_{x2}(x, 0) = J_z(x, 0) \quad \forall x$$
$$\Updownarrow \quad\quad\quad\quad\quad (4.69a)$$
$$\tilde{h}_{x1}(k_{xn}, 0) - \tilde{h}_{x2}(k_{xn}, 0) = \tilde{J}_z(k_{xn}, 0) \quad \forall n$$

$$h_{z2}(x, 0) - h_{z1}(x, 0) = J_x(x, 0) \quad \forall x$$
$$\Updownarrow \quad\quad\quad\quad\quad (4.69b)$$
$$\tilde{h}_{z2}(k_{xn}, 0) - \tilde{h}_{z1}(k_{xn}, 0) = \tilde{J}_x(k_{xn}, 0) \quad \forall n$$

with $J_z(x,0)$ and $J_x(x,0)$ non-zero only in the range $x = -W/2, x = W/2$. Combining the spectral definition (4.56a-c) for the spectral magnetic field with expressions (4.64a-d) yields in layers 1 and 2

$$\tilde{h}_{x1}(k_x, y) = [-\gamma j k_x \tilde{Y}_1^H(k_x) + j\omega\varepsilon_1 s_{01} \tilde{X}_1^E(k_x)] \cosh[s_{01}(y + H_1)] \quad (4.70a)$$

$$\tilde{h}_{z1}(k_x, y) = (k_1^2 + \gamma^2)\tilde{Y}_1^H(k_x) \cosh[s_{01}(y + H_1)] \quad (4.70b)$$

$$\tilde{h}_{x2}(k_x, y) = [-\gamma j k_x \tilde{Y}_2^H(k_x) + j\omega\varepsilon_2 s_{02} \tilde{X}_2^E(k_x)] \cosh[s_{02}(y - H)] \quad (4.70c)$$

$$\tilde{h}_{z2}(k_x, y) = (k_2^2 + \gamma^2)\tilde{Y}_2^H(k_x) \cosh[s_{02}(y - H)] \quad (4.70d)$$

Hence the boundary conditions (4.69a,b) are rewritten as

$$\begin{aligned}[-\gamma j k_x \tilde{Y}_1^H(k_x) + j\omega\varepsilon_1 s_{01} \tilde{X}_1^E(k_x)] \cosh(s_{01} H_1) \\ - [-\gamma j k_x \tilde{Y}_2^H(k_x) + j\omega\varepsilon_2 s_{02} \tilde{X}_2^E(k_x)] \cosh(s_{02} H) = \tilde{J}_z(k_x)\end{aligned} \quad (4.71)$$

$$\begin{aligned}(k_2^2 + \gamma^2)\tilde{Y}_2^H(k_x) \cosh(s_{02} H) - (k_1^2 + \gamma^2)\tilde{Y}_1^H(k_x) \cosh(s_{01} H_1) \\ = \tilde{J}_x(k_{xn}, 0)\end{aligned} \quad (4.72)$$

Equations (4.67), (4.68), (4.71), and (4.72) form a set of four linear equations in terms of the spectral coefficients $\tilde{X}_1^E(k_x)$, $\tilde{X}_2^E(k_x)$, $\tilde{Y}_1^H(k_x)$ and $\tilde{Y}_2^H$. This set can easily be solved for these coefficients, which are finally expressed as algebraic expressions of the spectral variable k_x and a linear combination of the spectral quantities $\tilde{J}_z(k_{xn}, 0)$ and $\tilde{J}_x(k_{xn}, 0)$:

$$\tilde{Y}_1^H(k_x) = \frac{[a_{22}\tilde{J}_z(k_{xn}, 0) - a_{21}\tilde{J}_x(k_{xn}, 0)]}{(a_{11}a_{22} - a_{12}a_{21})} \quad (4.73a)$$

$$\tilde{Y}_2^H(k_x) = \frac{[a_{11}\tilde{J}_z(k_{xn}, 0) - a_{12}\tilde{J}_x(k_{xn}, 0)]}{(a_{11}a_{22} - a_{12}a_{21})} \quad (4.73b)$$

$$\tilde{X}_1^E(k_x) = a_1\tilde{Y}_1^H(k_x) + a_2\tilde{Y}_2^H(k_x) \quad (4.73c)$$

$$\tilde{X}_2^E(k_x) = K\tilde{X}_1^E(k_x) \quad (4.73d)$$

with

$$K = -\frac{(k_1^2 + \gamma^2)\sinh(s_{01} H_1)}{(k_2^2 + \gamma^2)\sinh(s_{02} H)} \quad (4.73e)$$

$$a_1 = \frac{j\omega\mu_1 s_{01}\sinh(s_{01} H_1)}{-\gamma j k_x[\sinh(s_{01} H_1) + K\sinh(s_{02} H)]} \quad (4.73f)$$

$$a_2 = \frac{j\omega\mu_2 s_{02}\sinh(s_{02} H)}{-\gamma j k_x[\sinh(s_{01} H_1) + K\sinh(s_{02} H)]} \quad (4.73g)$$

$$\begin{aligned}a_{11} = [(-\gamma j k_x + j\omega\varepsilon_1 s_{01} a_1)\cosh(s_{01} H_1) \\ + j\omega\varepsilon_2 s_{02} a_1 K\cosh(s_{02} H)]\end{aligned} \quad (4.73h)$$

$$a_{12} = -[(-\gamma j k_x + j\omega\varepsilon_2 s_{02} K a_2)\cosh(s_{02}H)$$
$$+ j\omega\varepsilon_1 s_{01} a_2 K \cosh(s_{01}H_1)]$$
(4.73i)

$$a_{21} = -(k_1^2 + \gamma^2)\cosh(s_{01}H_1)$$
(4.73j)

$$a_{22} = (k_2^2 + \gamma^2)\cosh(s_{02}H)$$
(4.73k)

Hence the potentials and the fields are readily expressed as a linear combination of the spectral current density. The shape of the various components of the current density remains to be determined. As a first test, the transverse current density is neglected:

$$J_x(x,0) = 0 \quad \forall\, x \qquad \text{which yields} \qquad \tilde{J}_x(k_{xn},0) = 0 \quad \forall\, n$$

Because of the variational nature of the solution, only an approximation of the current density on the strip is needed. The error made on the current induces an error on the trial electric field (4.65a,b) and (4.66a,b), because this field is obtained from the potential (4.64a-d) whose spectral coefficients (4.73a-d) are linear combinations of the spectral current density. But, as a result of the variational character, this trial spectral field introduced in expressions (4.19a-e) does not induce a first-order error on the propagation constant obtained with (4.21). The longitudinal component of the current density is assumed to be constant on the strip, and imposed to be zero outside of the strip (unit step function):

$$J_z(z,0) = I_0/W \quad \text{for} \quad -W/2 < x < W/2$$
(4.74a)

$$J_z(z,0) = 0 \quad \text{for} \quad x < -W/2 \quad \text{or} \quad x > W/2$$
(4.74b)

The Fourier transform of this function is well known. It is

$$\tilde{J}_z(k_{xn},0) = I_0 2 \frac{\sin(k_{xn}W/2)}{(k_{xn}W)}$$
(4.74c)

The validity of the variational spectral equation (4.21) combined with the spectral fields derived from (4.64a-d) and satisfying boundary conditions (4.67) to (4.69a,b) is illustrated in Figure 4.5, where results obtained by Mittra and Itoh (circles) [4.27] are compared with results obtained with this formulation (solid line). They agree very well. It should be underlined, however, that the model of Mittra and Itoh solves the integral equation by Galerkin's method in the spectral domain. They report a solution at one frequency in 3 seconds on an IBM 360 computer. However, when the variational formulation is implemented on a regular PC, the calculation does not exceed a few seconds for the total of twenty frequency points involved in Figure 4.5. This comparison illustrates the validity of the variational formulation as well as its numerical efficiency.

Hence, the variational method can be used for on-line designs up to millimeter waves. As an example, the propagation constant of a millimeter wave shielded line was calculated up to 200 GHz [4.28]. Results are presented in Figure 4.6. The computation time is also of a few seconds on a regular PC.

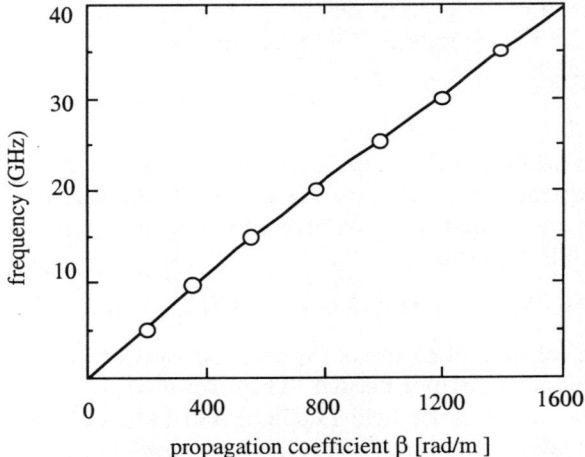

Fig. 4.5 *Propagation coefficient of shielded microstrip (Fig. 4.4) calculated with variational formulation (—) and by Mittra and Itoh [4.27] (o o o) (W = 1.27 mm, $H_1 = 11.43$ mm, $H = 1.27$ mm, $W_g = 12.7$ mm, $\varepsilon_{r1} = 1$, $\varepsilon_{r2} = 4.2$)*

4.4.5   Trial fields in terms of electric field components

The variational formulation (4.21) only involves the spectral electric fields components. Hence, using a suitable form for the electric field components and applying the conditions required by the variational formulation, it is possible to determine the unknown amplitude coefficients. This is applied to the open slot-line represented in Figure 4.7, with the definitions

x	horizontal coordinate, centered in middle of slot
y	vertical coordinate, centered in plane of interface containing the slot
H	thickness of substrate, called layer 2
W	width of slot
layer 1	infinite area corresponding to $y < 0$
layer 3	infinite area corresponding to $y > H$.

Only simple y-dependencies of the fields are of interest because of the integration along the y-axis in (4.19a-e). Hence the following expressions are chosen for the spectral trial electric field in the three layers:

$$\tilde{e}_{v1}(k_x, y) = \tilde{A}_v(k_x)e^{s_{01}y} \tag{4.75a}$$

$$\tilde{e}_{v2}(k_x, y) = \tilde{B}_v(k_x)\sinh(s_{02}y) + \tilde{C}_v(k_x)\cosh(s_{02}y) \tag{4.75b}$$

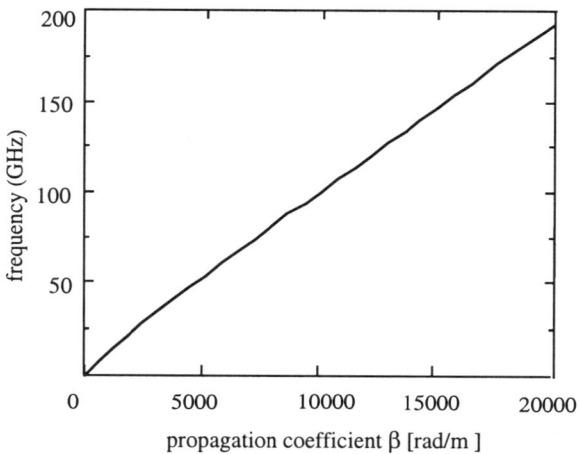

Fig. 4.6 *Propagation coefficient of shielded microstrip (Fig. 4.4) at millimeter waves, calculated with variational formulation (W = 0.5 mm, H₁ = 10.43 mm, H = 0.254 mm, Wg = 2.0 mm, εr1 = 1, εr2 = 25)*

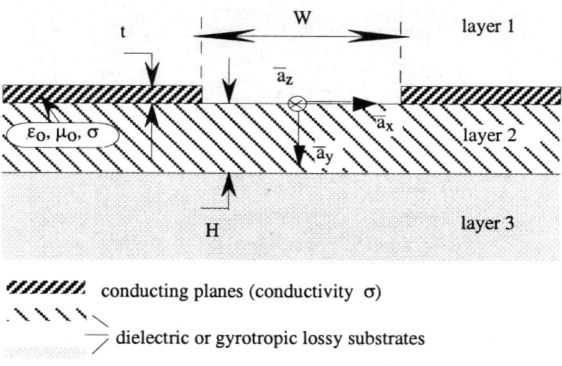

Fig. 4.7 *Geometry of open slot-line analyzed by using the variational principle*

$$\tilde{e}_{v3}(k_x, y) = \tilde{D}_v(k_x)e^{-s_{01}(y-H)} \tag{4.75c}$$

The continuity of the electric field at the interfaces is first imposed. Then, the spectral trial magnetic fields are deduced from Maxwell's equation (4.3) rewritten in the spectral domain and the continuity of the magnetic field at the interfaces is imposed. By doing so, six equations are obtained, in terms of the 14 unknowns, which are

$$\tilde{A}_v, \tilde{B}_v, \tilde{C}_v \text{ and } \tilde{D}_v \text{ where } v = x, y, z \quad (12 \text{ unknowns}) \tag{4.76a}$$

$$s_{01} \text{ and } s_{02} \quad (2 \text{ unknowns}) \tag{4.76b}$$

The variational character of equation (4.21) only holds when Maxwell's equation (4.4) is satisfied. Rewriting (4.4) in each layer in terms of the trial spectral electric field (4.75) and spectral magnetic field obtained from the spectral form of (4.3) provides a set of 9 equations which must be fulfilled for each value of y. It can easily be shown that this transforms the set of 9 equations into a homogeneous system of 12 equations to be solved for the 14 unknowns. With the particular choice

$$s_{01} = \sqrt{k_x^2 - \gamma^2 - k_1^2} \tag{4.77a}$$

$$s_{02} = \sqrt{k_x^2 - \gamma^2 - k_2^2} \tag{4.77b}$$

only four of them are independent. They are given by

$$s_{01}[jk_x\tilde{A}_y(k_x) - s_{01}\tilde{A}_x(k_x)] + \gamma[-jk_x\tilde{A}_z(k_x) - \gamma\tilde{A}_x(k_x)] = k_1^2\tilde{A}_x(k_x) \tag{4.78a}$$

$$-s_{01}[jk_x\tilde{D}_y(k_x) + s_{01}\tilde{D}_x(k_x)] + \gamma[-jk_x\tilde{D}_z(k_x) - \gamma\tilde{D}_x(k_x)] = k_1^2\tilde{D}_x(k_x) \tag{4.78b}$$

$$s_{02}[jk_x\tilde{B}_y(k_x) - s_{02}\tilde{C}_x(k_x)] - \gamma[jk_x\tilde{C}_z(k_x) + \gamma\tilde{C}_x(k_x)] = k_2^2\tilde{B}_x(k_x) \tag{4.78c}$$

$$s_{02}[jk_x\tilde{C}_y(k_x) - s_{02}\tilde{B}_x(k_x)] - \gamma[jk_x\tilde{B}_z(k_x) + \gamma\tilde{B}_x(k_x)] = k_2^2\tilde{B}_x(k_x) \tag{4.78d}$$

These four independent homogeneous equations relate the 14 unknowns. Hence, combining equations (4.77a,b) with the six equations (4.65a,b), (4.66 a,b), (4.67), and (4.68) resulting from boundary conditions, into the equations (4.78a-d) yields 10 equations in terms of the 12 spectral coefficients involved in (4.75a-c). Assuming a trial tangential electric field at the interface containing the slot, two additional equations may be used:

$$\tilde{A}_x(k_x) = \tilde{e}_x(k_x, 0) \tag{4.79}$$

$$\tilde{A}_z(k_x) = \tilde{e}_z(k_x, 0) \tag{4.80}$$

where $\tilde{e}_x(k_x, 0)$ and $\tilde{e}_z(k_x, 0)$ are the known trial electric field components at the interface. Finally, a system of 12 linear equations is obtained: the six equations resulting from boundary conditions, (4.78a-d), (4.79) and (4.80), to be solved for the 12 spectral coefficients $\tilde{A}_v, \tilde{B}_v, \tilde{C}_v, \tilde{D}_v$. As for the microstrip case treated in the former subsection, the spectral coefficients - hence the spectral trial fields - are related to the trial quantities at the interface containing a discontinuous conductive layer, namely for slot-like lines the tangential components of the electric field at this interface.

The shape of the electric field components at the interface remains to be determined. Because of the variational nature of the solution, only approximations of these components are needed. As a first test, the longitudinal component is neglected:

$$e_z(x, 0) = 0 \quad \forall x \quad \text{which yields} \quad \tilde{e}_z(k_x, 0) = 0 \tag{4.81}$$

The transverse component of the electric field is assumed to be constant across the slot, and imposed to be zero outside of the slot:

$$e_x(x, 0) = V_0/W \quad \text{for} \quad -W/2 < x < W/2 \tag{4.82a}$$

$$e_x(x, 0) = 0 \quad \text{for} \quad x < -W/2 \quad \text{or} \quad x > W/2 \tag{4.82b}$$

which yields

$$\tilde{e}_x(k_x, 0) = V_0 2 \frac{\sin(k_x W/2)}{(k_x W)} \tag{4.83}$$

Hence, the problem is totally characterized and can be solved.

4.4.6 Link with the Green's formalism

Two ways for obtaining trial fields, for strip-like and slot-like lines respectively, have been described in the spectral domain. It is underlined that both formulations, in terms of potentials or fields, respectively, can be used for either slot-like or strip-like structures. As an example, Table 4.1 compares results obtained with the second formulation for narrow and wide open slot-lines to results obtained by Janaswamy and Schaubert [4.29]. It appears that the difference is negligible. The formulation in terms of fields, however, offers some advantages when the influence of each component of the electric field has to be investigated.

Green's functions are present in the variational formalism, because the spectral fields are written as a function of spectral coefficients (4.54d) or (4.76a), expressed as linear combinations of the spectral current density or tangential electric field in the aperture. As a consequence, the spectral electric and magnetic fields can be written as linear algebraic combinations of the spectral current density for striplines, and of the spectral tangential electric field in the slot for slot-lines, respectively. Hence, in these cases the spectral domain formalism transforms the spatial integral definition of Green's

W/H	Frequency in GHz	(λ'/λ_0) [Jan.-Schaub.]	(λ'/λ_0) [Huynen]
1.335	2.0	0.8885	0.8890
	2.5	0.8834	0.8841
	3.0	0.879	0.8798
	3.5	0.875	0.8758
	4.0	0.871	0.8708
10.71	2.0	0.958	0.9581
	2.5	0.954	0.9545
	3.0	0.951	0.9515
	3.5	0.948	0.9488
	4.0	0.943	0.9460
	5.0	0.939	0.9409
	6.0	0.933	0.9360

Table 4.1 *Comparison of results obtained with variational formulation (Huynen) and by Janaswamy and Schaubert [4.29], for narrow and wide slots*

function [4.30][4.31] into a set of algebraic linear equations in the spectral domain. This algebraic formulation is a major advantage of the spectral domain method. As explained in Chapter 3, Green's functions are frequently involved as an operator in the linear integral equation being solved by the moment method. In this case, however, they are only used to find suitable trial fields to introduce into coefficients (4.19a-e) of the variational spectral equation (4.21). Green's functions appear in a quadratic form in equation (4.21) where squared powers of the fields are required.

For slot-like problems, the current density is obtained as the difference between the tangential magnetic fields in plane $y = 0$. Using the dependence between the electric field and the magnetic field given by Maxwell's equations, it is possible to express the electric field in the whole structure as a function of the tangential electric field in the slot, via the spectral coefficients, common to both fields. A new Green's function is then established between the electric fields in the whole space and in the slot.

Hence, for both formulations, potentials or fields, the final result is an expression of the electric field in terms of trial quantities at an interface with a discontinuous conductive layer. The trial quantities are the tangential current density flowing on the conducting layer for strip-like lines and the tangential electric field across the interface for slot-like lines. Since the trial fields satisfy Maxwell's equations, they differ from the exact (unknown) fields only by their shape. This is discussed in the next section.

4.5 Trial fields at discontinuous conductive interfaces

It has been shown that Mathieu functions (Appendix E), eigenfunctions of Laplace's equation in elliptical and hyperbolic coordinates, form a very efficient set of trial expressions for the tangential trial electric field components for slot-like structures [4.7]. In this section, the fundamentals of the derivation of these functions for slot-like structures are detailed. It will be shown that the same functions may be used to describe the current densities flowing on the strips in strip-like problems.

Due to the formulation developed in Section 4.4, the trial quantity reduces to components of the electric field or current density, tangential to the interface containing the discontinuous metallization. The analytical expression of the trial quantity is determined here by conformal mapping.

4.5.1 Conformal mapping

The change of coordinates

$$
\begin{aligned}
x &= a \cos u \cosh v \\
y &= a \sin u \sinh v
\end{aligned}
\tag{4.84}
$$

where $a = W/2$

transforms the (x, y) plane of Figures 4.1, 4.4 and 4.7 into the (u, v) plane. The configuration will be detailed when examining strip-like structures in Section 4.5.3. It is considered that the slot is placed in a homogenous medium of permittivity $(\varepsilon_1 + \varepsilon_2)/2$, which is the mean value of the permittivities of the two layers adjacent to the slot. The Helmoltz equations (4.51a,b) to be solved in the (x, y) plane for TE/TM modes can be transformed into the (u, v) plane as

$$
\nabla^2 \Phi^{H,E}(u, v) + (k_{eff}^2 + \gamma^2)(\frac{a^2}{2})(\cosh 2v - \cos 2u)\Phi^{H,E}(u, v) = 0
\tag{4.85}
$$

where $k_{eff}^2 = \omega^2 \mu_0 \frac{(\varepsilon_1 + \varepsilon_2)}{2}$

by using the following relations, derived from the change of coordinates:

$$
\begin{aligned}
\frac{\partial \Phi^{H,E}}{\partial u} &= -a \sin u \cosh v \frac{\partial \Phi^{H,E}}{\partial x} - a \cos u \sinh v \frac{\partial \Phi^{H,E}}{\partial y} \\
\frac{\partial \Phi^{H,E}}{\partial v} &= a \cos u \sinh v \frac{\partial \Phi^{H,E}}{\partial x} - a \sin u \cosh v \frac{\partial \Phi^{H,E}}{\partial y}
\end{aligned}
\tag{4.86}
$$

Letting $2q = (k_{eff}^2 + \gamma^2)(a^2/2)$ yields Mathieu's equation

$$
\nabla^2 \Phi^{H,E}(u, v) + 2q(\cosh 2v - \cos 2u)\Phi^{H,E}(u, v) = 0
\tag{4.87}
$$

whose general solutions are obtained by separating the variables:

$$
\Phi^{H,E}(u, v) = U^{H,E}(u)V^{H,E}(v)
\tag{4.88}
$$

The solutions of $U^{H,E}(u)$ are called Mathieu functions, while those of $V^{H,E}(v)$ are called the modified Mathieu functions. Only the solutions of $U^{H,E}(u)$ will be of interest here. Those solutions are classically noted $ce_n(q,u)$ and $se_n(q,u)$, respectively. They can be simply developed into cosine functions weighted by a power expansion of the q parameter as

$$ce_n(q,u) = \sum_{i=1}^{\infty} a_{ni}(q)\cos(iu) \tag{4.89}$$

$$se_m(q,u) = \sum_{i=1}^{\infty} b_{mi}(q)\sin(iu) \tag{4.90}$$

where n and m are arbitrary integers, except that $m \neq 0$. Suitable expressions for the coefficients $a_{ni}(q)$ and $b_{mi}(q)$ can be found in Appendix E and in [4.32].

4.5.2   Adequate trial components for slot-like structures

Because of this formulation, the trial quantity for slot-like structures reduces to components of the electric field, tangential to the interface between two dielectric layers adjacent to a discontinuous conductive interface. Due to the boundary condition, it must have the same x-variation across the slot on the two sides of the interface, and vanish on the metallization. The analytical expression of this trial tangential field is determined here from conformal mapping (4.84), for which it is considered that the slot is placed in a homogenous medium of permittivity $(\varepsilon_1 + \varepsilon_2)/2$, which is the mean value of the permittivities of the two layers adjacent to the slot.

Searching for a solution in plane $y = 0$ and for $-1 < x/a < 1$, the equivalent domain to be considered in the (u,v) plane is defined by

$$0 < u < \pi \quad \text{and} \quad v = 0$$

which yields

$$u = \arccos(\frac{2x}{W}) \tag{4.91}$$

For the slot-line in the homogeneous medium considered in Figure 4.8, the arccosine conformal mapping (4.91) transforms the conducting planes of the slot into a parallel plate waveguide (structure 3, Fig. 4.8).

The transverse component of the tangential electric field across the slot under TE assumption can be expressed as

$$e_x \approx -j\omega\mu_0 \left.\frac{\partial \Phi^H}{\partial y}\right|_{y=v=0} = \frac{j\omega\mu_0}{a\sin u} U^H(u) \left.\frac{dV^H}{dv}\right|_{v=0} \tag{4.92}$$

It can be seen that only the solutions for $U^{H,E}(u)$ are needed, because the x-dependence of the field across the slot is entirely determined by $U^{H,E}(u)$ at

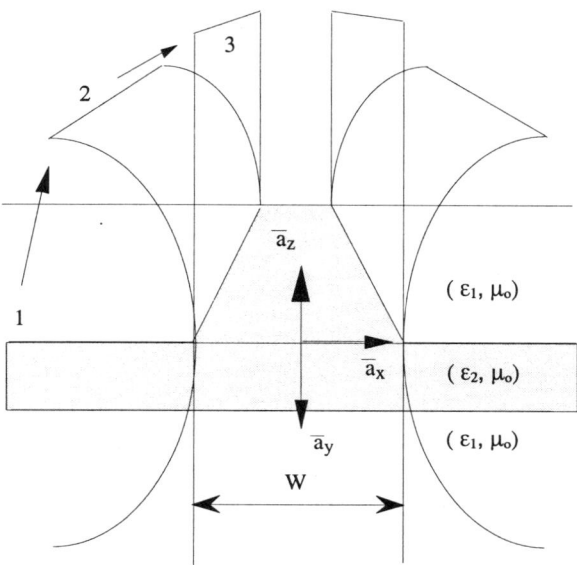

Fig. 4.8 *Transformation of conductors of slot-line when applying arccosine conformal mapping, from the conducting planes of the slot (1) to the parallel plate waveguide (3)*

$v = y = 0$. For $v = 0$, and to properly describe the e_x component in the slot, the potential Φ^H may not vanish in planes $u = 0$ and $u = \pi$. Hence, $se_n(q, u)$ are avoided here. The even (n even) and odd (n odd) x-dependencies of e_x across the slot can be expressed, using (4.89) to (4.92), as

$$
\begin{aligned}
e_{xn}(x, 0) &= \frac{K_x U^H(u)}{\sin u} \\
&= K_x \frac{ce_n[q, \arccos(2x/W)]}{\sqrt{1 - (2x/W)^2}} \\
&= K_x \sum_i a_{ni}(q) \frac{\cos[i \arccos(2x/W)]}{\sqrt{1 - (2x/W)^2}}
\end{aligned}
\tag{4.93}
$$

with i even for n even and i odd for n odd. It appears that this equation combine Mathieu functions with the correct singularity in the field at the edge of an infinitesimally thin metal sheet known to be of order $-1/2$, because of equation (4.91), due to the conformal mapping. Hence, one may expect a rapid convergence.

On the other hand, the longitudinal component e_z tangential to the interface is obtained from equations (4.52c) as

$$
\begin{aligned}
e_z &= (k^2 + \gamma^2)\Phi^E\big|_{y=v=0} \\
&= (k^2 + \gamma^2)U^E(u)V^E(v)
\end{aligned}
\tag{4.94}
$$

Similarly, for $v = 0$, the potential Φ^E has to vanish in planes $u = 0$ and $u = \pi$ to properly describe the e_z component in the slot. Hence, $\mathrm{ce}_i(q, u)$ have to be avoided here. The even (m odd) and odd (m even) x-dependencies of e_z across the slot are finally expressed, using (4.90), (4.91), and (4.94), as

$$
\begin{aligned}
e_{zm}(x,0) &= K_z U^E(u) \\
&= K_z\, \mathrm{se}_m[q, \arccos(2x/W)] \\
&= K_z \sum_i b_{mi}(q) \sin[i \arccos(2x/W)]
\end{aligned}
\tag{4.95}
$$

with i odd for m odd and i even for m even.

The reader recognizes in expansions (4.93) and (4.95) the Tchebycheff polynomial basis functions modified by an edge condition, used by some authors [4.33][4.34] as basis functions in Galerkin's procedure. Because of the physical point of view (conformal mapping of the structure) adopted here, the series expansion obtained is a very good formulation of the field across the slot: the coefficients affecting the basis functions are determined directly from physical considerations without solving any eigenvalue problem. The Fourier transform of any desired mode can be found as the Fourier transform of expressions (4.93) and (4.95) [4.35]:

$$
\tilde{e}_{xn}(k_x) = K_x \sum_i a_{ni}(q)(j)^i J_i(k_x W/2)
\tag{4.96a}
$$

$$
\tilde{e}_{zm}(k_x) = K_z \sum_i b_{mi}(q)(j)^{i-1}(i)\frac{J_i(k_x W/2)}{(k_x W/2)}
\tag{4.96b}
$$

where J_i is the Bessel function of first kind and order i. Any term $\tilde{e}_{xn}$ and $\tilde{e}_{zm}$ can be taken as a useful expression for higher order modes in slot-like structures.

The relationship between n and m is readily deduced from (4.52a-c) where it is shown that the x-dependence of the e_x component is proportional to the first x-derivative of the e_z component. Hence, if an even dependence of the e_x component is imposed by the choice of n even, e_z must have an odd dependence, which means that m is even. It is therefore concluded that n and m in expressions (4.93) to (4.96a,b) have the same even or odd parity.

For the dominant mode in the parallel plate waveguide (structure 3, Fig. 4.8), the x-component of the potential does not depend on the u-variable. Hence, a reasonable assumption for e_x in the slot is obtained by considering the image of expressions (4.89) and (4.90) taken for $n = 0$ and $m = 2$ when applying the arccosine conformal mapping, which yields (4.93) to (4.96a,b) rewritten for $n = 0$ and $m = 2$. Moreover, for moderate slot widths, the value of q remains negligible, so that only the first term in the serial expansions (4.93) to (4.96a,b) has to be considered. Hence the components of the dominant mode may be approximated by

$$
e_{x0} = \begin{cases} K_x \dfrac{1}{\sqrt{1 - (2x/W)^2}} & \text{if } 0 < |x| < \dfrac{W}{2} \\ 0 & \text{otherwise} \end{cases} \tag{4.97a}
$$

and

$$
e_{z2}(x,0) = K_z \sin\left[2 \arccos\left(\frac{2x}{W}\right)\right] \tag{4.97b}
$$

which respectively have the Fourier transforms [4.36]

$$
\tilde{e}_x(k_x) = K_x \frac{\pi W}{2} J_0\left(k_x \frac{W}{2}\right) \tag{4.97c}
$$

and

$$
\tilde{e}_z(k_x) = K_z j2 \frac{J_2(k_x W/2)}{(k_x W/2)} \tag{4.97d}
$$

It should be noted that the x-component of the electric field across the slot is easily related to an equivalent voltage V_0 across the line, which is obtained when integrating the e_x-field over the slot area $0 < |x| < W/2$. Hence, coefficient K_x is related to V_0 as

$$
K_x = \frac{2V_0}{\pi W} \tag{4.97e}
$$

The same dependence of the electric field can be used to describe coupled slot-lines (Fig. 4.9), by modifying the Fourier transform of the electric field to take into account the presence of two slots centered at planes $x = -(W_1 + S)/2$ and $x = (W_2 + S)/2$:

$$
\tilde{e}_x(k_x) = 2\pi[J_0(k_x W_1/2)e^{jk_x(W_1+S)/2} + C_v J_0(k_x W_2/2)e^{-jk_x(W_2+S)/2}] \tag{4.98}
$$

where C_v describes the ratio of voltages across the two slots of respective widths W_1 and W_2 and S is the spacing between the slots. In the case of symmetric lines, ratio C_v is made equal to -1 for odd coupled slot-lines or coplanar waveguide, and to $+1$ for even coupled slot-lines. In the case of asymmetric lines, ratio C_v is calculated by the Rayleigh-Ritz procedure, as mentioned for higher order modes in a next paragraph.

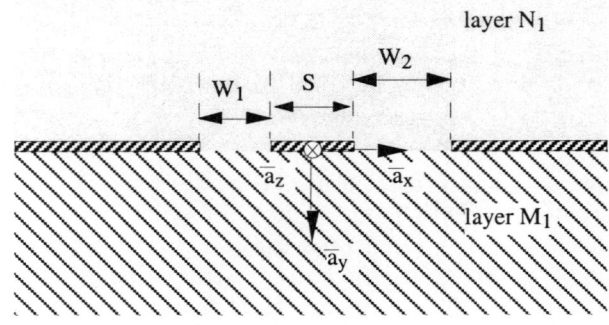

Fig. 4.9 *Geometry of coupled slot-lines for determining trial quantities*

General Mathieu functions (Appendix E), with no restriction on the value of q or m and n, are quite adequate functions as trial quantities for the higher order modes fields in slot-lines. The general expression for the transverse tangential electric field in the slot is therefore developed as a series expansion of Mathieu functions ce_n:

$$
e_x(x,0) = \begin{cases} \displaystyle\sum_{n=N_1}^{N_1+2N} R_n \frac{\mathrm{ce}_n[q, \arccos(\frac{2x}{W})]}{\sqrt{1 - (\frac{x}{W/2})^2}} & \text{if } 0 < |x| < \frac{W}{2} \\ 0 & \text{otherwise} \end{cases}
\tag{4.99a}
$$

while the longitudinal component is obtained from Mathieu functions se_m:

$$
e_z(x,0) = \begin{cases} \displaystyle\sum_{m=M_1}^{M_1+2M} Z_m\, \mathrm{se}_m[q, \arccos(\frac{2x}{W})] & \text{if } 0 < |x| < \frac{W}{2} \\ 0 & \text{otherwise} \end{cases}
\tag{4.99b}
$$

where R_n and Z_m are the unknown coefficients. The Fourier transform of those expressions are

$$
\tilde{e}_x(k_x) = \sum_{n=N_1}^{N_1+2N} R_n[\sum_{r=1}^{\infty} a_{nr}(q)(j)^r J_r(k_x W/2)]
\tag{4.99c}
$$

$$
\tilde{e}_z(k_x) = \sum_{m=M_1}^{M_1+2M} Z_m[\sum_{r=1}^{\infty} b_{mr}(q)(j)^{r-1} r \frac{J_r(k_x W/2)}{(k_x W/2)}]
\tag{4.99d}
$$

with

$$
q = (k_{eff}^2 + \gamma^2)\frac{W^2}{16}
\tag{4.99e}
$$

$$k^2_{eff} = \omega^2 \mu_0 \frac{(\varepsilon_1 + \varepsilon_2)}{2} \tag{4.99f}$$

The expression is valid for a single slot. For coupled slot-lines a formulation similar to (4.98) established for the dominant mode may be derived. The coefficients R_n and Z_m are obtained by applying the Rayleigh-Ritz procedure. The advantages of the formulation using Mathieu functions are twofold. First, the combination of the basis functions is known *a priori*, since q depends only on the geometrical and physical parameters of the line, and on the frequency. This is not the case for Galerkin's procedure where the coefficients weighting the basis functions are the eigenvectors associated with an eigenvalue problem. Secondly, the combination of basis functions is frequency-dependent, which ensures a dynamic formulation of the trial quantities counterbalancing the quasi-static hypothesis underlying the use of a conformal mapping. This is because the Helmoltz equation (4.85) being solved in the transformed domain involves the influence of frequency.

4.5.3 Adequate trial components for strip-like structures

Because of the formulation, the trial electric field for strip-like structures is related to the components of the current density flowing on strips in the interface between two dielectric medias. The analytical expression of these components is determined here by the same conformal mapping (4.84) as for slot-like lines, for which it is considered that the strip is placed in a homogenous medium of permittivity $(\varepsilon_1 + \varepsilon_2)/2$. The conducting strip in the homogeneous medium is shown on Figure 4.10. Solving (4.85) for the potential around the strip is equivalent to solving for the potential between two homofocal elliptical cylinders, when the inner cylinder has $b_i = 0$ and the outer cylinder goes to infinity.

The change of coordinates of (4.84) transforms the (x, y) plane of Figure 4.10a into the (u, v) plane. Looking for a solution in the whole space bounded by the two homofocal elliptical cylinders of parameters (a_i, b_i) and (a_e, b_e), the equivalent domain to be considered in the (u, v) plane is defined by

$$-\pi < u < \pi \quad \text{and} \quad \cosh^{-1}(a_i) < v < \cosh^{-1}(a_e)$$

When the inner cylinder is degenerate and the outer cylinder goes to infinity (Fig. 4.10b), the area between the two cylinders is transformed into a semi-infinite space bounded in the u-direction by the planes $u = \pm\pi$ (Fig. 4.10c):

$$-\pi < u < \pi \quad \text{and} \quad 0 < v < \infty \tag{4.100}$$

The strip area corresponds to $v = 0$, so that equation (4.91) still holds. The general solution (4.88) of Mathieu's equation (4.87) is chosen in the domain given in (4.100) so that $V^{H,E}(v)$ vanishes at infinity, taking advantage of the serial expansion of the modified Mathieu functions into hyperbolic functions of the v-variable [4.32]. Only solutions of $U^{H,E}(u)$ are considered here, because

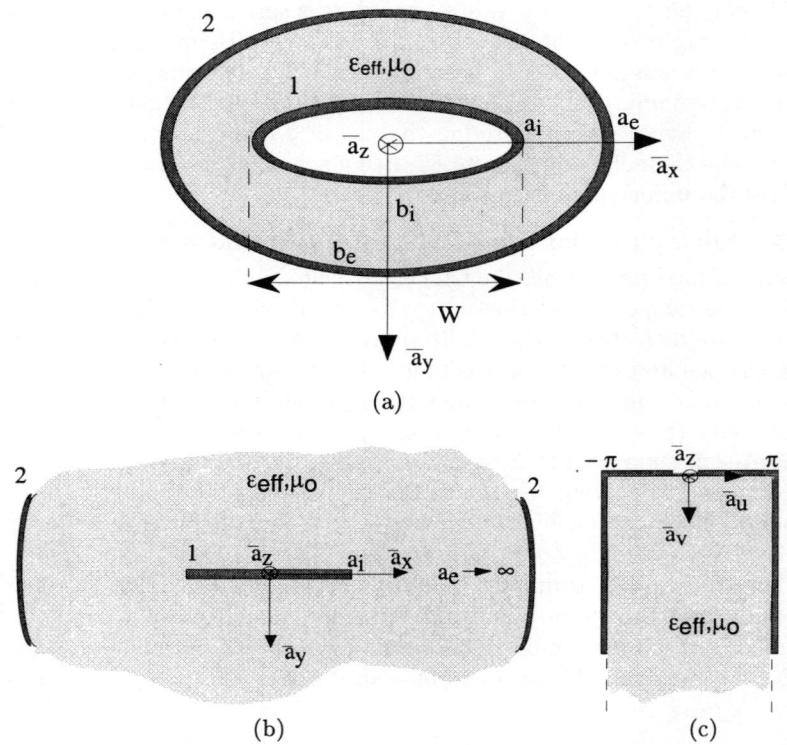

Fig. 4.10 *Transformation of conductors of microstrip when applying arccosine conformal mapping (a) original elliptical cylinders; (b) microstrip configuration; (c) equivalent structure in transformed domain*

one is looking for the surface charge density on the strip, which is proportional to the e_y-component of the field, normal to the strip at plane $y = v = 0$. The component of the electric field normal to the strip under the TM assumption can be expressed (4.52) as

$$e_y \approx -\gamma \frac{\partial \Phi^E}{\partial y}\bigg|_{y=v=0} = \frac{\gamma}{a \sin u} U^E(u) \frac{dV^E}{dv}\bigg|_{v=0} \tag{4.101}$$

From this e_y-component, the even (n even) and odd (n odd) x-dependence of the surface charge density ρ_s using (4.89), (4.91), and (4.101) is

$$
\begin{aligned}
\rho_{sn}(x, 0) &= e_{yn}(x, 0) \\
&= \frac{K_s U^E(u)}{\sin u} \\
&= K_s \frac{ce_n[q, \arccos(2x/W)]}{\sqrt{1 - (2x/W)^2}} \\
&= K_s \sum_i a_{ni}(q) \frac{\cos[i \arccos(2x/W)]}{\sqrt{1 - (2x/W)^2}}
\end{aligned}
\tag{4.102}
$$

with i even for n even and i odd for n odd.

Following the static approximation of Denlinger [4.37], the longitudinal current density on the strip is related to the surface charge density by

$$j_z = \frac{\omega \rho_s}{\beta} \tag{4.103a}$$

Hence, the even and odd x-dependencies of the longitudinal current density on the strip are obtained as

$$j_{zn}(x, 0) = K_s' \sum_i a_{ni}(q) \frac{\cos[i \arccos(2x/W)]}{\sqrt{1 - (2x/W)^2}} \tag{4.103b}$$

As for slot-like structures, the shape of the transverse current component is deduced from the relationship between the potential and the h_z component responsible for the j_x current. Equations (4.53a-c) yield indeed

$$
\begin{aligned}
j_x &= h_{z1} - h_{z2} \\
&= (k^2 + \gamma^2) \left[\Phi_1^H - \Phi_2^H \right]\big|_{y=v=0} \\
&= (k^2 + \gamma^2) \left[U_2^H(u)V_2^H(0) - U_1^H(u)V_1^H(0) \right]\big|_{y=v=0}
\end{aligned}
\tag{4.104}
$$

$U^H(u)$ is a Mathieu function, because it is a solution of (4.87). Hence, j_x has the form

$$
\begin{aligned}
j_{xm}(x, 0) &= K_s'' U^H(u) \\
&= K_s'' \, se_m[q, \arccos(2x/W)] \\
&= K_s'' \sum_i b_{mi}(q) \sin[i \arccos(2x/W)]
\end{aligned}
\tag{4.105}
$$

with i odd for m odd and i even for m even. Because at $v = 0$, j_{xm} has to vanish at planes $u = 0$ and $u = \pi$ to describe properly the finite width of the strip, all $ce_i(q, u)$ are avoided in (4.105). The Fourier transform of the current densities (4.103b) and (4.105) are of course similar to those of (4.96a,b):

$$\tilde{j}_{zn}(k_x) = K'_s \sum_i a_{ni}(q)(j)^i J_i(k_x W/2) \tag{4.106a}$$

$$\tilde{j}_{xm}(k_x) = K''_z \sum_i b_{mi}(q)(j)^{i-1}(i) \frac{J_i(k_x W/2)}{(k_x W/2)} \tag{4.106b}$$

Similarly, the relation between n and m is obtained as for slot-like lines by inspection of equations (4.53a-c), which imply that the x-dependence of the h_x-component is proportional to the first x-derivative of h_z. The j_z-component is related to the transverse tangential magnetic field by

$$j_z = h_{x1} - h_{x2} \tag{4.107}$$

and equation (4.104) provides a similar relationship between j_x and h_z. It can therefore be concluded that j_x must have an even/odd x-dependence to preserve the odd/even dependence of j_z, so that m and n must have the same even or odd parity. Any term $\tilde{j}_{xn}$ and $\tilde{j}_{zm}$ can be taken as a useful expression for higher-order modes in strip-like structures. Hence, it has been demonstrated that Mathieu functions used as trial expressions for the tangential e_x and e_z-components are also adequate trial expressions for the current densities j_z and j_x, respectively.

As for slot-lines, the potential corresponding to the dominant mode in the parallel plate structure of Figure 4.10 does not depend on the u-variable. Hence, by virtue of (4.101) to (4.103b), a reasonable assumption for j_z on the strip is the image of expressions (4.89) and (4.90) taken for $n = 0$ and $m = 2$ when applying the conformal mapping of (4.84), namely (4.103b) with $n = 0$ and $m = 2$. The approximations made for moderate slot width are readily rewritten for moderate strip width, which yields the following approximations for the components of the dominant mode

$$j_z(x,0) = \begin{cases} K'_s \dfrac{1}{\sqrt{1 - (2x/W)^2}} & \text{if } 0 < |x| < \frac{W}{2} \\ 0 & \text{elsewhere} \end{cases} \tag{4.108a}$$

and

$$j_x(x,0) = K''_s \sin[2\arccos(\frac{2x}{W})] \tag{4.108b}$$

which, respectively, have as Fourier transform

$$\tilde{j}_z(k_x) = K'_s \frac{\pi W}{2} J_0(k_x W/2) \tag{4.108c}$$

and

$$\tilde{j}_x(k_x) = K_s'' 2j \frac{J_2(k_x W/2)}{(k_x W/2)} \tag{4.108d}$$

Of course, the z-component of the current density can be related to the total current I_0 flowing on the strip by integrating j_z over the strip area, which yields

$$K_s' = 2\frac{I_0}{\pi W} \tag{4.108e}$$

As for slot-like lines, the dependence of the longitudinal current density can be used to describe coupled strip-lines (Fig. 4.11), by modifying the Fourier transform of the longitudinal current to take into account the presence of two strips, centered at planes $x = -(W_1 + S)/2$ and $x = (W_2 + S)/2$, respectively:

$$\tilde{j}_z(k_x) = 2\pi[J_0(k_x W_1/2)e^{jk_x(W_1+S)/2} + C_i J_0(k_x W_2/2)e^{-jk_x(W_2+S)/2}] \tag{4.109}$$

where C_i describes the ratio of longitudinal currents flowing on the two strips of respective widths W_1 and W_2 and S is the spacing between the strips. In the case of symmetric lines, the ratio C_i is made equal to -1 for odd coupled strip lines, and to $+1$ for even coupled strip-lines. In the case of asymmetric lines, the ratio C_i is calculated by the Rayleigh-Ritz procedure, as mentioned before for slot-like lines.

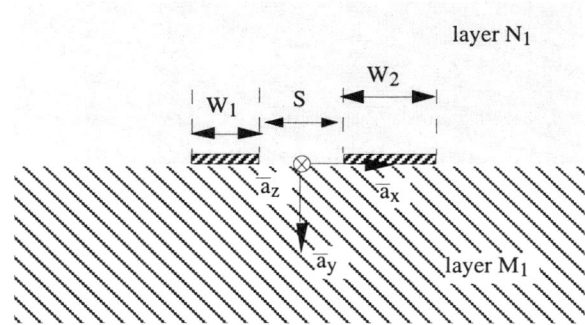

Fig. 4.11 *Geometry of coupled microstrip lines for determining trial quantities*

It has been seen that the trial expressions obtained for the j_x- and j_z-components flowing on a strip are similar to those obtained respectively for the trial e_z- and e_x-components in slot-like structures. The general expressions for the trial current densities on a strip for higher-order modes are therefore readily deduced from equations (4.99a,b) as

$$j_z(x,0) = \begin{cases} \displaystyle\sum_{n=N_1}^{N_1+2N} S_n \frac{\text{ce}_n[q, \arccos(\frac{2x}{W})]}{\sqrt{1-(\frac{x}{W/2})^2}} & \text{if } 0 < |x| < \frac{W}{2} \\ 0 & \text{elsewhere} \end{cases} \tag{4.110a}$$

and

$$j_x(x,0) = \sum_{m=M_1}^{M_1+2M} T_m \, \text{se}_m[q, \arccos(\frac{2x}{W})] \tag{4.110b}$$

where S_n and T_m are the unknown coefficients. The Fourier transform of these expressions are:

$$\tilde{j}_z(k_x) = \sum_{n=N_1}^{N_1+2N} S_n[\sum_{r=1}^{\infty} a_{nr}(q)(j)^r J_r(k_x W/2)] \tag{4.110c}$$

$$\tilde{j}_x(k_x) = \sum_{m=M_1}^{M_1+2M} T_m[\sum_{r=1}^{\infty} b_{mr}(q)(j)^{r-1} r \frac{J_r(k_x W/2)}{(k_x W/2)}] \tag{4.110d}$$

with q and k_{eff}^2 given by (4.99e,f). For coupled strip-lines, it can be modified as for the dominant mode. The advantages resulting from the *a priori* expression of the coefficient q have been mentioned above and are still valid. Similarly, coefficients S_n and T_m are obtained from Rayleigh-Ritz procedure, which will be applied in the following section.

4.5.4   Rayleigh-Ritz procedure for general variational formulation

Since a variational principle is the basis of the calculation, the Rayleigh-Ritz procedure [4.1] and [4.38] can be applied to obtain the values of coefficients R_n and Z_m in (4.99a,c), or S_n and T_m in (4.110a-d). This method simply makes use of the variational nature of equation (4.21): because it is variational, coefficients R_n are determined by imposing the condition that the value of γ extracted from (4.21) is extremum.

When evanescent modes are considered, namely when γ is real, equation (4.21) simplifies because in this case the calculation of (4.19b,c) shows that one has $\tilde{B}_i = \tilde{C}_i$, and can be rewritten as

$$\gamma^2 = \frac{\left(\sum_i \tilde{D}_i - \omega^2 \sum_i \tilde{E}_i\right)}{\sum_i \tilde{A}_i} \tag{4.111}$$

For convenience it is assumed that the trial fields are described by the series expansion

$$\tilde{e}_x(k_x) = \sum_{n=N_1}^{N_1+2N} R_n \tilde{e}_{xn}(k_x) \tag{4.112}$$

Expression (4.112) is a simplified notation of the series expansion (4.99c) where the left-hand side of expression (4.96a) has been used. For slot-like structures the integrand of each term of (4.111) is proportional to $|\tilde{e}_x(k_x)|^2$ when $\tilde{e}_z(k_x)$ is neglected, so that each term can be rewritten when $N = 1$ as

$$\tilde{Y}_i = \sum_{m=1}^{2} \sum_{n=1}^{2} R_m R_n \tilde{Y}_{imn} \tag{4.113}$$

with $\tilde{Y}_i = \tilde{A}_i, \tilde{B}_i, \tilde{C}_i, \tilde{D}_i$, and $\tilde{E}_i$, and where $\tilde{Y}_{imn}$ is calculated as in (4.19a-e), with $|\tilde{e}_x(k_x)|^2$ replaced by $[\tilde{e}_{xm}(k_x)\tilde{e}_{xn}(k_x)]^*$. Entering (4.113) into (4.111) yields

$$\gamma^2 = \frac{R_1^2 \tilde{N}_{11} + 2R_1 R_2 \tilde{N}_{12} + R_2^2 \tilde{N}_{22}}{R_1^2 \tilde{D}_{11} + 2R_1 R_2 \tilde{D}_{12} + R_2^2 \tilde{D}_{22}} \tag{4.114a}$$

with

$$\tilde{N}_{mn} = \sum_i \tilde{D}_{imn} - \omega^2 \sum_i \tilde{E}_{imn} \tag{4.114b}$$

$$\tilde{D}_{mn} = \sum_i \tilde{A}_{imn} \tag{4.114c}$$

Taking advantage of the variational nature of (4.111) and (4.114a), the Rayleigh-Ritz procedure imposes that the solution is extremum. Normalizing the numerator and denominator of (4.114a) to R_1^2 and defining $X = R_2/R_1$ yields

$$\gamma^2 = \frac{\tilde{N}_{11} + 2X\tilde{N}_{12} + X^2 \tilde{N}_{22}}{\tilde{D}_{11} + 2X\tilde{D}_{12} + X^2 \tilde{D}_{22}} \tag{4.115}$$

Taking then the first derivative of the right-hand side of this equation with respect to X, the second-order equation in (R_2/R_1) is obtained:

$$\begin{aligned}(2X\tilde{N}_{22} + 2\tilde{N}_{12})(\tilde{D}_{11} + 2X\tilde{D}_{12} + X^2 \tilde{D}_{22}) \\ - (2X\tilde{D}_{22} + 2\tilde{D}_{12})(\tilde{N}_{11} + 2X\tilde{N}_{12} + X^2 \tilde{N}_{22}) = 0\end{aligned} \tag{4.116}$$

Only the solution for $X = (R_2/R_1)$ yielding a positive value for γ^2 is used.

It must be emphasized that the use of the Rayleigh-Ritz procedure for more than one term in development (4.99a-d) requires the solution of a determinantal equation, which leads to a set of linear equations in terms of the ratios R_n/R_{N_1}. In this case, the method results in solving a system of equations related to an eigenvalue problem [4.1], having about the same complexity as the Galerkin's procedure. Mathieu functions, however, are efficient enough to provide accurate trial fields when expansion (4.99) is limited to $N = 0$. This will be illustrated in the next subsection.

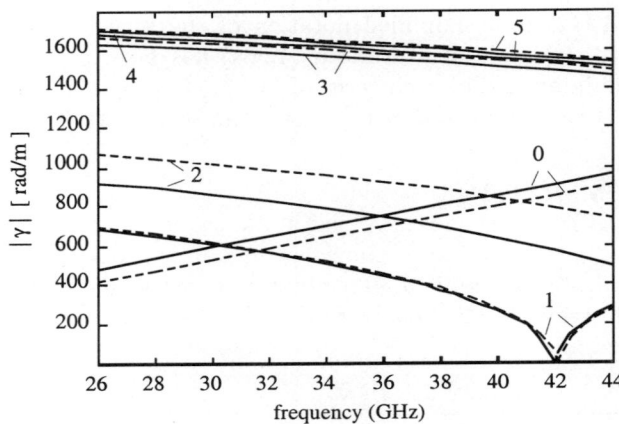

Fig. 4.12 *Real and imaginary parts of propagation constant calculated with $N = 0$ for finline (Hélard et al. [4.33]), for two different slot widths (—— 1 mm, - - - 2 mm) (numbering refers to order of mode, 0 denoting dominant propagating mode)*

4.5.5   Efficiency of Mathieu functions as trial components

Mathieu functions have been used as trial fields in [4.7] for calculating higher-order modes in a shielded slot-like line, namely a finline operating in the Ku-band. The efficiency of the method combined with the series expansion of (4.99c,d) limited to only one Mathieu function ($N = 0$) is highlighted when calculating the propagation constant of several higher-order modes of the finline analyzed by Hélard et al. in Figure 3 of [4.33]. The advantage of the formulation of (4.99c,d) limited to one term, however, is that the combination of the basis functions is known *a priori*. This is not the case for the results presented at Figure 3 of Hélard et al., where Galerkin's procedure is used with at least two basis functions and requires the solution of a determinantal eigenvalue equation.

Figure 4.12 shows the real (α) and imaginary (β) part of the propagation constant in the $\gamma - \omega$ plane for the dominant mode (curve 0), and for the first five higher order modes (curves 1 to 5), calculated with only one term in (4.99c,d) in the range 26 to 44 GHz, for two different slots as considered by Hélard et al. The comparison with their results shows a very good agreement, and the behavior of the dominant and first higher-order modes is correctly predicted by our model. The reduced mathematical complexity of expression (4.99a-d) combined with the explicit formulation (4.21) is obvious: it does not require the solution of an eigenvalue problem to obtain the amplitude of the basis functions and the value of γ which sets the determinantal equation equal to zero.

4.5.6 Ratio between longitudinal and transverse components

It has been shown that one Mathieu function is sufficient for describing the shape of any component of the trial quantity. The ratio between the magnitudes of the transverse and longitudinal components, however, remains unknown. For all the results presented up to now, only the transverse component of the trial electric field was considered for slot-like lines, while for strip-like lines the transverse current density was neglected. It has been observed, however, that both the transverse and longitudinal components are important when modeling higher-order modes in open lines. The knowledge of these modes is necessary for modeling planar discontinuities, for example when using a mode matching technique. For microstrip lines, it is possible to deduce the characteristics of higher-order modes from a planar equivalent waveguide [4.39]-[4.41] modeling the microstrip. For slot-like lines, the moment method is usually applied to determine the ratio between the transverse and the longitudinal components.

It will now be shown that a fairly good estimate of this ratio can be obtained by imposing an additional boundary condition on the electric field at the dielectric interface. This is of prime interest, since the variational equation involves the electric field. One simply imposes the continuity of the mean value of the displacement field at the interface between two dielectric layers. For the slot-line of the previous section, this is equivalent to imposing

$$\varepsilon_2 \tilde{e}_{y2}(k_x, H) = \varepsilon_1 \tilde{e}_{y3}(k_x, H) \tag{4.117}$$

Because the case is lossless, we do not need to care about the phase, so that the previous condition is equivalent to imposing the continuity of the squared magnitude of the displacement field

$$|\varepsilon_2 \tilde{e}_{y2}(k_x, H)|^2 = |\varepsilon_1 \tilde{e}_{y3}(k_x, H)|^2 \tag{4.118}$$

Integrating over the whole k_x-spectrum yields:

$$\int_{-\infty}^{\infty} |\varepsilon_2 \tilde{e}_{y2}(k_x, H)|^2 dk_x = \int_{-\infty}^{\infty} |\varepsilon_1 \tilde{e}_{y3}(k_x, H)|^2 dk_x \tag{4.119}$$

Since it has been shown that the trial fields in each layer are linear combinations of the components of the trial quantity, a simple analytical expression for the ratio R_n/Z_m may be obtained from (4.119).

Few explicit results are found in the literature for higher-order modes on open slot-like lines. They are usually computed for shielded lines. To our knowledge, only Fedorov *et al.* [4.42] provide such results for wide slots. Figure 4.13 compares their results (symbols □ and o), for the first higher-order mode, with those obtained by the variational formulation (4.21) (lines), used with one Mathieu function and both $\tilde{e}_x$- and $\tilde{e}_z$-components in the slot. The comparison is made for two configurations on the same substrate corresponding to two different slot widths, $W = 4$ mm and 8 mm. The agreement between the two results is excellent for each of the two slot widths considered. It must be emphasized, however, that Fedorov *et al.* use the Galerkin's procedure and several basis functions, so that the order of the system to be solved is important. Since the formulation used here is explicit and analytical, no tedious iterations are necessary, except for the integration along the k_x-axis. So, the variational method is an efficient tool for calculating planar discontinuities by mode-matching methods.

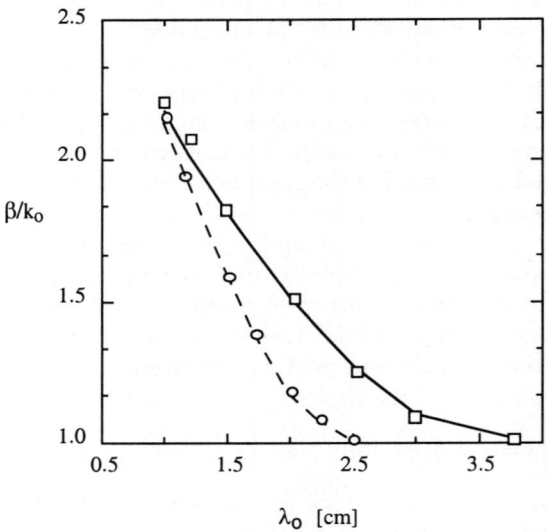

Fig. 4.13 *Propagation coefficient for first higher-order mode of open slot-line (Fig. 4.7, with $H = 1$ mm, $\varepsilon_{r1} = \varepsilon_{r2} = 1, \varepsilon_{r3} = 9$), calculated respectively with variational principle (4.21) (lines) and by Fedorov* et al. *[4.42] (symbols), for two different slot widths (□ □ □ : $W = 8$ mm; o o o: $W = 4$ mm)*

4.6   Resulting advantages of variational behavior

To start the calculations, trials have to be introduced for the complex propagation constant as well as for the fields. The choice made for the trials, especially for the fields, may strongly affect the accuracy of the result, and the computation time.

1. The unknown value γ appears in variational equations (4.2) and (4.21) which are solved for γ. It may also appear in the spatial or spectral formulation of the trial fields, as shown for instance by (4.55a,b), (4.56a,b) and (4.78a-d). A reasonable initial value for γ, noted γ_0, is introduced in the expression of the fields.

2. Applying the boundary conditions to the trial fields results in a set of homogeneous linear equations relating the spectral coefficients (4.54d) or (4.76a) to the tangential electric field across the slot for slot-like lines, and to the current densities flowing on the strip for strip-like lines. Only an approximation of these quantities is necessary because of the variational character.

3. The integration over the k_x spectrum for open lines, or the infinite summation for shielded lines, is stopped when the convergence is considered as satisfactory; that is when the addition of the last term of the summation or of the contribution of the last subinterval of integration does not significantly modify the solution of (4.21). Because of the linearity of the Fourier transform, any truncation of the x-Fourier transform of the fields yields an error over the field in the spatial domain. However, due to the variational formulation, this error may also exist in the trial field used in (4.19a-e), without significantly affecting the result.

The variational nature of equation (4.2) yields rapidly convergent results once a trial field has been chosen. It can also improve the quality of the approximation made on the trial field. Three major advantages of variational formulation (4.2) are summarized here. Each of them is detailed in the following subsections.

1. The calculation of the propagation constant does not require solving either a determinantal equation or a high number of iterations, as in other procedures [4.1],[4.5],[4.14] The propagation constant is simply obtained by solving a second-order equation, with an approximate value of γ inserted in the trial fields if necessary.

2. The tangential spectral quantity required at an interface containing a discontinuous conductive layer can be approximated with a strongly reduced number of basis functions. These are best derived by a conformal mapping and adequate boundary conditions at the interface. The Rayleigh-Ritz procedure may be applied in case of asymmetric coupled lines for which the actual ratio between the spectral quantities on the two lines is unknown.

3. The integration or series expansion is limited to a small number of terms because the integrands of expressions (4.19a-e) for calculating the coefficients of the second-order equation (4.2) or (4.21) are rapidly decreasing

functions of the spectral variable k_x, as will be shown in a further subsection. Moreover, the propagation constant is explicit in equation (4.21). It is calculated at each step of the summation, which directly indicates when the convergence is considered satisfactory, that is when the addition of one more term does not modify the value of the solution.

4.6.1 Error due to approximation made on γ in trial fields

A suitable approximation of γ is necessary in the expression of the trial fields. It is denoted $\gamma_t = \alpha_t + j\beta_t$ in the following discussion. A good choice for γ_t is [4.7]

$$\gamma_t = j\beta_t = j\sqrt{\frac{(\varepsilon_1 + \varepsilon_2)}{2}} k_0 \qquad (4.120)$$

The solution of variational equation (4.21), obtained using approximation (4.120), is introduced as a new value for γ_t in the spectral fields, which yields a new solution for the equation. The quality of the approximation is then evaluated, and improved if necessary. Iterations are stopped when the difference between γ and γ_t is found negligible. Since the formulation is variational with respect to the field, the error introduced by an approximation made on the value γ_t involved in the field expression is only of the second-order, which strongly reduces the number of iterations. This is illustrated in Figure 4.14 for a lossless slot-line on a dielectric substrate [4.4]. The solid line depicts the solution of equation (4.21) obtained for a wide range of values of the imaginary part β_t. Since there are no losses, purely imaginary γ and γ_t are considered for the solution of (4.21) and for the approximation for the field are considered. Starting with the value provided by (4.120) for β_t in the trial fields (A in Fig. 4.14), the first solution (denoted by β_0) of (4.21) is point B. This value is then used as a new approximation, denoted by β_{t1}, for the trial fields (point C). After solving again (4.21), it leads to an updated value for β, denoted by β_1 (point D). It is observed that this solution is within 0.067 % of the final solution, which is reached when $\beta_n = \beta_{n-1} = \beta_{tn} = 527$ rad/m (point E). This explains why the results of Figures 4.3, 4.5, and 4.6, as well as other results presented earlier [4.7], have necessitated no more that 3 iterations ($n = 2$) to obtain a final value of β within a 0.05 % uncertainty.

The convergence scheme may thus be formalized as follows.

1. Equation (4.21) is rewritten as

$$\gamma^2 \sum_i \tilde{A}_i[\tilde{\bar{u}}_t] + \gamma \sum_i \tilde{B}_i[\tilde{\bar{u}}_t] - \sum_i \tilde{D}_i[\tilde{\bar{u}}_t] + \omega^2 \sum_i \varepsilon_i \mu_i \tilde{E}_i[\tilde{\bar{u}}_t] = 0 \qquad (4.121)$$

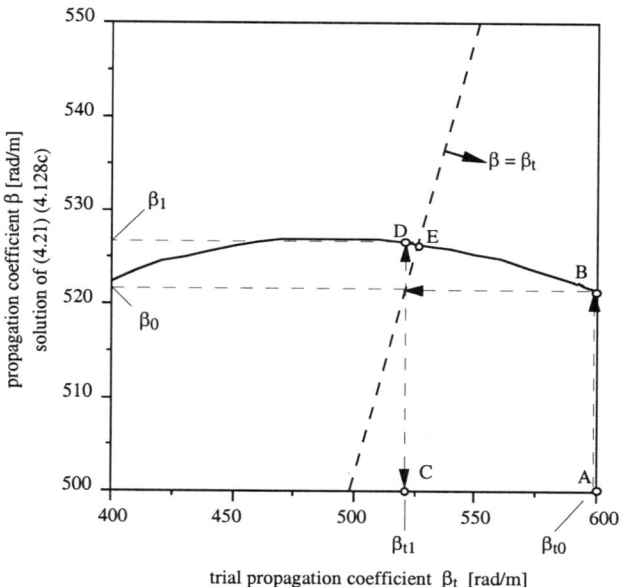

Fig. 4.14 *Convergence properties of variational principle for lossless open slot-line*

where $\tilde{\bar{u}}_t = \tilde{\bar{e}}_i(\gamma_t)$. It has the general solution

$$\gamma = \alpha + j\beta = \frac{-\sum_i \tilde{B}_i[\tilde{\bar{u}}_t]}{2\sum_i \tilde{A}_i[\tilde{\bar{u}}_t]}$$

$$\pm \frac{\sqrt{\left\{\sum_i \tilde{B}_i[\tilde{\bar{u}}_t]\right\}^2 - 4\sum_i \tilde{A}_i[\tilde{\bar{u}}_t]\left\{\omega^2 \sum_i \varepsilon_i\mu_i \tilde{E}_i[\tilde{\bar{u}}_t] - \sum_i \tilde{D}_i[\tilde{\bar{u}}_t]\right\}}}{2\sum_i \tilde{A}_i[\tilde{\bar{u}}_t]}$$

$$(4.122)$$

For a lossless case, in (4.121) one has $\gamma = j\beta$ and $\gamma_t = j\beta_t$. The imaginary part of solution (4.122) is represented by the solid curve in Figure 4.14.
2. The solution at iteration n is obtained by solving (4.21) with β_{tn} equal to the solution obtained at iteration $n - 1$, that is with

$$\beta_{tn} = \beta_{n-1} \qquad\qquad (4.123)$$

in the coefficients of (4.121) and in (4.122). For $n = 0$, β_{t0} is given by (4.120).
3. The convergence is obtained when the relative difference between results

after two successive iterations is lower than a given value x, that is

$$\left| \frac{\beta_n - \beta_{n-1}}{\beta_n} \right| < x\% \tag{4.124}$$

which, by virtue of (4.123), yields approximately

$$\beta_n = \beta_{n-1} = \beta_{tn} \tag{4.125}$$

With convergence criterion (4.124), the final solution is located at the intersection between the solid curve depicting solution (4.122) as a function of the trial value γ_t and the linear function

$$\gamma(\gamma_t) = \gamma_t \tag{4.126}$$

represented by the dashed line in Figure 4.14.

It is underlined that the variational behavior of (4.21) is related to a stationarity concept. Indeed starting with an approximate value β_{t0} of about 600 rad/m (point A) corresponding to an error of about 14 % on β, point B yields a value of 522 for β_0 corresponding to an error of about 1 % on β. Hence, the first-order error is reduced to a second-order one. So, a very limited number of iterations is necessary. It has also to be observed that the solution defined by (4.124) is located exactly at the extremum value of the curve representing the solution (4.122).

When lossy layers are considered, a most important feature of the variational character is that the solution obtained with equation (4.21) is complex, corresponding to a stationary value for both the real and imaginary parts of the propagation constant. This is observed at Figure 4.15, where the real (a) and imaginary (b) parts of the solution for a coplanar waveguide lying on a doped semiconductor substrate is represented as a function of a wide range of values for (α_t, β_t) (α_t in the range 100-300 Np/m, β_t in the range 300-700 rad/m). The resistivity of the substrate, due to the presence of carriers in the doped semiconductor, is introduced in the variational formulation via a frequency-dependent loss tangent associated with each layer and yielding the imaginary part of its relative permittivity. A frequency-dependent complex value of the permittivity is thus considered, in the trial fields as well. The real and imaginary parts of solution (4.122) written as a function of the real and imaginary parts of the trial γ_t are each represented by a surface above the plane (α_t, β_t). A number of features may be pointed out in Figure 4.15.

First, the *variational behavior* of the solution is obvious: the real and imaginary parts of the solution remain quite constant over a very wide range of values for both the real and imaginary parts of γ_t. For a relative variation of 300 % for α_t and of 230 % for β_t, the variations of the real part α and of the imaginary part β of the solution from (4.122) remain within 10 % and 5 %, respectively.

Next, in the case of iterations carried out to improve the trial fields and hence the results on γ, *the convergence is quite immediate*. As a first step,

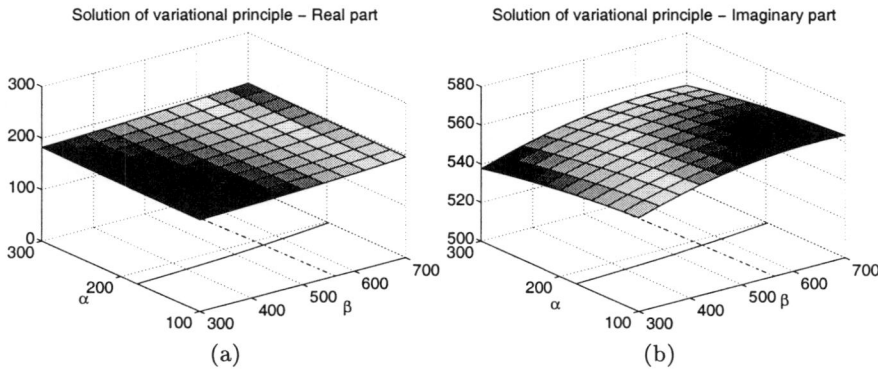

Fig. 4.15 *Variational solution as a function of real and imaginary parts of trial propagation constant introduced in trial fields for a CPW (20 Ω cm resistivity substrate) (a) real part; (b) imaginary part*

α_t is kept constant. Again solving equation (4.121) yields a new value for β noted β_0. The process is repeated n times with α_t kept constant, until the imaginary part of the solution (4.122) becomes equal to the imaginary part of the trial β_t, that is when criterion (4.124) is satisfied. Practically, as said before, iterations are stopped when the relative difference between results obtained after two successive iterations is lower than a given value x, that is

$$\left| \frac{\beta_n - \beta_{n-1}}{\beta_n} \right| < x\% \tag{4.124}$$

Because of the very small variation of the solution β with respect to β_t which is observed, (Fig. 4.15b), a very limited number of iterations is necessary. This number of iterations is noted N_I. The convergence condition (4.124) written for every value of α_t forms the intersection of the surface representing the imaginary part of solution (4.122) (Fig. 4.15b) with plane $\beta(\alpha_t, \beta_t) = \beta_t$. This intersection is obtained for the set of values in plane (α_t, β_t) which are represented by the dashed curve. Using in a second step a similar convergence criterion for the real part of solution (4.122), that is, for a fixed value of β_t

$$\alpha_n = \alpha_{n-1} = \alpha_{tn} \tag{4.127}$$

and writing the convergence condition (4.127) for every value of β_t yields the solid curve in Fig. 4.15. It corresponds to the intersection between the surface representing the real part of the solution (4.122) (Fig. 4.15a) and plane $\alpha(\alpha_t, \beta_t) = \alpha_t$. The solid and dashed curves are each parallel to a coordinate axis. This property expresses the fact that the real and imaginary parts of the

solution of (4.122), obtained after some iterations, become independent of the
imaginary and real parts of the trial propagation constant introduced in the
trial fields, respectively. Hence, iterating on the imaginary part while fixing
the real part at an arbitrary value, then iterating on the real part while retain-
ing the solution found for the imaginary part, converges towards a solution
located at the intersection between the solid and dashed curves (Fig. 4.15).
Moreover, at each step of the iterative process, the real and imaginary parts
of the trial can be simultaneously updated without affecting the convergence,
because of the orthogonality of the two curves. Hence, *when new iterations are
necessary, updating at the same iteration the values of α_t and β_t also reduces
the numerical complexity without significantly damaging the convergence.* The
convergence scheme is finally equivalent to solving equation (4.121) n times
with, at iteration n, γ_n obtained as solution of (4.121)

$$\gamma_{tn} = \gamma_{n-1} \tag{4.128a}$$

which means

$$\alpha_{tn} = \alpha_{n-1} \tag{4.128b}$$

$$\beta_{tn} = \beta_{n-1} \tag{4.128c}$$

The convergence is obtained when the two conditions (4.124) and (4.127)
are satisfied at the same iteration. The number of iterations is very limited,
because of the flatness of the surfaces representing both the real and imaginary
parts of the solution.

It has been observed that the two dashed and solid curves intersect each
other, so that a solution exists. The same property is found for gyrotropic
layers. The formulation is also efficient in the case of leaky planar lines or slow-
wave devices. The propagation constants of these lines indeed exhibit both
real and imaginary parts, even in the case of lossless media, and a solution
is difficult to obtain from an implicit method, as mentioned by Das [4.15].
Moreover, the results in Figure 4.15 are computed for a *high-loss* case, and
the solution found exhibits this feature, because the real and imaginary parts
of the solution are of the same order. Such a case is thus particularly suitable
for comparing the efficiency of the variational solution (4.122) with that of
the implicit Galerkin's procedure. This will be presented in the next section.

4.6.2   Influence of shape of trial fields at interfaces

The formulation is variational with respect to the electric field introduced in
equation (4.2). This has particular applications in the case of coupled lines,
where the ratio between the trial quantities considered for each line has to be
determined. A ratio of voltages C_v has been defined for coupled slot-like lines
(4.98), while a ratio of currents C_i is considered for strip-like lines (4.109).
Figure 4.16 shows, for lossless coupled symmetrical slot-lines having a width
ratio W_2/W_1 equal to 1, the value of β obtained from equation (4.21) as a

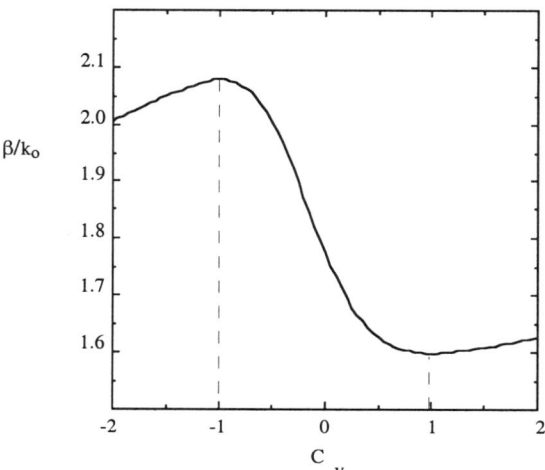

Fig. 4.16 *Propagation coefficient of coupled slot-lines (Fig. 4.9) calculated with variational formulation for different values of ratio between trial electric fields in the two slots* $(W_1 = W_2 = 1$ *mm,* $H = 1.2$ *mm)*

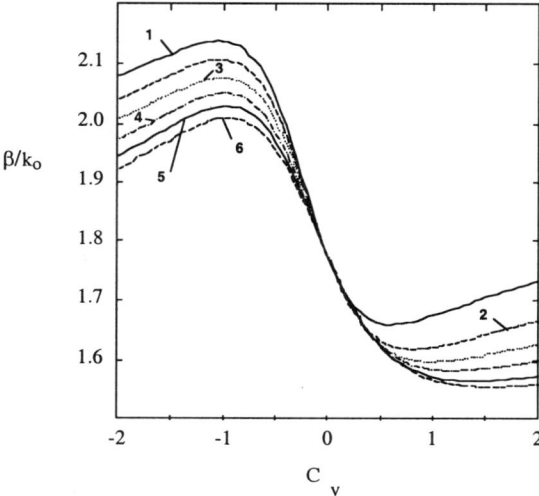

Fig. 4.17 *Propagation coefficient of lossless coupled slot-lines calculated with variational formulation as a function of ratio between trial electric fields in the two slots, with ratio* $W_2/W_1 = 0.5$ *(1),* *0.75 (2),* *1.0 (3),* *1.25 (4),* *1.5 (5), and 2.0 (6)*

function of ratio C_v. It is observed that two extrema of β occur, respectively a maximum for $C_v = -1$, and a minimum for $C_v = +1$. This confirms that equation (4.2), rewritten in the spectral domain as (4.21), is variational with respect to the electric field, since the solution exhibits extrema for values of C_v which are indeed the actual ratios of the transverse tangential electric fields in this symmetrical structure. Hence, the application of the Rayleigh-Ritz procedure in this case will provide the actual ratios, since they correspond to extrema of the solution.

In the case of asymmetric lines, the ratio C_v or C_i cannot be determined *a priori*, because of the lack of x-symmetry of the tangential electric field in the structure. However, according to the theory of asymmetric coupled lines [4.43], there are two solutions for β, corresponding to two values of C_v or C_i. The first is associated with the in-phase behavior of the trial quantity on the two lines (C_v, C_i positive) and is called the π-mode. The second is associated with the out-of-phase behavior of the trial quantity (C_v, C_i negative) and is called the C-mode. One can expect that the Rayleigh-Ritz procedure in the case of asymmetric lines will determine the actual value of β by searching the values of C_v which render β extremum. Figure 4.17 shows the variation of the solution of (4.21) as a function of C_v for coupled slot lines having different widths (W_2/W_1 varying between 0.5 and 2.0). In this asymmetric case, extrema now occur for values of C_v which vary with ratio W_2/W_1. Applying the Rayleigh-Ritz procedure, as explained for the higher-order modes in a preceding section, yields the value of C_v which maximizes the solution of (4.21), as well as the value of this extremum.

The variational procedure is applied to the lossless asymmetrical coupled slot-like line previously studied with Galerkin's method by Kitlinski and Janisacz [4.44]. Results are compared in Tables 4.2 (C-mode) and 4.3 (π-mode) for the case $W_2/W_1 = 2$. The values of the normalized propagation constant obtained with this variational formulation (Huynen) are compared to results obtained by Kitlinski *et al.* over the frequency range 2-20 GHz. The corresponding values obtained for the ratio C_v are also given. The agreement between the two methods is better than one percent over the whole frequency range.

4.6.3   Truncation error of the integration over spectral domain

Figure 4.18 shows a typical variation of the integrands of the coefficients defined by (4.19) as a function of the spectral variable k_x, for a slot-line at x-band. As explained in Section 4, these integrands are similar for laterally-open or - shielded slot-lines, because the spectral fields are the same for a given value of k_x. If the line has lateral shielding, the integration is simply replaced by a summation over discrete values of the integrands corresponding to discrete values of k_x. The figure shows that all the integrands are rapidly decreasing as the spectral variable k_x increases. In this case it indicates clearly that the integration or summation can be stopped when k_x equals about 8000 m^{-1}. If the slot-line is shielded at least in the x-direction, and for W_g taken 10

C-mode			
Frequency (GHz)	C_v [Huynen]	(λ_0/λ') [Kit.-Jan.]	(λ_0/λ') [Huynen]
2	-0.975	1.970	1.9436
4	-0.975	1.960	1.9542
6	-0.975	1.978	1.9680
8	-0.925	1.985	1.9842
10	-0.925	2.007	2.0031
12	-0.925	2.025	2.0200
14	-0.925	2.040	2.0376
16	-0.925	2.066	2.0604
18	-0.925	2.093	2.0829
20	-0.925	2.120	2.1000

Table 4.2 *Comparison of results obtained with this formulation (Huynen) and by Kitlinski and Janisacz [4.44] for C-mode of asymmetrical coupled slot-lines*

π-mode			
Frequency (GHz)	C_v [Huynen]	(λ_0/λ') [Kit.-Jan.]	(λ_0/λ') [Huynen]
2	1.700	1.330	1.2350
4	1.725	1.370	1.3650
6	1.725	1.425	1.4226
8	1.775	1.470	1.4731
10	1.775	1.540	1.5447
12	1.725	1.580	1.5838
14	1.825	1.640	1.6349
16	1.725	1.690	1.6810
18	1.775	1.730	1.7255
20	1.875	1.775	1.7709

Table 4.3 *Comparison of results obtained with this formulation (Huynen) and by Kitlinski and Janisacz [4.44] for π-mode of asymmetrical coupled slot-lines*

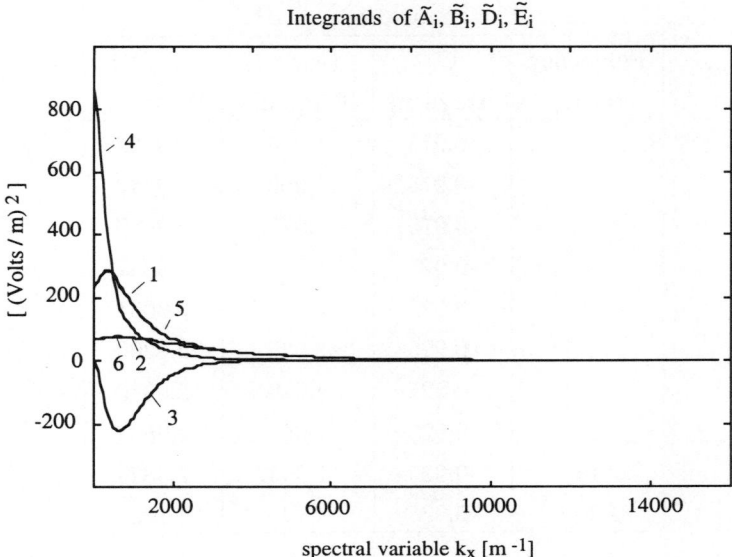

Fig. 4.18 *Integrands of set of the coefficients (4.19) for a slot-line (RTduroid 6010, thickness 0.635 mm, manufacturer's relative dielectric constant 10.8, slot width = 0.390 mm); curve 1: $k_0^2\tilde{A}_1$; curve 2: $k_0^2\tilde{A}_3$; curve 3 : $\tilde{C}_1 + \tilde{C}_3$; curve 4: $\tilde{D}_1 + \tilde{D}_3$; curve 5: $k_0^2\tilde{E}_1$ (indistinguishable from curve 1); curve 6: $k_0^2\tilde{E}_3$ (indistinguishable from curve 2)*

times larger than the slot width ($W = 0.390$ mm in the present case), then the maximum number n_{max} of terms needed for the summation is obtained from relationship $k_{xn} = 2n_{max}\pi/W_g$ which yields

$$n_{max} = \frac{8000 \cdot 10 \cdot 0.39\ 10^{-3}}{2\pi} = 5 \tag{4.129}$$

If the slot-line is open, then the Romberg integration method [4.45] is successfully applied to the integrands, because a large integration step size is sufficient for most part of the integration range (interval 2000-8000 in this case). Hence, only a few evaluations of the integrands are needed in this interval. Only the interval 0-2000 has to be strongly discretized by the Romberg algorithm to take into account the presence of a maximum in the integrands. A reduced number of evaluations of the integrands is needed when they have a monotonic behavior in some subintervals, which ensures a rapid convergence of the Romberg algorithm in these subintervals. Hence, the integration along k_x can be performed with a very few number of terms in the case of shielded lines, or very few steps in the case of open lines. This is a major advantage

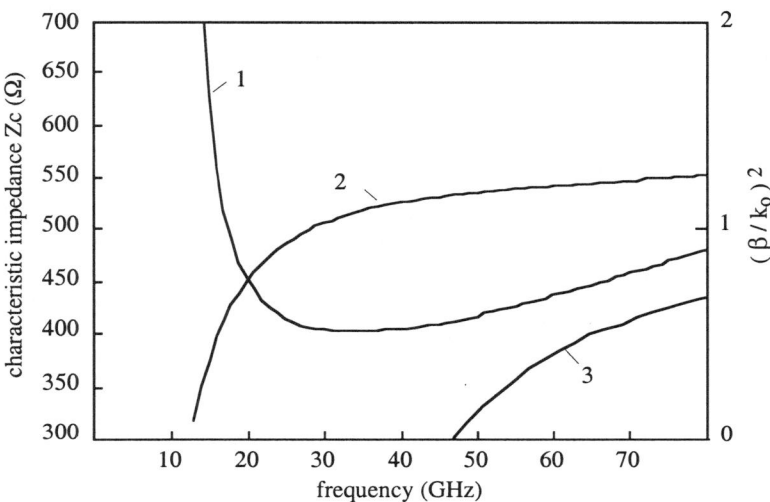

Fig. 4.19 *Characteristic impedance of the dominant mode (curve 1), and effective dielectric constant of the dominant mode (curve 2) and first higher-order mode (curve 3) of finline with 20 spectral terms*

of the formulation: the spectral fields, hence the integrands of (4.19), have to be evaluated for a very limited number of values for k_x.

Figure 4.19 shows results obtained when calculating the parameters of the dominant and first higher-order mode of the shielded line analyzed by Schmidt and Itoh (Fig. 3b of [4.46]). They report that 200 spectral terms and appropriate basis functions led to a convergence within 0.2 % for the ratio $(\beta/k_0)^2$. Using the variational approach, however, the characteristic impedance (curve 1) and the propagation constant (curve 2) of the dominant mode are calculated with equation (4.21) while expression (4.111) is used for the propagation constant of the first higher-order mode (curve 3). The number of spectral terms used in (4.19) never exceeds 20. Hence, the numerical complexity is strongly reduced, although the comparison with Figure 3b of [4.46] shows such a perfect agreement that it is impossible to see the difference between the two results on a figure of normal size.

Applying Galerkin's procedure in the spectral domain results in a determinantal equation involving γ and the spectral variable k_x, which is therefore an implicit equation for γ. As reported by Jansen [4.14], a typical number of spectral terms necessary to obtain a good convergence is between 100 and 500 terms for each calculation. Moreover, a number of evaluations of the determinantal equation is necessary to find the roots. According to Jansen, a typical number of evaluations is 10. As a rule of thumb, the numerical complexity of Galerkin's procedure can be evaluated by multiplying the number of terms by the number of evaluations, the result being of the order of a few thousands. Since no explicit formulation of γ exists in this case, the tedious procedure for finding the root of the determinantal equation requires that a number of integrations over the whole k_x-spectrum have to be performed each time the determinantal equation is solved. Hence, to evaluate the improvement in accuracy when increasing the integration domain, the determinantal equation has to be solved each time the integration domain is changed. Moreover, the number of terms necessary may vary with the geometry of the structure and the physical parameters of the layers. Hence, an *a priori* knowledge of an upper limit for the integration is not available. For this reason, some authors [4.6] choose an arbitrary very large limiting value for the integration, to ensure sufficiently accurate results for a wide variety of topologies. They also assume a virtual lateral shielding placed at a distance of about 10 times the slot width and transform the integration process into a discrete summation [4.5]. Choosing a large limit of integration, they increase the computation time for most of the cases of practical interest. On the contrary, the explicit variational formulation directly evaluates the effect of a modification of the size of the integration domain on the result: the integration can be stopped when the convergence is found to be satisfactory, which minimizes the computation time for each specific topology.

4.7 Characterization of the calculation effort

All the results presented in this chapter were obtained on a 25 MHz IBM PC 80386. The propagation constant (a), attenuation (b), and characteristic impedance (c) of a slot-line are represented in Figure 4.20, in the frequency range 4-24 GHz. At each frequency, the calculation is stopped when the values of the three parameters computed from the solution of (4.21) each remain within 0.1 % , when the size of the integration domain is increased. Figure 4.20d shows that this result is achieved in less than 1 second for each frequency point. It illustrates that *on-line* results can be obtained with this explicit variational formulation in the case of lossy multilayered planar transmission lines.

In the following discussion, results provided by the variational principle are compared with those from other methods, in terms of number of iterations, accuracy of the results obtained, and computation time, taking into account the computational environment.

4.7.1 Comparison with Spectral Galerkin's procedure

As already discussed in Chapter 3, the Spectral Domain Approach (SDA) uses the Galerkin's procedure in the Fourier-transform domain in order to find the propagation constant. A review of this technique is presented in [4.14] and [4.47]. The SDA can be tedious for on-line designs of planar lines [4.14]. When compared with variational equation (4.121), Galerkin's procedure searches for

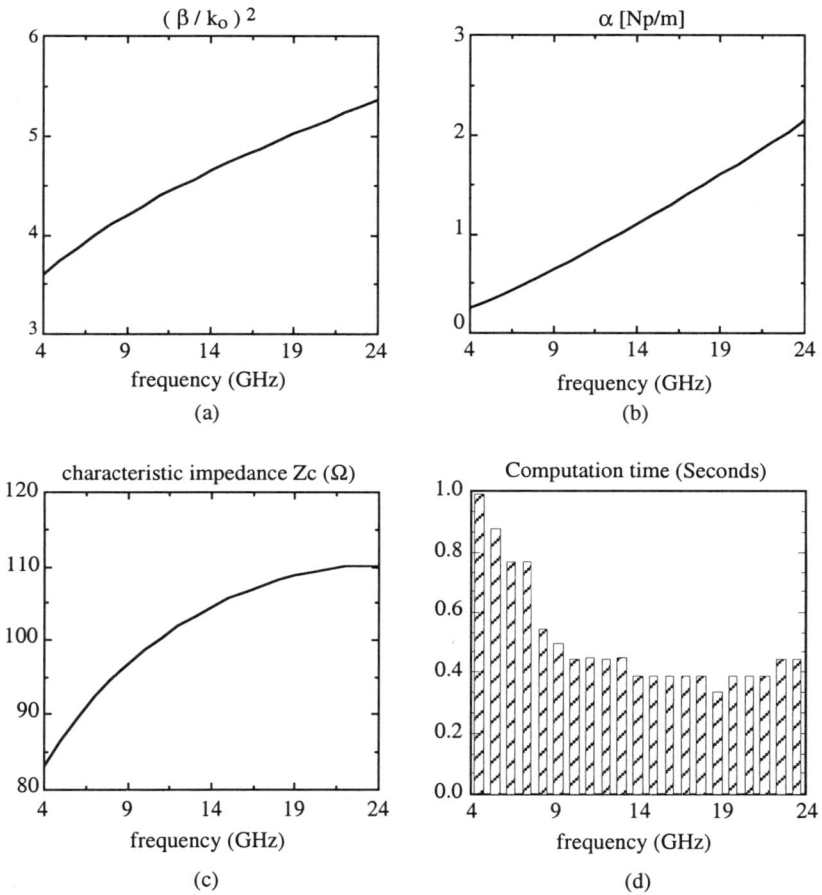

Fig. 4.20 *Calculation of line parameters of slot-line (a) effective dielectric constant; (b) attenuation coefficient; (c) characteristic impedance, (d) computation time at each frequency*

the complex root of the following implicit determinantal equation, obtained in Chapter 3 (3.60):

$$\det(\tilde{\bar{\bar{\mathcal{L}}}}_{st}(\gamma)) = 0 \qquad \text{for strip-like problems} \qquad (4.130a)$$

$$\det(\tilde{\bar{\bar{\mathcal{L}}}}_{sl}(\gamma)) = 0 \qquad \text{for slot-like problems} \qquad (4.130b)$$

Solving the second-order variational equation for γ differs from Galerkin's procedure. The general solution (4.122) of the equation is indeed known *a priori* while Galerkin's method replaces the equation by the implicit determinantal equation (4.130a,b) to be solved for γ. Solving this equation requires the evaluation of the determinantal equation several times, and may be very difficult, as mentioned by a number of authors [4.48]. This is because the method provides no information about the behavior of the implicit equation (4.130a,b) with respect to γ. In particular spurious solutions may occur [4.11]. Solving (4.130a,b) requires specific numerical techniques, especially when losses are present and evanescent modes are considered [4.12] and [4.15]. For these reasons, the method is usually limited to lossless applications on dielectric multilayered media, as mentioned by Jansen [4.14]. Under low-loss assumptions, Das and Pozar [4.15] propose to use "the spectral-domain perturbation method" but, as they mention, it is only valid for small loss applications. Such a method is not suited for general uses. Indeed Figure 4.15 shows values of the attenuation and propagation coefficients having the same order of magnitude. When using the SDA, the number N_S of spectral terms necessary to obtain a good accuracy for each calculation of the determinant is typically between 100 and 500 [4.12]. It has already been commented that a number of evaluations of the determinant are necessary to find the complex roots. A typical number of evaluations is $N_E = 10$. The numerical complexity of SDA can be quantified by $N_S \times N_E$, which is of the order of a few thousands [4.12].

On the other hand, for the variational equation (4.21), the coefficients $\tilde{A}_i$, $\tilde{B}_i$, $\tilde{D}_i$, and $\tilde{E}_i$ are obtained by summing N_S spectral terms, while N_E has to be replaced by N_I, the number of iterations necessary to obtain a satisfactory convergence on γ. More precisely, N_I is defined as the number of iterations, carried out simultaneously on the real and imaginary parts of (4.21) or (4.121), which are necessary to obtain successive values within x % for both the real and imaginary parts of the solution:

$$\left| \frac{\alpha_n - \alpha_{n-1}}{\alpha_n} \right| < x\% \tag{4.131a}$$

$$\left| \frac{\beta_n - \beta_{n-1}}{\beta_n} \right| < x\% \tag{4.131b}$$

Figure 4.21 shows several parameters comparing the efficiency of the variational principle (VP) (4.21), with that of the SDA using Galerkin's procedure, on a coplanar waveguide up to 20 GHz. Both methods are applied to the same low-resistivity Silicon-on-Insulator (SOI) structure as shown in Figure 4.15 (resistivity 20 Ωcm). When computing the propagation coefficient, the iteration procedure is stopped when the convergence to $x = 0.5$ % is simultaneously obtained on both the real and imaginary parts of the solution. Figure 4.21a shows that VP (-) agrees very well with SDA (o). From Figure 4.21b,d, it is concluded that the spectral variational principle is much more efficient than SDA. It needs far fewer spectral terms and iterations per frequency point: *the numerical complexity of the SDA ($N_S \times N_E$) is between 8000 and 12000 whilst that of the VP ($N_S \times N_I$) is between 100 and 200.*

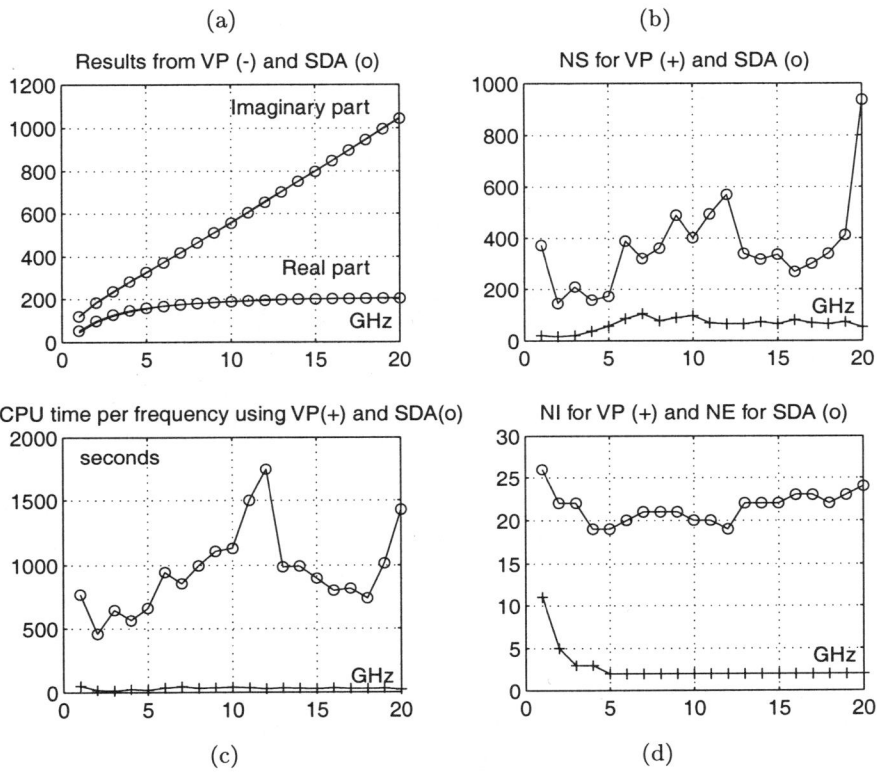

Fig. 4.21 *Comparison of the efficiency of the variational formulation (VP) and of the SDA for a CPW transmission line (standard 20 Ωcm resistivity) (a) real and imaginary parts of the propagation constant; (b) number of spectral terms needed; (c) CPU times; (d) number of iterations for VP (+) and number of evaluations of the determinantal equation for the SDA (o)*

The very low values of N_I are explained by the properties of the variational solution, pointed out in the comments dealing with Figure 4.15. The reduced number of spectral terms N_S is due first to the variational nature of the solution, which ensures that truncation effects on the spectral field have only a small influence on the result. Secondly, it is also due to the fact that a ratio of spectral terms is used in the variational expression (4.122), so that errors made on terms involved in the numerator and denominator vanish. It has to be underlined that the results provided by (4.21) and by Galerkin's method, respectively, have been obtained using the same computer and programming language (Matlab 4.0). As a consequence, the comparison between CPU times is valid (Figure 4.21c). Hence, the variational principle (4.21) is much less time-consuming: it is less than 50 s for VP, while it is of the order of 1,000 s for the SDA using Galerkin.

4.7.2 Comparison with finite-elements solvers

The High Frequency Structure Simulator (HFSS) from Hewlett-Packard is a 3-D electromagnetic wave simulator, computing electromagnetic fields and scattering parameters of a 3-D structure, geometrically and physically defined by the user and enclosed in a box. The ports of the equivalent N-port are specified as limiting plane surfaces of the box. The user has to specify all the materials inside the box, and boundary conditions on its surfaces. The 3-D mesh to be solved by the finite-element method is built from a 2-D port analysis: the vector wave equation is first solved in 2-D for each port using a triangular adaptive mesh (iterative scheme) and yields the transmission line parameters of the considered mode in each port. Hence, the 2-D port analysis provides the transmission line parameters of any structure having the same transverse section as the port.

Figure 4.22 compares results obtained from Version 3.0 of HFSS and from the variational spectral form, for the same coplanar waveguide shown in Figures 4.15 and 4.21, whose transmission line parameters have been measured [4.49]. The VP transmission line parameters (- - -) agree perfectly with the experiment (—), while the results obtained with HFSS (o) are less accurate. HFSS seems unable to take into account materials having high-loss frequency dependent constitutive parameters. Furthermore, a 2-D computation with HFSS takes about 7 minutes per frequency point, whereas only a few seconds per frequency are needed using (4.21) on the same computer. Hence, the variational principle (4.21) proves to be more efficient for high-loss structures than this example of a popular 3-D tool.

Similar results are observed on strip lines, as shown in Figure 4.23. The variational principle (4.21) and the finite-element method are applied to the shielded microstrip line investigated both in Section 4.4 and by Itoh and Mittra [4.5],[4.26]. Attention is drawn to the fact that the structure is now lossy. Figure 4.23a,b shows the effective dielectric constant and attenuation coefficient, obtained using the variational principle (solid line) and a finite-element algorithm (circled line). In these figures, it is observed that even after refining the mesh from 180 elements to 448 elements (crossed line) discrepancies remain between VP and FEM values. Figure 4.23c illustrates once more the very limited numerical complexity of the VP in terms of number of spectral terms and iterations, while Figure 4.23d compares the computation time for FEM (in minutes) with that of the VP (in seconds). Applying FEM code to the structure of Figure 4.4 requires about 3 and 9 minutes per frequency for a 180 and 448 element-mesh respectively, while only a few seconds are necessary when using the VP on the same computer. The CPU time in FEM is crucially influenced by the number of elements of the mesh and by the efficiency of the eigenvalue solver required by the FEM algorithm, which in turn strongly affects the accuracy of the solution obtained [4.50].

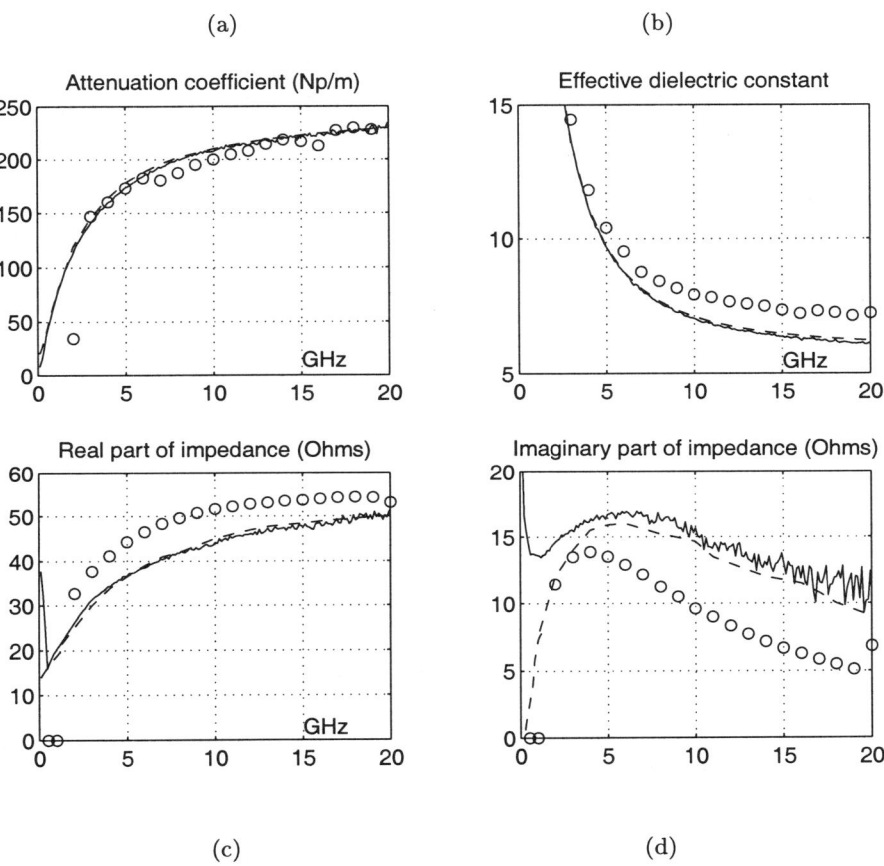

Fig. 4.22 *Measured results (——), simulated with (VP) (- - -), and simulated with HFSS (o) for transmission line parameters of multilayered lossy CPW (a) attenuation coefficient; (b) effective dielectric constant; (c) real part of impedance; (d) imaginary part of impedance (Reproduced from Proc. NUM-ELEC'97 [4.49])*

4.8   Summary

A general variational formulation for planar multilayered lossy transmission lines has been established in this chapter. The formulation is very general. It has a number of advantages, related to its variational character and the explicit formulation. It can be used in both the spatial and the spectral domain, depending upon the effects it is sought to highlight. In combination with conformal mapping, it drastically reduces the complexity and the time of numerical computation. It leads to rapidly convergent results even when

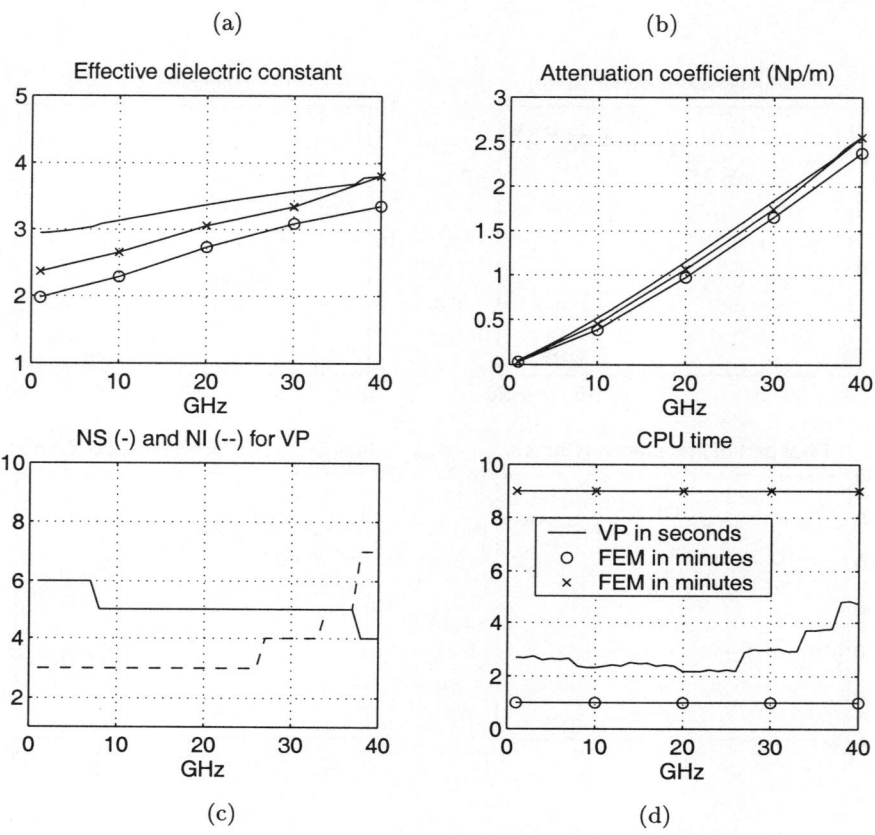

Fig. 4.23 *Comparative results using variational principle (VP) and finite-element method (FEM) for shielded microstrip (a) effective dielectric constant; (b) attenuation coefficient; (c) number of spectral terms (N_S) and of iterations (N_I) required by VP; (d) CPU time in seconds for VP (—), and in minutes for FEM*

higher-order modes are considered. Mathieu functions are shown to be very efficient expressions for trial fields of the dominant and the higher-order modes in slots. The calculation is fast: it is performed on-line on a regular PC. Results obtained on open and shielded lines have been successfully compared with previously published data. The method is general enough to accomodate gyrotropic lossy substrates. The method has been validated by actual measurements on various types of lines like slot-lines, coupled slot-lines, finlines, and coplanar waveguides, in which YIG-layers may be included. It has also been validated on various structures involving lines and transitions between them. These experimental results are presented in the next chapter.

4.9 References

[4.1] J. Schwinger, D. S. Saxon, *Discontinuities in Waveguides - Notes on Lectures by J. Schwinger*. London: Gordon and Breach, 1968.

[4.2] V.H. Rumsey, "Reaction concept in electromagnetic theory", *Phys. Rev.*, vol. 94, pp. 1483-1491, June 1954, and vol. 95, p. 1705, Sept. 1954.

[4.3] A.D. Berk, "Variational principles for electromagnetic resonators and waveguides", *IEEE Trans. Antennas Propagat.*, vol. AP-4, no. 2, pp. 104-111, Apr. 1956.

[4.4] I. Huynen, B. Stockbroeckx, "Variational principles compete with numerical iterative methods for analyzing distributed electromagnetic structures", *Annals of Telecommunications*, vol. 53, no. 3-4, pp. 95-103, March-April 1998.

[4.5] T. Itoh, *Numerical Techniques for Microwave and Millimeter Wave Passive Structures*. New York: John Wiley&Sons, 1989, Ch. 5.

[4.6] S.S. Bedair, I. Wolff, "Fast, accurate and simple approximate analytic formulas for calculating the parameters of supported coplanar waveguides for (M)MIC's", *IEEE Trans. Microwave Theory Tech.*, vol. MTT-40, no. 1, pp. 41-48, Jan. 1992.

[4.7] I. Huynen, D. Vanhoenacker-Janvier, A. Vander Vorst, "Spectral domain form of new variational expression for very fast calculation of multilayered lossy planar line parameters", *IEEE Trans. Microwave Theory Tech.*, vol. 42, no. 11, pp. 2099-2106, Nov. 1994.

[4.8] J.B. Knorr, P. Shayda, "Millimeter wave finline characteristics", *IEEE Trans. Microwave Theory Tech.*, vol. MTT-28, no. 7, pp. 737-743, July 1980.

[4.9] I. Huynen, A. Vander Vorst, "A new variational formulation, applicable to shielded and open multilayered transmission lines with gyrotropic non-hermitian lossy media and lossless conductors", *IEEE Trans. Microwave Theory Tech.*, vol. 42, no. 11, pp. pp. 2107-2111, Nov. 1994.

[4.10] E. Yamashita, "Variational method for the analysis of microstrip-like transmission lines", *IEEE Trans. Microwave Theory Tech.*, vol. MTT-16, no. 8, pp. 529-535, Aug. 1968.

[4.11] M. Essaaidi, M. Boussouis, "Comparative study of the Galerkin and the least squares boundary residual method for the analysis of microwave integrated circuits", *Microwave and Opt. Technol. Lett.*, vol. 7, no. 3, pp. 141-146, Mar. 1992.

[4.12] F.L. Mesa, M. Horno, "Quasi-TEM and full wave approach for coplanar multistrip lines including gyromagnetic media longitudinally magnetized", *Microwave and Opt. Technol. Lett.*, vol. 4, no. 12, pp. 531-534, Nov. 1991.

[4.13] J.B. Davies in T. Itoh, *Numerical Techniques for Microwave and Millimeter Wave Passive Structures*. New York: John Wiley&Sons, 1989, Ch. 2, p. 91.

[4.14] R.H. Jansen, "The spectral domain approach for microwave integrated circuits", *IEEE Trans. Microwave Theory Tech.*, vol. MTT-33, no. 10, pp. 1043-1056, Oct. 1985.

[4.15] N.K. Das, D.M. Pozar, "Full-wave spectral-domain computation of material, radiation, and guided wave losses in infinite multilayered printed transmission lines", *IEEE Trans. Microwave Theory Tech.*, vol. MTT-39, no. 1, pp. 54-63, Jan. 1991.

[4.16] D.C. Champeney, *A Handbook of Fourier Theorems*. Cambridge: Cambridge University Press, 1987.

[4.17] J.P. Parekh, K.W. Chang, H.S. Tuan, "Propagation characteristics of magnetostatic waves", *Circuits, Systems and Signal Proceedings*, vol. 4, no. 1-2, pp. 9-38, 1985.

[4.18] K. Yashiro, S. Ohkawa, "A new development of an equivalent circuit model for magnetostatic forward volume wave transducers", *IEEE Trans. Microwave Theory Tech.*, vol. MTT-36, no. 6, pp. 952-960, June 1988.

[4.19] R.I. Joseph, E. Schlomann, "Demagnetizing field in nonellipsoidal bodies", *J. Applied Physics*, vol. 36, no. 5, pp. 1579-1593, May 1965.

[4.20] P. Kabos, V.S. Stalmachov, *Magnetostatic Waves and their Applications*. London: Chapman&Hall, 1994.

[4.21] S.N. Bajpai, "Insertion losses of electronically variable magnetostatic wave delay lines", *IEEE Trans. Microwave Theory Tech.*, vol. MTT-37, no. 10, pp. 1529-1535, Oct. 1989.

[4.22] I.W. O'Keefe, R.W. Paterson, "Magnetostatic surface waves propagation in finite samples", *J. App. Phys.*, vol. 49, pp. 4886-4895, 1978.

[4.23] J.D. Adam, S.N. Bajpai, "Magnetostatic forward volume wave propagation in YIG strips", *IEEE Trans. Magnetics*, vol. MAG-18, no. 6, pp. 1598-1600, Nov. 1982.

[4.24] P.R. Emtage, "Interaction of magnetostatic waves with a current", *J. App. Phys.*, vol. 49, no. 8, pp. 4475-4484, Aug. 1978.

[4.25] R.E. Collin, *Field Theory of Guided Waves*. New York: McGraw-Hill, 1960, Ch. 1.

[4.26] T. Itoh, R. Mittra, "Spectral-domain approach for calculating the dispersion characteristics of microstrip lines", *IEEE Trans. Microwave Theory Tech.*, vol. MTT-21, no. 7, pp. 496-499, July 1973.

[4.27] R. Mittra, T. Itoh, "A new technique for the analysis of the dispersion characteristics of microstrip lines", *IEEE Trans. Microwave Theory Tech.*, vol. MTT-19, no. 1, pp. 47-56, Jan. 1971.

[4.28] I. Huynen, *Modelling planar circuits at centimeter and millimeter wavelengths by using a new variational principle*, Ph.D. dissertation. Louvain-la-Neuve, Belgium: Microwaves UCL, 1994.

[4.29] R. Janaswamy, D.H. Schaubert, "Dispersion characteristics for wide slot-lines on low-permittivities substrates", *IEEE Trans. Microwave Theory Tech.*, vol. MTT-33, no. 8, pp. 723-726, Aug. 1985.

[4.30] R.E. Collin, *Field Theory of Guided Waves*. New York: McGraw-Hill, 1960, Ch. 2.

[4.31] R.E. Collin, *Field Theory of Guided Waves*, 2nd ed. New York: IEEE Press, 1991, Ch. 4.

[4.32] N. McLachlan, *Theory and Applications of Mathieu Functions*. Oxford, UK: Oxford University Press, 1951.

[4.33] M. Hélard, J. Citerne, O. Picon, V. Fouad Hanna, "Theoretical and experimental investigation of finline discontinuities", *IEEE Trans. Microwave Theory Tech.*, vol. MTT-33, no. 10, pp. 994-1003, Oct. 1985.

[4.34] E.B. El-Sharawy, R. W. Jackson, "Coplanar waveguide and slot line on magnetic substrates: analysis and experiment", *IEEE Trans. Microwave Theory Tech.*, vol. MTT-36, no. 6, pp. 1071-1079, June 1988.

[4.35] G.N. Watson, *A Treatise on the Theory of Bessel Functions*. Cambridge: Cambridge University Press, 1966.

[4.36] M. Abramowitz, I. Stegun, *Handbook of Mathematical Functions*. New York: Dover Publications, 1965.

[4.37] E.J. Denlinger, "A frequency dependent solution for microstrip transmission lines", *IEEE Trans. Microwave Theory Tech.*, vol. MTT-19, no. 1, pp. 30-39, Jan. 1971.

[4.38] R.F. Harrington, *Time-Harmonic Electromagnetic Fields*. New York: McGraw-Hill, 1961, Ch. 7.

[4.39] E. Kuhn, "A mode matching method for solving fields problems in waveguide and resonator circuits", *Arch. Elek. Übertragung.*, vol. AEÜ-27, no. 12, pp. 511-518, Dec. 1973.

[4.40] G. Kompa, "Dispersion measurements of the first two higher-order modes in open microstrip", *Arch. Elek. Übertragung.*, vol. AEÜ-27, no. 4, pp. 182-184, Apr. 1973.

[4.41] R. Mehran, "The frequency-dependent scattering matrix of microstrip right angle bends", *Arch. Elek. Übertragung.*, vol. AEÜ-29, no. 11, pp. 454-460, Nov. 1975.

[4.42] A.N. Fedorov, N.N. Levina, N.A. Khametova, "Some results of a numerical investigation of slot and strip-slot lines", *Radio Engin. and Electron. Physics (USA)*, vol. 28, no. 7, pp. 42-49, July 1983.

[4.43] I. Huynen, "Modélisation d'interférences sur circuits microruban hyperfréquences", *Revue E*, 107, no. 3, pp. 40-45, Mar. 1991.

[4.44] M. Kitlinski, B. Janiczak, "Dispersion characteristics of asymmetric coupled slot lines on dielectric substrates", *Electronic Lett.*, vol. 19, no. 3, pp. 91-92, Feb. 1983.

[4.45] G. Dahlquist, A. Bjorck, *Numerical Methods*. New Jersey: Prentice-Hall, 1974, Ch. 7.

[4.46] L.P. Schmidt, T. Itoh, "Spectral domain analysis of dominant and higher order modes in finlines", *IEEE Trans. Microwave Theory Tech.*, vol. MTT-28, no. 9, pp. 981-985, Sept. 1980.

[4.47] T. Itoh (Editor), *Numerical Techniques for Microwave and Millimeter Wave Passive Structures*. New York: John Wiley&Sons, 1989, Ch. 2 and 5.

[4.48] J.B. Davies in T. Itoh, *Numerical Techniques for Microwave and Millimeter Wave Passive Structures*. New York: John Wiley&Sons, 1989, Ch. 2.

[4.49] I. Huynen, B. Stockbroeckx, "Variational principles compete with numerical iterative methods for analyzing distributed electromagnetic structures", *Proceedings of the NUMELEC'97 Conference*, Lyon, France, pp. 76-77, 19-21 March 1997.

[4.50] Z. Raida, "Comparative efficiency of semi-analytical and finite-element methods for analyzing shielded planar lines", *Internal Report*, Microwaves Lab., UCL, Feb. 1997.

CHAPTER 5

Applications

5.1 Transmission lines

5.1.1 Transmission line parameters

Transmission line theory usually uses three parameters to describe the propagation along a guiding structure [5.1]: the propagation constant β taking into account the velocity of propagation, the attenuation coefficient α describing the attenuation along the line, and the characteristic impedance Z_c, the ratio between the forward voltage and current waves. The impedance is usually chosen in such a way that the power flow on the equivalent line equals the power flow on the physical guiding structure, obtained by integrating the Poynting vector over the cross-section of the structure. Defining the z-axis as propagation axis, the transmission line equations are

$$V(z) = V_+ e^{-\gamma z} + V_- e^{+\gamma z} \tag{5.1a}$$

$$I(z) = \frac{1}{Z_c} \left[V_+ e^{-\gamma z} - V_- e^{+\gamma z} \right] \tag{5.1b}$$

where we have defined γ as in (4.1):

$$\gamma = \alpha + j\beta \tag{5.1c}$$

The power flow along the z-axis is expressed as

$$P(z) = \frac{1}{2} \operatorname{Re} \left[V(z) I(z)^* \right] = P_0 e^{-2\alpha z} \tag{5.2a}$$

with the definition

$$P_0 = \frac{|V_+|^2}{2Z_c} = \frac{|V_0|^2}{2Z_c} = \frac{Z_c |I_0|^2}{2} \tag{5.2b}$$

Requiring P_0 to equal the electromagnetic power flow yields:

$$P_0 = \frac{1}{2} \int_S \left(\overline{e} \times \overline{h}^* \right) \cdot \overline{dS} \tag{5.3}$$

5.1.1.1   Propagation constant

In Chapter 4 we derived spatial and spectral forms of an explicit variational principle for the propagation constant γ. They consist of second-order equations for γ

$$-\gamma^2 \sum_i A_i - \gamma \sum_i (B_i - C_i) + \sum_i D_i - \omega^2 \sum_i E_i = 0 \qquad (4.17\text{a})$$

in the spatial domain, and

$$-\gamma^2 \sum_i \tilde{A}_i - \gamma \sum_i (\tilde{B}_i - \tilde{C}_i) + \sum_i \tilde{D}_i - \omega^2 \sum_i \tilde{E}_i = 0 \qquad (4.21)$$

in the spectral domain, equivalent to (4.2), where the spatial or spectral coefficients are functions of the spatial or spectral transverse dependencies of the trial electric field components, as shown by expressions (4.17b-f) and (4.19a-e), respectively.

5.1.1.2   Power flow

A simple calculation directly shows that P_0 (5.3) can be expressed as a function of coefficients involved in the variational equations of (4.17a-f) and (4.21):

$$P_0 = \frac{-1}{j\omega\mu_0} \left[\gamma \sum_i A_i - \sum_i C_i \right]^* \qquad \text{for spatial trial fields} \qquad (5.4\text{a})$$

$$= \frac{-1}{j\omega\mu_0} \left[\gamma \sum_i \tilde{A}_i - \sum_i \tilde{C}_i \right]^* \qquad \text{for spectral trial fields} \qquad (5.4\text{b})$$

where the coefficients A_i, C_i, $\tilde{A}_i$, $\tilde{C}_i$ are defined, respectively, by expressions (4.17b,d) and (4.19a,c).

Once the propagation constant has been calculated by (4.17a) or (4.21), we directly obtain the value of P_0 without any additional computation, since (5.4a,b) only uses coefficients already calculated for obtaining γ. This is another major advantage of our formulation. The power lost per unit length on the equivalent line is then obtained as the first derivative of the power flow along the z-axis:

$$P_{diss} = -\frac{\partial P(z)}{\partial z} = 2\alpha P_0 e^{-2\alpha z} \qquad (5.5)$$

5.1.1.3   Attenuation coefficient

A number of phenomena contribute to a non-zero value for the attenuation coefficient α. Two of them have already been considered when using (4.17a).

A.   Evanescent modes

These are readily obtained by solving the variational equation (4.17a), since this equation may yield a complex solution, depending on the choice of trial fields. If the trial fields associated with a mode are properly chosen, then the behavior of the propagation constant γ will give the correct evanescent behavior. This has already been illustrated in Chapter 4 where higher-order mode propagation constants were calculated for shielded lines below cut-off, yielding a real value for the propagation constant γ (Figs. 4.12 and 4.19).

B.   Losses in the layers

As has been shown in Chapter 4, substrate losses are calculated by using complex expressions for $\bar{\bar{\varepsilon}}$ and $\bar{\bar{\mu}}$ in the variational equation, yielding a complex solution for equations (4.2) and (4.21):

$$\gamma = \alpha_s + j\beta \qquad \text{where } \alpha_s \text{ takes into account substrate losses.}$$

Hence the substrate loss constant is directly obtained from the solution of (4.2) calculated when $\bar{\bar{\varepsilon}}$ and $\bar{\bar{\mu}}$ are lossy. This has already been illustrated for a gyrotropic YIG-film, where real and imaginary parts of the propagation constant were found simultaneously.

Two other categories of losses may also be present on planar lines, namely conductor losses and radiation losses. The modeling of conductor losses is performed as follows. Since equation (4.17a) is variational in the absence of conductor losses, these are calculated by using a perturbational expression, based on a skin-effect formulation [5.2]. Indeed in Chapter 4, we have developed expressions for trial fields, assuming that the thickness of the perfectly conducting layer is negligible, so that surface current densities are flowing on these layers. The perturbational approach that will now be used is valid when the conductors are good, although not perfect, in other words under a low loss assumption. The perturbational approach considers that the value of the tangential magnetic field at the conducting interfaces under low loss does not differ much from its lossless value. This lossless value may be combined with a skin-effect formulation to obtain an estimation of the conductor losses. The skin-effect formulation has already been used for planar lines analyzed in the spatial domain by Rozzi et al. [5.3]. In [5.4] we presented a new spectral domain formulation of the skin effect, taking into account the variation along the x- and z-axis of the magnetic field. It is summarized below.

A good conductor is characterized by the condition

$$\sigma >> \omega\varepsilon_0 \tag{5.6}$$

Under this assumption, Maxwell's equations result in diffusion equations in terms of spectral electric field, magnetic field or current density. They have as a general spectral solution

$$\tilde{\bar{x}}(k_x, y) = \tilde{\bar{x}}(k_x, 0)\, e^{-Ky} \tag{5.7}$$

where $K = \sqrt{j\omega\mu_0\sigma}$
$\quad\quad x = e, h, j$ for the electric field, magnetic field, and current density, respectively.

By virtue of the perturbational approach, $\tilde{\bar{h}}(k_x, 0)$ is equal to the magnetic field at the conducting interface assuming losslessness. Hence it can be calculated as shown in Chapter 4. The surface impedance Z_m of the conductor is related to the tangential electric field at the surface of the conductor to the tangential magnetic field as:

$$\tilde{\bar{e}}(k_x, 0) = Z_m \tilde{\bar{h}}(k_x, 0) \tag{5.8a}$$

where

$$Z_m = \sqrt{\frac{j\omega\mu_0}{\sigma}} \tag{5.8b}$$

Using Parseval's relations (Appendix C), the power dissipated on the two sides of any conductive layer of thickness t is obtained as a function of the spectral fields as

$$\begin{aligned} P_{diss} &= e^{-2\alpha z} \int_{-\infty}^{+\infty} \int_0^t \tilde{\bar{e}}(x, y) \overline{\tilde{j}^*}(x, y) \, dy \, dx \\ &= e^{-2\alpha z} \int_{-\infty}^{+\infty} \int_0^t \tilde{\bar{e}}(k_x, y) \overline{\tilde{j}^*}(k_x, y) \, dy \, dk_x \\ &= \frac{\sigma |Z_m|^2}{2} (I_{N1} + I_{M1}) e^{-2\alpha z} \end{aligned}$$

with

$$I_i = \int_{-\infty}^{+\infty} \int_0^t \left(|\tilde{h}_{xi}(k_x, 0)|^2 + |\tilde{h}_{zi}(k_x, 0)|^2 \right) e^{-2\,\text{Re}(K)y} \, dy \, dk_x \tag{5.9}$$

where the subscript $i = N1, M1$ refers to the two layers adjacent to the conductor, according to a previous definition (Chapter 4, Fig. 4.1).

The attenuation constant due to conductor losses is then obtained from (5.5) as

$$\alpha_c = \frac{P_{diss}|_{\alpha=0}}{2P_0|_{\alpha=0}} \tag{5.10}$$

It should be noted that (5.10) is equivalent to (5.5) rewritten using the perturbational approach, that is for P_{diss}, P and P_0 calculated for $\alpha = 0$ using (5.9), (5.2a) and (5.4) respectively.

An important comment has to be made about the shape of the field used for the skin-depth effect. This field is calculated not only without losses but also for zero-thickness conductors: it is not the actual field for the lossless finite thickness case. However, the error made on the conductor losses due to a finite thickness of conducting layer is assumed to be small. Some authors have studied theoretically [5.5]-[5.7] and experimentally [5.8]-[5.10] the limitations of existing modeling techniques when a finite thickness is considered. In fact, as mentioned by Itoh [5.11], the simplicity of the spectral domain technique related to a Green's formalism fails when the zero-thickness assumption is not valid. For these reasons, the zero-thickness assumption is usually maintained but corrected by a perturbational calculation of conductor losses, based on a skin-effect formulation which involves the thickness of the conductor.

Hence the total attenuation coefficient is the sum of two terms; α_s due to the substrate, obtained by the variational procedure, and α_c due to the conductors, obtained by a skin-depth perturbational formulation.

5.1.1.4 Characteristic impedance

When the guiding structure supports a pure TEM propagation mode, a unique correspondence is found between the current and voltage of the equivalent line and the physical magnetic and electric fields existing in the structure [5.2] and [5.12]. In the case of TEM lines, the characteristic impedance is equal to the wave impedance, which depends only on the constitutive parameters of the homogeneous medium surrounding the conductors of the TEM line.

The definition of a characteristic impedance for a multilayered guiding structure (Fig. 4.1), supporting TE or TM modes is not unique, since a wave impedance is defined for each TE or TM mode considered and also for each of the layers considered. We only may use a relation between V_+ and Z_c when the power flow P_0 is assumed to be known.

Hence we have to relate the electromagnetic fields in (5.3) to the voltage or current waves of the equivalent transmission line. Using (5.2b) we relate V_+ to either a current or a voltage, as:

$$Z_{cPV} = \frac{V_0^2}{2P_0} \qquad \text{Power-Voltage definition} \qquad (5.11)$$

$$Z_{cPI} = \frac{2P_0}{I_0^2} \qquad \text{Power-Current definition} \qquad (5.12)$$

A. Characteristic impedance for slot-like lines

As many authors [5.13]-[5.15], we choose for the characteristic impedance of the dominant mode of slot-like lines the definition (5.11). It is easy to relate the transverse component of the trial quantity defined in Chapter 4 for slot-like lines to a voltage concept, by integration over the finite area of the slot. The normalization factor for the trial transverse electric field across the slot is chosen such that its integral across the slot is equal to the voltage V_0 arbitrarily imposed across the slot. As a consequence, the spectral fields and hence the spectral quantities in (4.19a-e), are proportional to V_0. When introduced into (5.4), the power flow P_0 becomes proportional to the squared power of V_0, which renders expression (5.11) for the impedance independent of V_0.

This illustrates another advantage of the method presented here: no additional computation is necessary to calculate the characteristic impedance of the structure, because the power flow and hence the impedance, is directly expressed as a function of the coefficients already used in the calculation of the propagation constant.

B.   Characteristic impedance for strip-like lines

The longitudinal component of the current density defined as the trial quantity for strip-like problems is related to the total current I_0 flowing on the strip, using definition (5.12). This current is simply obtained as the integration of the longitudinal current density over the strip area. Some authors [5.16], however, propose another definition of the characteristic impedance, based on a power-voltage definition, where the voltage is obtained from the line integration of the electric field between the center of the strip and the ground plane. Jansen and Kirschning have pointed out in [5.17] that the power current definition has various theoretical advantages, because it is related to a current, hence to a magnetic field, which is not responsible for the dispersion due to the dielectric interface. As a consequence, the behavior of a calculation based on a power-current definition will be less frequency-dependent than that with a power-voltage definition, as observed from results presented in [5.16]. Hence, the power-current definition of the characteristic impedance will better agree with the pure TEM definition of Z_c.

C.   Other definitions for the characteristic impedance

It is important to note that the problem of a correct definition of characteristic impedance remains unsolved, and is still the subject of discussions in the literature. Williams, for example, attempts to present a uniform definition for planar lines, based on causality considerations [5.18]. Also, measuring this parameter is particularly difficult [5.19]. It requires specific calibration techniques, such as those developed by the National Institute of Standards and Technology (NIST) [5.20].

On the other hand, to the best of our knowledge, it is not proven that expression (5.4) combined with either definition (5.11) or (5.12) yields a variational expression for the characteristic impedance and hence that the resulting value of Z_c is correct to the second-order.

We will nevertheless illustrate in this chapter that definitions (5.11) (5.12) are very well validated by experiments carried out on a number of configurations, because they are calculated using the solution of variational principle (4.17a) and (4.21) for the propagation constant involved in their expressions [5.21].

5.1.2   Microstrip

The validity of the variational approach (4.21) is first tested on a quasi-TEM microstrip line, etched on a dielectric substrate with a complex permittivity $2.36(1 - j.0038)$. The line width is 1.5 mm and has been designed to present a 50 Ω characteristic impedance on the substrate considered in this example . Figure 5.1a shows the losses per unit length, obtained as the real part of the complex propagation constant γ, while Figure 5.1b shows the real part of the complex effective dielectric constant, obtained from the imaginary part β of γ as

$$\varepsilon_{eff} = \left(\frac{c_0\beta}{\omega}\right)^2 \tag{5.13}$$

Solid lines are for parameters extracted from measurements, while dashed ones show values of parameters calculated using the variational principle (4.21).

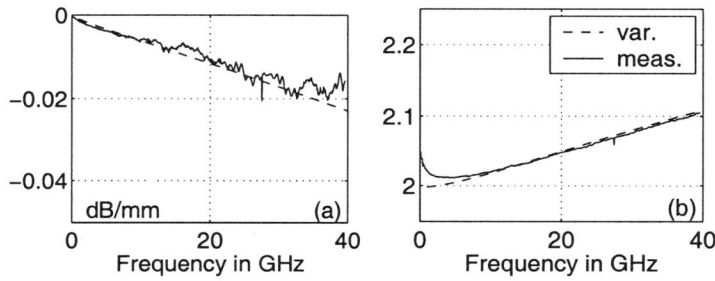

Fig. 5.1 *Transmission parameters of microstrip line etched on a substrate plate of thickness 0.508 mm and relative permittivity $2.36(1 - j0.0038)$ (a) losses; (b) effective dielectric constant; solid lines are for measurements, dashed ones for variational principle (4.21)*

A good agreement is obtained for both losses and effective dielectric constant. The measurement method is accurate and will be described in Section 5.5.

5.1.3 Slot-line

Next, the validation is extended to a slot-line, whose dominant mode is of TE-type. The line considered is etched on the same substrate as for the microstrip in Figure 5.1, but its width is designed to be 0.37 mm in order to obtain a good matching with the slot-to-microstrip transitions used for the measurement. Such transitions will be described in Section 5.2. After measurements and calibration, the effects of the microstrip transition to the slot-line are removed, and the experimental values for the transmission parameters presented in Figure 5.2 are only those of the slot-line. Again, a good agreement is observed for both losses and effective dielectric constant. This illustrates that our formulation is efficient for dynamic (non quasi-TEM) situations. It has to be noted, however, that the measuring frequency range is limited by the bandwidth of the microstrip-to-slot-line transition (2-25 GHz). The out-of-band mismatch is too high and the calibration procedure can no longer compensate for the transition.

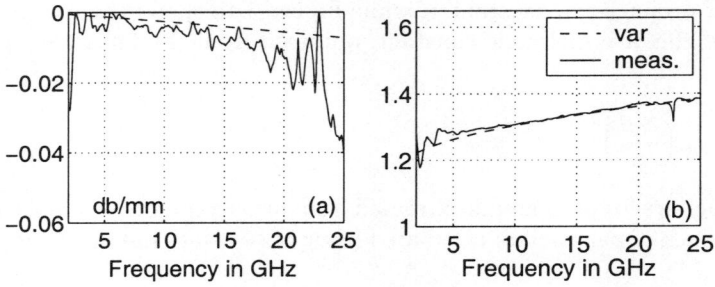

Fig. 5.2 *Transmission parameters of slot-line etched on a substrate plate of thickness 0.508 mm and relative permittivity 2.36(1 − j0.0038) (a) losses; (b) effective dielectric constant; solid lines are for measurements, dashed ones for variational principle (4.21)*

5.1.4   Coplanar waveguide

As another example, Figure 5.3 compares the measured and calculated terms of the scattering matrix of a coplanar waveguide impedance step. The simulation for this structure involves the attenuation coefficient, the propagation constant, and the characteristic impedance of the lines. Only the dominant modes on the two lines are considered. The characteristic impedance of the access line is taken as reference impedance for the scattering matrix of the junction, as the measurement technique uses a de-embedding algorithm which moves the reference planes for junctions between the two lines of different width. It can be seen in Fig. (5.2a,b) that a good agreement is obtained for both the reflection and transmission parameters. Hence, calculated impedance agrees with measurement. Measured and computed phases also agree very well, especially in the case of transmission coefficient (Fig. 5.3d). This validates the calculated propagation constant. The calculated dielectric and conductor losses also agree with the measured ones: calculated and measured transmission curves are indeed in coincidence when the electric length of the step line is a multiple of a half wavelength. The degradation of the measured scattering terms around 18 GHz for the coplanar line is attributed to surface wave losses due to a mismatch of the coaxial-to-coplanar launcher used for de-embedding.

5.1.5   Finline

Finally, a validation is performed for a boxed (or shielded) line, namely a finline. It can be viewed as a slot-line on a dielectric substrate, surrounded by a waveguide enclosure. Hence, the variational principle (4.21) can be used, but the spectral integrals yielding the coefficients of its second-order equation have to be replaced by discrete summations over spectral variable k_x. The slot-line inside the waveguide is etched on a substrate having a thickness and permit-

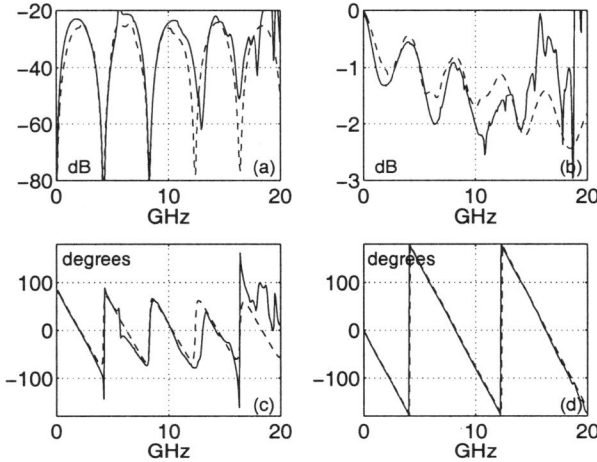

Fig. 5.3 *Modeled (- -) and measured (—) scattering terms of coplanar waveguide step (a) magnitude of reflection coefficient; (b) magnitude of transmission coefficient; (c) phase of reflection coefficient; (d) phase of transmission coefficient*

tivity different from Figure 5.2; its width is designed to be 0.45 mm in order to obtain a good matching with the slot-to-waveguide transitions used for the measurement. The particular measurement method is described in Section 4.5 [5.22]. Figure 5.4 shows results obtained for the effective dielectric constant. Again, a very good agreement is observed between the experimental values (lines) and the calculations using the variational principle (4.21) (circles).

5.1.6 Planar lines on lossy semiconductor substrates

Planar transmission lines are frequently used as interconnects on multilayered low-resistivity substrates in high-frequency MMICs, such as bulk silicon wafers, doped areas in gallium arsenide (GaAs), or photo-illuminated layers in optoelectronics. These substrates are modeled by using an equivalent loss tangent, obtained from their resistivity ρ:

$$j\omega\varepsilon = j\omega\varepsilon_0\varepsilon_r + 1/\rho \tag{5.14a}$$

$$= j\omega\varepsilon_0\varepsilon_r\left(1 - \frac{j}{\rho\omega\varepsilon_0\varepsilon_r}\right) \tag{5.14b}$$

$$= j\omega\varepsilon_0\varepsilon_r\left(1 - j\tan\delta_{equ}\right) \tag{5.14c}$$

Figure 4.22 compares the transmission line parameters obtained from the variational spectral forms (4.21), (5.4) and (5.11), with measured transmission line parameters, for the same coplanar waveguide as in Figures 4.15 and

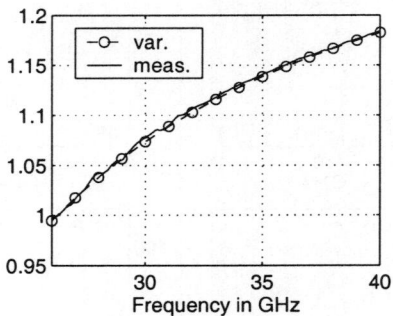

Fig. 5.4 *Effective dielectric constant of finline etched on a substrate plate of thickness 0.272 mm and relative permittivity 2.22(1 − j0.0018), waveguide section 2a² with a = 3.556 mm; solid lines are for measurements, symbols o for the variational principle (4.21)*

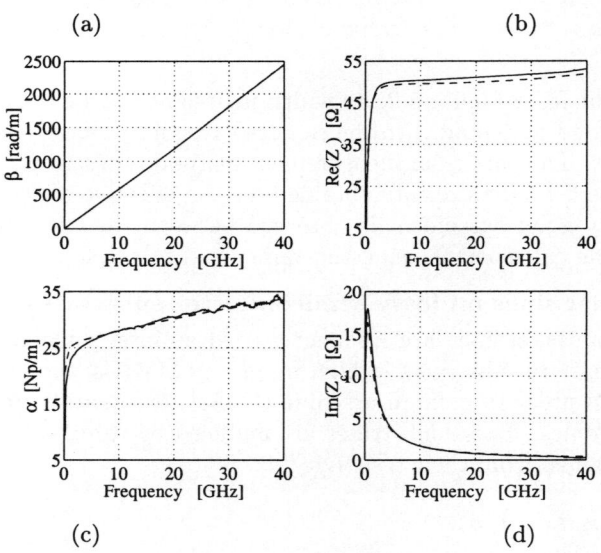

Fig. 5.5 *Results measured (solid), simulated with (VP) (- -), for transmission line parameters of microstrip line on low-resistivity silicon substrate of resistivity 225 Ωcm (a) propagation constant; (b) real part of impedance; (c) attenuation coefficient; (d) imaginary part of impedance*

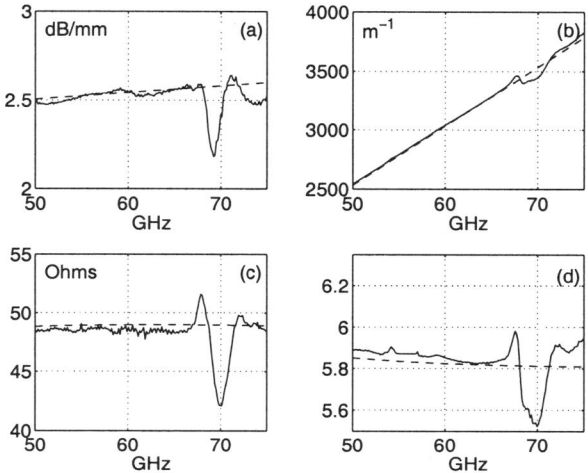

Fig. 5.6 *Results measured (solid), simulated with (VP) (- -), for transmission line parameters of coplanar waveguide on low-resistivity SOI substrate of resistivity 20 Ω cm (a) attenuation coefficient; (b) propagation constant; (c) real part of impedance; (d) effective dielectric constant*

4.21. The VP transmission line parameters (- - -) perfectly agree with experimental results (—). It should be observed that propagation on high-loss substrates is highly-dispersive in the low frequency range. A slow-wave mode of propagation occurs, typical of integrated circuits (ICs) on low resistivity substrates.

The variational principle (4.21) yields a similar agreement in the case of stripline topologies on low-resistivity substrates. As an example, Figure 5.5 shows the transmission line parameters of a microstrip line on a lossy silicon substrate, using (4.21). Again, perfect agreement occurs between the VP transmission line parameters (- -) and measured ones (—) [5.23].

At higher frequencies, substrate losses have no effect on dispersion, but the variational modeling still holds, as shown in Figure 5.6. Values of attenuation, effective dielectric constant and characteristic impedance extracted from measurements agree with simulated ones (Fig. 5.6a,c,d) in the millimeter wave frequency range (50-75 GHz). The peaks observed in the measurement around 70 GHz are due to a problem during calibration and not to the device.

5.1.7 Planar lines on magnetic nanostructured substrates

Variational formulas established for perturbation problems are particularly efficient for modeling the behavior of magnetically-nanostructured planar circuits and devices over a wide frequency range, as shown in Figure 5.7. The device consists of an array of magnetic metallic nanowires embedded in a

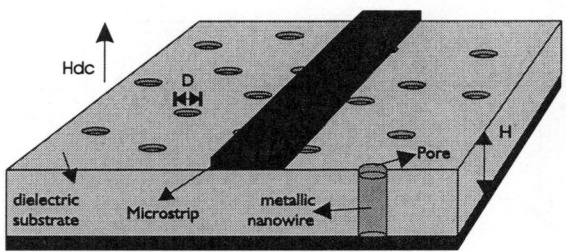

Fig. 5.7 *Microstrip line on nanostructured substrate, including nanowires (Reproduced from Proc. PIERS'2000 [5.25])*

porous dielectric substrate of thickness H [5.24][5.25]. The metallic wires are perpendicular to the dielectric plate, all with diameter D. Typical values of D are in the nanometer range, *i.e.* between 50 and 500 nm. This substrate was first developed for microstrip applications: a microstrip is deposited on the top side of the composite substrate, while the bottom is metallized to serve as ground plane. The magnetic wire is made of cobalt, nickel or iron.

The experimental results presented in Figure 5.8 (solid lines) show that the device has interesting tuning properties at microwave frequencies. Transmission measurements between the input and output of microstrip exhibit a stopband behavior in a certain frequency range, which is shifted to higher frequencies when a DC-magnetic field is applied parallel to the nickel nanowires (Fig. 5.8a,c). A similar shifting behavior is obtained with the other ferromagnetic metals considered (cobalt and iron), but at fixed or zero DC-field the center frequency of stopband differs with the kind of wire material.

We have successfully related these experimental relationships to ferrimagnetic theory [5.24][5.25]. This theory predicts a resonance behavior at microwaves for ferrimagnetic materials, tunable when applying a DC-magnetic field. Hence, ferromagnetic nanowires act like thin ferrite implants in the dielectric substrate, and can be treated by variational principles particularized for material perturbation problems. In Chapter 2, a formulation correct to second-order is obtained for the shift induced on the propagation constant, in the case of material perturbation:

$$j(\beta_1 - \beta) \approx \frac{j\omega \int_S \{\overline{e}^* \cdot (\Delta\overline{\overline{\varepsilon}} \cdot \overline{e}) + \overline{h}^* \cdot (\Delta\overline{\overline{\mu}} \cdot \overline{h})\}\, dS}{\int_S \overline{h}^* \cdot (\overline{a}_z \times \overline{e})\, dS - \int_S \overline{e}^* \cdot (\overline{a}_z \times \overline{h})\, dS} \qquad (2.94b)$$

with $\Delta\overline{\overline{\varepsilon}} = \overline{\overline{\varepsilon}}_1 - \overline{\overline{\varepsilon}}$
$\qquad \Delta\overline{\overline{\mu}} = \overline{\overline{\mu}}_1 - \overline{\overline{\mu}}$

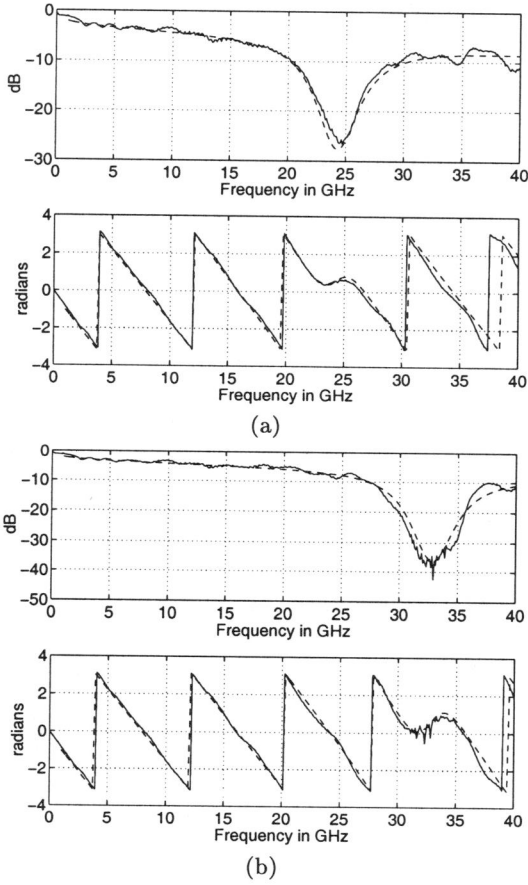

Fig. 5.8 *Transmission model (2.94a) (dashed line) and experiment(solid line) on microstrip on nanostructured substrate (D = 500 nm, H = 20 μm), with cobalt nanowires; magnitude [dB] and phase [radians] for (a) $H_{dc} = 350$ Oe; (b) $H_{dc} = 3650$ Oe*

Based on the ferromagnetic resonance hypothesis, a full analytic model is derived using the variational principle (2.94a,b). We take as trial fields the electric and magnetic fields between the two conductors of a parallel plate transmission line, assuming that a uniform current density is flowing on each conductor. This assumption is valid for a wide microstrip line, that is for geometries with high W/H ratios (typical values for Fig. 5.7: $W/H = 20$). The permeability tensor $\overline{\overline{\mu}}_1$ is taken equal to the gyrotropic tensor modeling the ferromagnetic resonance effect, as introduced in Appendix D. Other permeabilities and permittivities are scalar. Assuming that the propagation constant in the absence of nanowires is calculated exactly, formula (2.94a) yields the propagation constant of the microstrip line represented in Figure 5.7. The dashed curves in Figure 5.8 are calculated with the resulting model, obtained by computing the transmission factor $e^{-\gamma L}$ where L is the length of the microstrip. An excellent agreement with experiment is observed, for both magnitude and phase and for each value of magnetic field considered.

5.2   Junctions

5.2.1   Microstrip-to-slot-line transition

Simple transitions between slot lines and measuring equipment are necessary for testing and designing slot-line circuits. The first attempt to measure slot-line characteristics was made by Cohn [5.26], followed by Mariani *et al.* [5.27] and Robinson and Allen [5.28]. They used a direct transition between a coaxial cable and the slot-line. The drawbacks of this transition are the poor agreement between modeling and experiment, and the difficulty of manufacturing repeatable transitions. The transition has, however, been used by Lee [5.29] to check the validity of his slot-line impedance model.

The microstrip-to-slot-line transition shown in Figure 5.9 offers an elegant way to measure and characterize slot-like devices. The frequency range over which this characterization is efficient is related to the bandwidth of the microstrip-to-slot-line transition. Thus we need an accurate model for this transition, in order to optimize its design for wideband applications. This transition was introduced by Cohn [5.26] and used by Mariani *et al.* [5.27], and Robinson and Allen [5.28]. It has been used for the propagation constant measurements presented in the previous section.

We model the transition as two transmission lines ended by quarter-wavelength stubs (Fig. 5.9a) and interconnected via a coupling network (Fig. 5.9b). The network represents the coupling area shown in Figure 5.9c. It consists of:

1. a transformer of ratio [5.5],[5.26], where n is a function of frequency: it depends on the configuration of the magnetic fields around the strip and slot, whose shape is a function of wavelength and hence of frequency (Fig. 5.9c)

2. two fringing capacitances C_1 modeling the interaction of the electric field between the strip and slot in the coupling area

3. an equivalent capacitance C_2 describing the electric field across the slot in coupling area

4. a series inductance L modeling the strip crossing the slot in coupling area.

5.2.1.1   Inductance

The value of L is taken equal to the internal inductance of a wire of length equal to the width W_s of the slot in the coupling area:

$$L = \frac{10^{-7}\, W_s}{2} \quad \text{[H]}$$

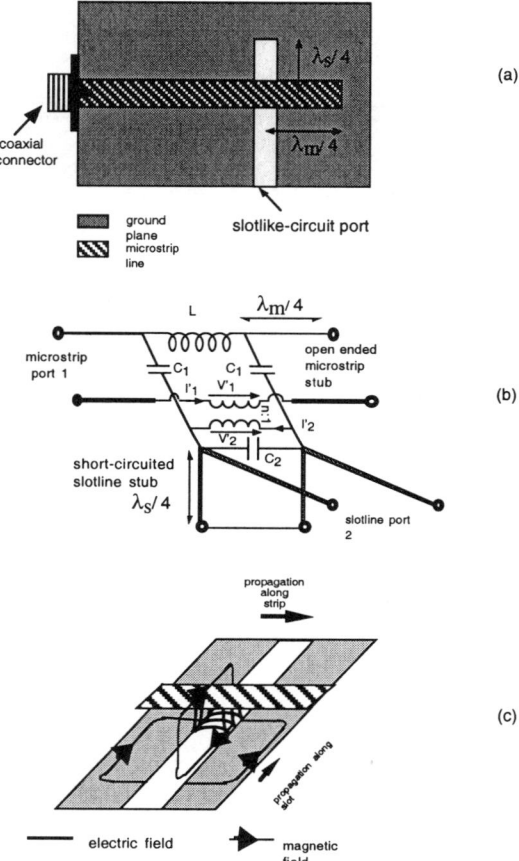

Fig. 5.9 *Microstrip-to-slot-line transition (a) topology; (b) equivalent circuit for modeling coupling area; (c) coupling area*

5.2.1.2 Capacitances C_1 and C_2

We deduce the equivalent capacitances C_1 and C_2 from the line parameters of the mode propagating along the piece of conductor-backed slot-line (Fig. 5.10). It is obvious that the electric field behavior in this structure induces a capacitive effect.

5.2.1.3 The ratio n of the transformer

Knorr [5.30] proposes the following formulas for the transformer ratio n:

$$V_1' = nV_2' \tag{5.15a}$$

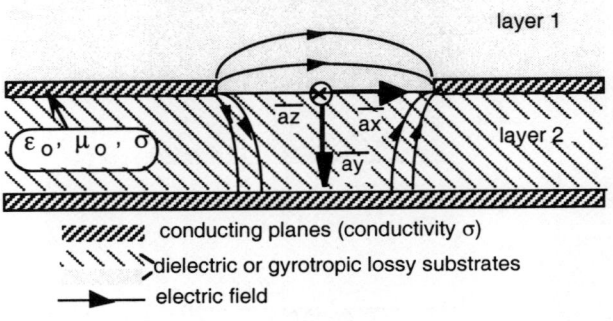

Fig. 5.10 *Conductor-backed slot-line*

$$I_2' = nI_1' \tag{5.15b}$$

where

$$n = \frac{1}{\cos(\alpha) - \frac{\sin\alpha}{\tan q}} \tag{5.15c}$$

$$\alpha = \frac{2Huf}{3} \tag{5.15d}$$

$$q = \alpha + \arctan\left(\frac{u}{v}\right) \tag{5.15e}$$

$$v = \sqrt{\varepsilon_{eff\, s} - 1} \tag{5.15f}$$

$$u = \sqrt{\varepsilon_r - \varepsilon_{eff\, s}} \tag{5.15g}$$

Since the slot-line is conductor-backed in the coupling area, we improve Knorr's formula by introducing into v and u the effective dielectric constant of the conductor-backed slot-line of Figure 5.10 computed using variational principle (4.21) instead of the value corresponding to a simple slot-line.

5.2.1.4   VSWR of junction

The input impedances of the quarter-wavelength stubs are

$$Z_s = Z_{cs} \tanh(\gamma_s L_s) \tag{5.16a}$$

for the slot stub, where L_s is the effective length of the stub, taking end effects into account, and γ_s the propagation constant of slot, and

$$Z_m = Z_{cm} \coth(\gamma_m L_m) \tag{5.16b}$$

for the microstrip stub, where L_m is the effective length of the stub, taking end effects into account, and γ_m is the propagation constant of microstrip. Z_s and

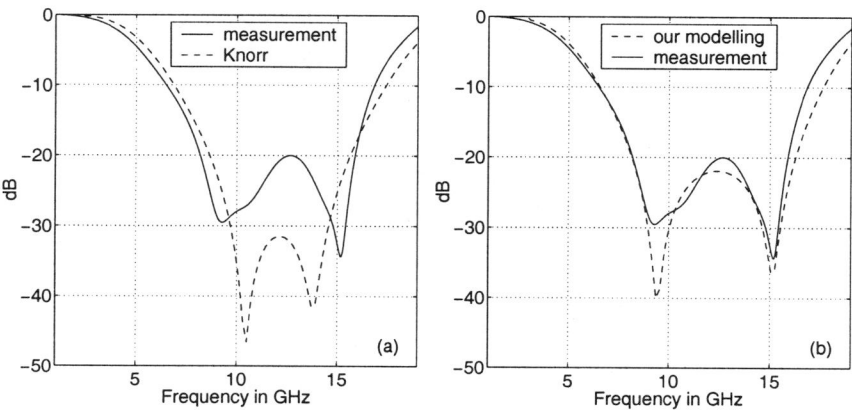

Fig. 5.11 *Characterization of the microstrip-to-slot-line transition (a) comparison between Knorr's model and measured reflection coefficient; (b) comparison between conductor-backed slot-line model and measured reflection coefficient*

Z_m are calculated using (5.11) and (5.12) for the characteristic impedances and variational principle (4.21) for the complex propagation constants. Using the intermediate variables

$$Y = j\omega C_2 + \frac{1}{Z_s} + \frac{1}{Z_{cs}} \tag{5.17a}$$

$$A_1 = 1 - \frac{2}{\omega^2 L C_1} \tag{5.17b}$$

$$A_2 = \frac{1 - nA_1}{\frac{YA_1}{n} + \frac{1}{nj\omega L}} \tag{5.17c}$$

$$A_3 = Z_m + (1 - n + \frac{2Y}{j\omega C_1})A_2 + \frac{2n}{j\omega C_1} \tag{5.17d}$$

the VSWR of the equivalent circuit of Figure 5.9b is defined as

$$VSWR = \frac{1 + \Gamma_1}{1 - \Gamma_1} \tag{5.18a}$$

where

$$\Gamma_1 = \frac{A_3 - Z_{cm}}{A_3 + Z_{cm}} \tag{5.18b}$$

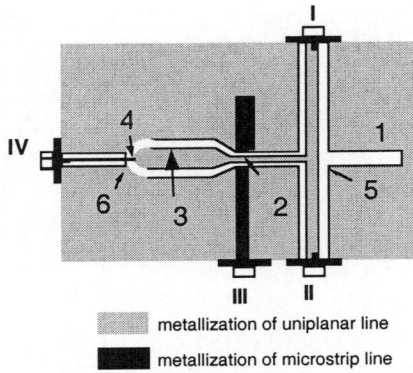

Fig. 5.12 *Topology of planar hybrid T-junction*

5.2.1.5   Results

Figure 5.11a compares our measurement of the reflection coefficient of a microstrip-to-slot-line junction and Knorr's model evaluated with only a transformer ratio. It is observed that the bandwidth of the measured and simulated reflection coefficients do not match well. A significant improvement is obtained when introducing the modified value for the transformer ratio, capacitances C_1 and C_2 and inductance L, as observed at Figure 5.11b. In particular, the location of the zeros and the magnitude of the reflection coefficient in the bandpass agree well. The observed improvement is due to accurate modeling of the effective dielectric constant formula (5.15a-g).

5.2.2   Planar hybrid T-junction

Figure 5.12 shows a planar hybrid T-junction. It has the following phase-shift properties:

1. When port IV is fed, the waves at ports I and II are in phase.

2. When port III is fed, the waves at ports I and II are out-of-phase.

The structure of Figure 5.12 has been designed at 25 GHz by using the frequency-dependent models developed for:

a. propagation coefficient and impedance of single (1) and coupled (3) slot-lines, including dielectric and ohmic losses

b. uniplanar junctions (4)(5)

c. microstrip-to-coupled slot-line junction (2).

All junctions were optimized to be as wide-band as possible. Figure 5.13 shows the phases measured at the different ports. A good behavior is observed, since paths I-IV and II-IV are in phase (Fig. 5.13a,b), while paths I-III and II-III are out-of-phase (Fig. 5.13c). This illustrates that variational models are accurate, especially for the line parameters and the slot-line-to-microstrip transition.

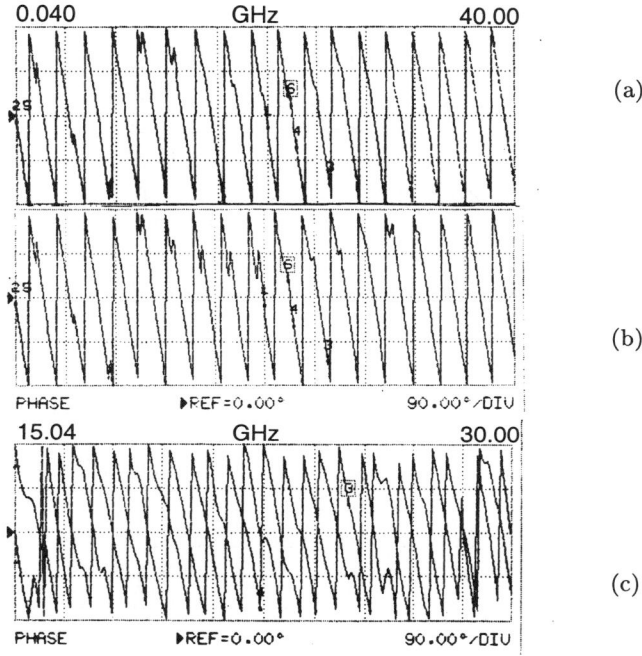

Fig. 5.13 *Phase measurements on planar hybrid T-junction (a) phase of path I-IV; (b) phase of path II-IV; (c) phases of paths I-III and II-III*

5.3 Gyrotropic devices

5.3.1 State-of-the-art

Because of the increasing development of wideband communication systems, there is a growing need for tunable components compatible with planar integrated configurations. Such elements are used as phase-shifters or tunable resonators and filters in oscillators and frequency synthesizers, for on-board and automotive systems operating at microwave and millimetre-wave frequencies.

For this reason, research involving planar lines on anisotropic or gyrotropic substrates for the 20 past years has concentrated on two areas: the first is modeling and designing planar lines on magnetic substrates that may serve as ferrite phase shifters or planar circulators. The second is designing magnetostatic wave (MSW, Appendix D) devices, including YIG-films operating as frequency-tunable devices in frequency synthesizers, channel-filters, delay lines, and tuned oscillators. In the first category, the DC-biasing magnetic field is usually kept constant, while tuned devices use the variation of DC-field to modify the operating frequency.

5.3.1.1   Planar lines on magnetic and anisotropic substrates

Various configurations have been investigated for planar phase shifters. In all cases, the propagation constant has to be calculated. Wen [5.31] proposed using a coplanar waveguide supporting a piece of YIG-film and experimentally tested the phase-shift properties of the structure. No attempt to calculate this phase-shift was made. The same year, Robinson and Allen [5.28] experimentally tested a slot-line phase shifter in the same configuration as Wen. A number of quasi-TEM modeling methods have been developed, based on variational formulas for inductance or capacitance per unit length. Massé and Pucel [5.32][5.33] defined an equivalent filling factor for the effective permeability of the line and calculated the capacitance per unit length of a microstrip on a magnetic substrate to deduce its dispersion and phase behavior. Other configurations supporting quasi-TEM modes on magnetic substrates were investigated with a similar approach by Kitazawa [5.34] and Horno et al. [5.35].

On the other hand, lines supporting non-TEM modes are usually analyzed by the moment method, and the implicit Galerkin procedure is used to find the dispersion of the gyrotropic layer. This was done by Jackson for slot-lines and coplanar waveguides [5.36], and by Mesa et al. [5.37] for coplanar multistrip lines. Usually, the magnetostatic range is never taken into account when calculating propagation constants, and the losses of the layer are neglected [5.38].

5.3.1.2   YIG-tuned planar MSW devices

For many years, gyrotropic passive planar components have been used as wide-band tunable resonators or phase-shifters for space qualified YIG-tuned oscillators. They compare advantageously with YIG-sphere resonators and filters, because of their planar geometry and two-dimensional coupling mechanism, yielding a reduction of mass and size. The desired effect results from a judicious combination of planar transmission lines with planar gyrotropic YIG-ferrite films which operate in their magnetostatic wave frequency range [5.39]. Various configurations are summarized in Figure 5.14. They all consist of planar YIG-ferrite layers cut as resonators and coupled to planar transducers etched on a dielectric substrate. The magnet poles generate a DC-magnetic field. Taking advantage of the ferrite nature of the film, the resonant frequency of the structure is varied by changing the applied DC-magnetic field. The most significant part of the theoretical work performed on planar MSW devices deals with MSW delay-lines (Fig. 5.14f). It consists of a planar YIG-film epitaxially grown on a Gadolinium-Gallium-Garnet (GGG) crystal, coupled to microstrip transducers. The insertion loss and delay between ports 1 and 2 are varied with frequency by changing the applied DC-magnetic field. The model is based on the following assumptions:

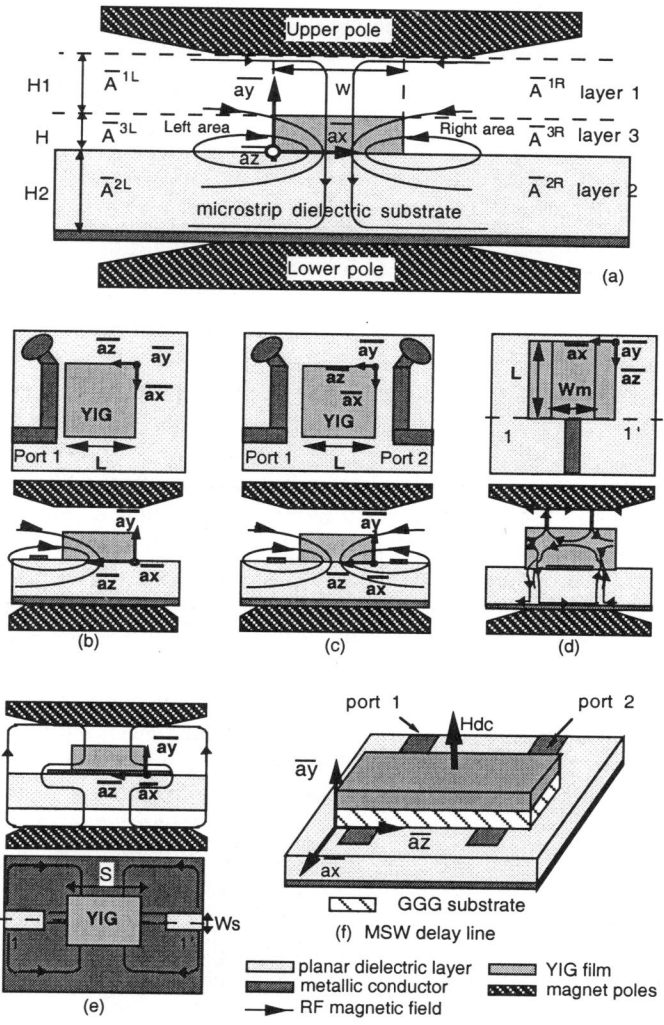

Fig. 5.14 *Configurations for YIG-tuned devices (a) geometry of YIG-film for MSFVW; (b) one-port undercoupled YIG-to-microstrip configuration; (c) two-port undercoupled YIG-to-microstrip configuration; (d) one-port overcoupled YIG-to-microstrip configuration; (e) MSFVW two-port undercoupled YIG-to-slot configuration; (f) MSFVW delay-line*

1. magnetostatic approximation
2. non-linear effects are neglected
3. uniformity of fields along the length of conduction strips
4. good conductors are assumed and thickness of strips is small
5. finite length (along z-axis) of YIG-film is not considered
6. non-uniformity of DC-magnetic field is not considered.

The theoretical model evaluates the part of power which is "radiated" from the microstrip into the YIG-film. The result is an equivalent radiation impedance loading the microstrip transducer. It is obtained by integrating over the continuous spectrum of the various propagating modes along the z-axis in the YIG-film. As a consequence, tedious integrations in the complex plane using the residue method are necessary [5.40]. This technique has been applied by a number of authors [5.41]-[5.44]. Then, a rough evaluation of the power transmitted between ports 1 and 2 is obtained.

To our knowledge, no attempt has been made to model the various YIG-resonator configurations for magnetostatic forward volume waves (MSFVW) depicted in Figure 5.14b-e, used for YIG-tuned oscillators. In particular, the configurations for which no physical contact exists between the planar transducer and the YIG-film have never been modeled. Also, the structures of Figure 5.14 cannot be easily analyzed by conventional variational principles [5.45] or transmission line analysis techniques, since these methods only are valid provided the present media are either isotropic or lossless.

The following section illustrates the efficiency of various variational formulations that we have developed for planar multilayered lossy gyrotropic structures used for MSW devices. The approach departs from the implicit full-wave methods for modeling multilayered magnetic planar lines, and from the radiation impedance concept, because we consider the resonator as a resonant transmission line whose parameters are calculated from explicit variational formulations. Hence, this approach advantageously compares with implicit methods since on-line results can be obtained with a regular PC in a few seconds. The formulation includes any spatial non-uniformity of the non-Hermitian tensor describing the gyrotropic layer. It can be used both in the spatial and the spectral domains. This will be illustrated when calculating an undercoupled MSW-resonator modeled with the spatial domain formulation, and an overcoupled MSW-resonator modeled with the spectral domain formulation. In both cases, trial potentials and fields are derived under the magnetostatic assumption (Appendix D, Chapter 4).

5.3.2 Undercoupled topologies

The configurations depicted in Figure 5.14 can be divided into two classes: under- and overcoupled topologies, depending on the proximity between the YIG sample and the coupling planar transducer.

Configurations b,c,d involve microstrip transducers. A schematic representation of the magnetic field patterns inside the YIG-film is shown for each configuration in Figure 5.14b-d. These should be compared with the field representation in an isolated YIG-film of finite width (Fig. 5.14a). As we see, configurations b and c, for which the microstrip has no contact with the YIG-layer (side-coupling), exhibit a magnetic field pattern which slightly differs from the pattern of an isolated film. As a consequence, the field configuration in the YIG-film can be approximated by that of the isolated YIG-film of Chapter 4. We model the film as an equivalent transmission line (Fig. 5.15a,b) along the z-axis (Fig. 5.14a), with forward and reverse propagation constants computed from MSW assumption as follows. In Chapter 4, the Perfect Magnetic Wall (PMW) assumption was used to derive the dispersion relationship for MSW inside the planar YIG-film. We have shown in [5.46] that a more suitable description, taking into account the non-uniform demagnetizing effect (Appendix D), is obtained when trial magnetostatic potential (4.33) inside the YIG-film is replaced by the following:

$$\Psi_3^{YIG}(x,y,z) = \left[A \sin k_x x + B \cos k_x x\right] Y_3(y) e^{-\gamma z} \quad \text{for } 0 \leq x \leq W \quad (5.19a)$$

$$\Psi_3^{L}(x,y,z) = L e^{+\gamma_n x} Y_3(y) e^{-\gamma z} \qquad\qquad \text{for } x \leq 0 \qquad (5.19b)$$

$$\Psi_3^{R}(x,y,z) = R e^{+\gamma_n(x-W)} Y_3(y) e^{-\gamma z} \qquad \text{for } x \geq W \qquad (5.19c)$$

where W is the width of the YIG-film and Y_3 the y-dependence (Chapter 4). The x-dependence of the potential is left unknown in (5.19a-c) in order to allow use of a more rigorous boundary condition than the PMW-condition. On the left- and right-sides of the YIG-film in layer 3 the potential is assumed to exponentially decrease from the edges (5.19b,c). Imposing boundary conditions at planes $x = 0$ and $x = W$ yields a relationship between k_x and γ, involving values at the edge of the film because of the non-uniform demagnetizing effect, and an expression for the ratio B/A:

$$\tan k_x W = \frac{-2\gamma_n \mu_0 \mu_{xx,3} k_x}{(\mu_0 \gamma_n)^2 - \left(\mu_{zx,3}\gamma\right)^2 - \left(\mu_{xx,3} k_x\right)^2} \tag{5.20}$$

$$B/A = \frac{\mu_{xx,3} k_x}{\gamma \mu_{zx,3} + \gamma_n \mu_0} \tag{5.21}$$

Assuming that k_x is the unknown in (5.20), the k_x-solution is independent of the sign of γ, that is from the forward or reverse nature of propagation. However, this is not the case for coefficient B whose value differs with the sign of γ, as shown by (5.21). Hence, we have theoretically demonstrated that when taking into account an edge effect related to a non-uniform demagnetizing effect for a finite-width YIG-film, it is impossible, with the same pattern of field, to obtain identical forward and reverse propagation coefficients for a unique physical value of k_x. This is only possible for isotropic layers, having no zx tensor components.

Having obtained these dependencies, the various integrands of coefficients of equation (4.2) are expressed using (4.34a-c) and (4.36a-c) for the trial

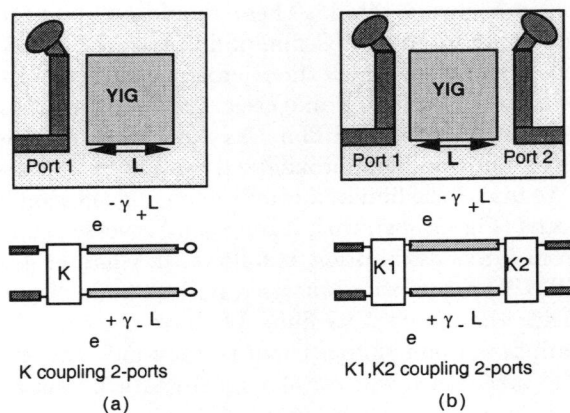

Fig. 5.15 *Equivalent circuits modeling various YIG-devices topologies (a) one-port undercoupled YIG-to-microstrip configuration; (b) two-port undercoupled YIG-to-microstrip configuration*

electric field, derived from the magnetostatic potential. The x-dependence of the integrands is the product of the x-dependence of the inverted non-uniform permeability tensor with simple sinusoidal functions. This integral is evaluated numerically by a simple trapezoidal rule algorithm. Coefficients A, B, C, and E in each layer are summed and equation (4.2) is solved for γ.

The final result consists of two complex propagation constants noted γ_f and γ_r, associated respectively with a forward and reverse propagation direction. They are obtained when solving equation (4.2) with the set of values (k_x, B) from (5.20) (5.21) which best matches the outside source field on the microstrip transducer.

Figure 5.16 (configuration of Fig. 5.14a) shows the results obtained for γ_f and γ_r. A significant difference is observed between the forward (solid) and reverse (dashed) propagation coefficients at the same frequency. Hence non-reciprocal effects are possible. This is in contradiction with the assertions found in literature [5.47][5.48], where the propagation of magnetostatic forward volume waves (MSFVW) is stated to be reciprocal. This is because the analysis methods found in literature are based on the PMW-assumption, and do not take into account the edge effect due to the demagnetizing effect. In our formulation, we involve the constitutive parameters $\mu_{xx,3}$ and $\mu_{zx,3}$ in calculating the k_x-constant.

Undercoupled resonators are hence described by an equivalent transmission line modeling the YIG-film, having different forward and reverse propagation constants, and coupled to microstrip-transducers by coupling networks K (Fig. 5.15). The z-dependence of the magnetostatic magnetic fields inside

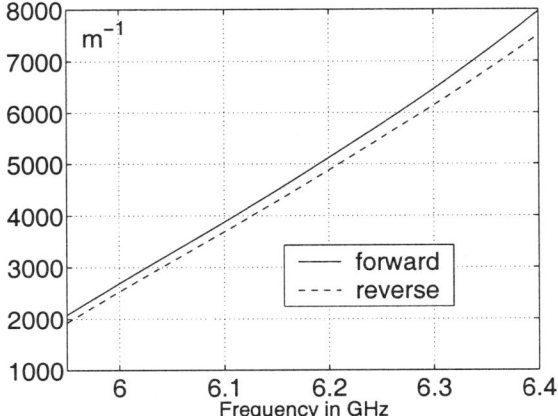

Fig. 5.16 *Forward (solid line) and reverse (dashed line) propagation constants in YIG-film calculated with variational principle using non-uniform internal DC-field and rigorous boundary conditions at edge of film*

the YIG-film thus has the general form:

$$Z_{x,y}(z) = V_f e^{-\gamma_f z} - V_r e^{+\gamma_r z} \qquad \text{for } x\text{- and } y\text{-components} \qquad (5.22a)$$

$$Z_z(z) = \gamma_f V_f e^{-\gamma_f z} + \gamma_r V_r e^{+\gamma_r z} \qquad \text{for } z\text{-component} \qquad (5.22b)$$

5.3.3 One-port undercoupled topology

For a one-port configuration, the calculation of the input impedance is sufficient. We consider port 1 (Fig. 5.14b). From Poynting's theorem, the one-port equivalent impedance Z_{r1} of the resonator is related to the energy contained in volume V [5.49]:

$$\frac{1}{2}|V_1|^2 \left(\frac{1}{Z_{r1}}\right)^* = 2j\omega \int_V \left(\overline{B} \cdot \overline{H}^* - \overline{E} \cdot \varepsilon^* \overline{E}^*\right) dV \qquad (5.23)$$

The fields in (5.23) are derived from the magnetostatic assumptions (Chapter 4), but with a z-dependence (5.22a,b) and transverse dependencies (5.19a-c) for the magnetostatic potential in the YIG-film. The boundary conditions at $z = 0$ and $z = -L$ are approximated by a PMW, together with an exponentially-decaying B_{z2} field in the left and right areas outside the YIG-sample, yielding

$$\frac{V_r}{V_f} = e^{(\gamma_f + \gamma_r)L} \qquad (5.24)$$

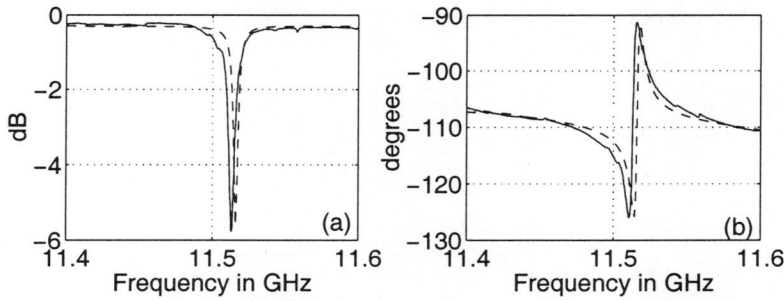

Fig. 5.17 *Comparison between modeled (dashed) and measured (solid) reflection coefficient of one-port undercoupled microstrip-to-YIG configuration (a) magnitude; (b) phase*

and the voltage induced on the microstrip line

$$V_1 = -j\omega \int_0^W \int_{-H_2}^0 B_{z2}^L(x, y, S + L + W_{mic}/2)\, dy\, dx \qquad (5.25)$$

where W_{mic} is the width of microstrip transducer and S the spacer between transducer and YIG-film. Equation (5.24) combined with field descriptions (5.19a-c) to (5.22a,b) yield electric trial fields under magnetostatic assumption, with a z-dependence derived from (5.22a,b). These trials are introduced in expression (5.23) for the input impedance. The modeled reflection coefficient (dashed) is compared with the measured one (solid) in Figure 5.17. Model and experiment agree very well. This demonstrates the efficiency of the formulation for lossy gyrotropic media. As expected, the configuration is undercoupled, as can be seen from the phase of reflection coefficient (Fig. 5.17b): the phase variation does not exceed 180 degrees in the resonant range.

5.3.4 Two-port undercoupled topology

The scattering matrix of the two-port configuration is obtained from its impedance matrix. We define two impedance matrices $\overline{\overline{Z}}_r$ and $\overline{\overline{Z}}_m$ associated respectively with the YIG-resonator and the coupled microstrip transducers (without YIG-sample). Terms Z_{11r} and Z_{22r} have already been calculated:

$$Z_{11r} = Z_{r1} \qquad (5.26a)$$

$$Z_{22r} = Z_{r2} \qquad (5.26b)$$

The two mutual terms are deduced from the self-impedances as follows. Looking at port 1 (Fig. 5.14c), the total ratio V_2/V_1 derived from (5.22b) and (5.24) must preserve the phase delay from one port to the other, so that the mean delay factor $\mathrm{Im}\, 0.5(\gamma_f + \gamma_r)L$ is added:

$$\frac{V_2}{V_1} = e^{0.5\,\mathrm{Im}(\gamma_f+\gamma_r)L}\frac{\gamma_f e^{\gamma_f L} + \gamma_r e^{-\gamma_r L}}{\gamma_f e^{-\gamma_f L} + \gamma_r e^{\gamma_r L}} \qquad (5.27)$$

From ratio V_2/V_1 we easily deduce the coupled terms of the impedance matrix of the resonator:

$$Z_{21r} = \frac{V_2}{V_1}Z_{r1} \qquad (5.28a)$$

$$Z_{12r} = \frac{Z_{r1}}{V_2/V_1} \qquad (5.28b)$$

The impedance matrix of the YIG-microstrip transducer configuration is the sum of the two impedance matrices

$$\overline{\overline{Z}} = \overline{\overline{Z}}_r + \overline{\overline{Z}}_m \qquad (5.29)$$

where $\overline{\overline{Z}}_m$ is the impedance of the coupled microstrip transducers without the YIG-film. Defining a reference impedance $\overline{\overline{Z}}_R$, the scattering matrix $\overline{\overline{S}}$ is derived from the impedance matrix $\overline{\overline{Z}}$.

We validate the model by designing a two-port microstrip-to-YIG MS-FVW Stray-Edge Resonator (SER) (Fig. 5.14c). The phases of the reflection terms, not shown here, exhibit an undercoupled behavior similar to Figure 5.17b. Hence, the modal propagation constants are calculated without taking into account the vicinity effect of the transducers, because of the weak coupling. Due to demagnetizing effects, the internal field is not uniform along the width of the sample, and the permeability tensor is not uniform over the cross-section. As a consequence, the two forward and reverse solutions computed for the propagation constant are applicable. The measured and modeled magnitudes of both the reverse and forward transmission terms show a good agreement (Fig. 5.18a,b). However, Figure 5.18c,d highlight a significant non-reciprocal effect: the modeled phase (dashed line) of the reverse and forward transmission terms are markedly different. This disparity has also been confirmed experimentally (solid lines) [5.46],[5.50][5.51]. It should be noted that the designs of tunable oscillators using MSFVW SERs found in the literature use one-port configurations [5.52][5.53].

5.3.5 One-port overcoupled topology

A large strip located between the YIG-sample and the dielectric layer strongly modifies the pattern of the magnetic field, since the current flowing on the microstrip forces the magnetic field to surround the strip. Hence, the magnetostatic fields in the resonator have to be calculated in the presence of the conducting strip. We expect the resonator in configuration Figure 5.14d to be overcoupled.

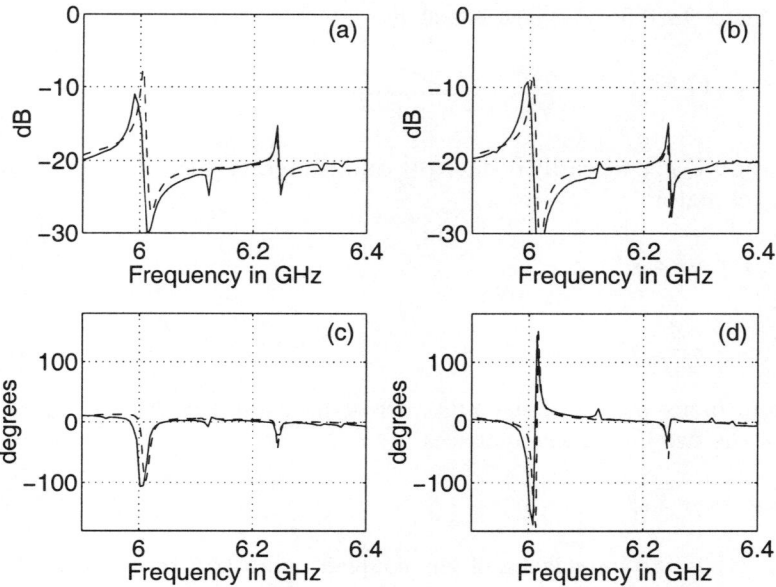

Fig. 5.18 *Comparison between modeled (dashed) and measured (solid) trans-mission scattering terms of two-port undercoupled microstrip-to-YIG config-uration at 6 GHz (a) magnitude of reverse transmission; (b) magnitude of forward transmission; (c) phase of reverse transmission; (d) phase of forward transmission*

Two models have been tested for the one-port overcoupled topology. The first is the transmission-line approach depicted in Figure 5.19, showing a top view of the topology (a), its equivalent circuit (b) and the cross-section of its equivalent transmisson line (c). One simply calculates the propagation constant and characteristic impedance of a boxed microstrip line in the pres-ence of a YIG-layer and with lateral perfect magnetic walls. Hence, the input impedance of the line and the resulting reflection coefficient are obtained, for a particular reactive load at the end of the microstrip transducer. The cal-culations use the variational principle (4.21) in the spectral domain, because of the presence of a strip conductor of finite extent in the close vicinity of the YIG-layer. As a consequence, trial spectral fields are derived from the spectral magnetostatic potential solution of a Fourier-transformed magneto-static equation [5.54]. The spectral field components are related to surface current densities on the strip, as previously done in Chapter 4 for the boxed microstrip line. This enables the computation the trial spectral field in the whole transverse section.

The second approach uses the fact that the following expression for the

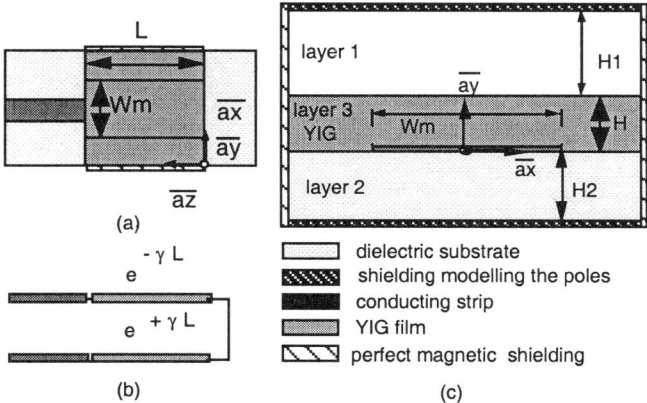

Fig. 5.19 *Equivalent transmission line model of one-port overcoupled YIG-to-microstrip configuration (a) top view; (b) equivalent circuit; (c) cross-section in $z = L/2$ of equivalent line*

impedance, similar to (5.23), is based on squared powers of fields, hence on energy, so that it can be expected that its right-hand side exhibits a stationary behavior:

$$\frac{1}{2}Z_{r1}|I_1|^2 = 2j\omega \int_V \left(\overline{B} \cdot \overline{H}^* - \overline{E} \cdot \varepsilon^* \overline{E}^* \right) dV \tag{5.30}$$

Davies [5.55] mentions the use of the stationarity of the right-hand of equation (5.30) to deduce some variational expressions for the resonant frequency of cavities. Collin [5.56] makes use of the same stationarity assumption to deduce stationary formulas for the equivalent reactive load of dielectric obstacles in waveguides. We have proven [5.54] the stationarity of the right-hand side of equation (5.30) under magnetostatic wave assumption when the volume of integration contains gyrotropic lossy media. We applied it to the top-overcoupled configuration of Figure 5.14d, modeled as an equivalent cavity instead of a transmission line. Basically, the trial transverse magnetostatic field configurations are identical for the two approaches, but their integration is made in a transverse section only for transmission line formulation (4.21) while it is computed in the whole cavity volume for the energetic formulations (5.30). Hence, the total trial field for the second approach involves a trial longitudinal z-dependence similar to (5.24) but with $\gamma_f = \gamma_r$ and PMW-boundary conditions at $z = 0$ and $z = -L$. Figure 5.20a,b compares the two variational approaches, *i.e.* the equivalent transmission line modeled by (4.21) (solid), and the cavity model using energetic formula (5.30) (dashed). Both formulations predict the overcoupled behavior which is experimentally observed [5.54]. The calculated magnitudes of reflection coefficient differ by less than 0.5 dB.

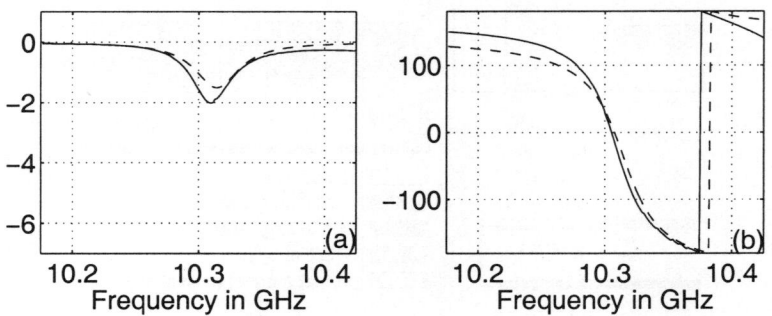

Fig. 5.20 *Comparison between reflection coefficient of one-port top overcoupled modeled either by energetic approach (- -) or transmission line approach (—) (a) modeled magnitude; (b) modeled phase*

5.4 Optoelectronic devices

Multilayered p-i-n photonic devices are developed for ultra wideband optoelectronic applications, with several advantages. First, ultra-large intrinsic bandwidths can be obtained by reducing the thickness of the intrinsic layer and hence the carrier transit times inside the device. Secondly, suitable values for the thickness and dielectric constant of some layers can be selected in order to confine the optical beam in the vicinity of the absorbing intrinsic layer, and to guide it in the case of travelling wave operation. Finally, high-doping levels are imposed on some layers to obtain the desired opto-electrical (O/E) conversion, implying that the propagation mechanism is usually of a slow wave type. This means that even for very short guiding structures the propagation effects are important at microwave frequencies.

Several papers have been published about modeling slow wave propagation in multilayered semiconductor transmission lines. To our best knowledge they deal with metal-insulator-semiconductor structures studied by Guckel *et al.* [5.57], or Schottky structures on Si or GaAs semiconducting layers, as presented by Jäger [5.58]-[5.60]. The methods found in literature for analyzing such structures (quasi-TEM or spectral domain analysis, mode-matching method) assume that the semiconductor is divided into several homogeneous layers of *infinite extent* [5.61]-[5.63].

This section presents a variational approach for modeling the transmission line behavior in Travelling Wave Photodetectors (TWPDs). Excellent reviews of the TWPD concept are given in [5.61] and [5.64]. Waveguide photodetectors are devices where the electrodes collecting the photogenerated current act as a transmission line. Hence, the photogenerated carriers act as a distributed radio-frequency (RF) source of current inducing voltages and currents travelling on the electrodes towards their ends. TWPDs are waveguide photodetectors in which the optical signal is guided along the length of the device as well as the RF photogenerated signal, so that the RF modulating envelope of the optical carrier travels towards the matched output of the device at the group velocity, undergoing no reflections. Also, the trade-off between maximal internal efficiency and transit-time bandwidth limitation is avoided, because the directions of optical beam propagation and carrier drift are orthogonal. Under such conditions, and when the group velocity of the optical signal matches the velocity of the microwave signal, the bandwidth of TWPDs is limited only by the finite carrier transit times, and by microwave losses and the resulting unwanted mismatch of its electrical ports [5.65]. Hence, it is of prime interest to have accurate transmission line models in order to optimize the TWPD bandwidth.

5.4.1 Topologies

Most of TWPDs presented in literature use mesa-type p-i-n structures (strip-like transmission line as in Fig. 5.21a), in which the electrodes collecting the microwave photocurrents lay on a multilayered substrate with each layer having a uniform carrier concentration. Hence, both the optical and RF propagation can be modeled by simple equivalent transmission lines having a TEM behavior. In [5.66] a coplanar waveguide TWPD topology is proposed for Silicon-on-Insulator (SOI) technology, where CPW electrodes lay on P and N diffusion areas (Fig. 5.21b). Both mesa and CPW topologies presented in Figure 5.21 exhibit two particular features: some of the layers have a significant conductivity due to the high doping, responsible for the slow-wave phenomena, and some of the layers are not homogeneous or of finite extent along the x-axis. The variational approach proposed in this section takes these two features into account simultaneously. The efficiency of the variational principle (4.21) for slow wave phenomena has already been illustrated in Section 5.1, where CPW and microstrips on low resistivity substrates have been successfully modeled in the slow wave range.

5.4.2 Modeling mesa p-i-n structures

The geometry of the transverse section of the mesa p-i-n photodetector is shown in Figure 5.21a. Microwave and optical propagation occurs along the z-axis. The p-i-n junction is reverse biased with an adequate saturating voltage between top strip conductor and bottom grounding plane (Au), in such a manner that the intrinsic layer (third layer) is fully depleted, with an equivalent conductivity equal to zero. The RF-modulated part of the optical power intensity induces an RF-variation of the photogenerated current in the intrinsic area, which is collected at the top and bottom electrodes and propagates on the equivalent RF-transmission line. The optical beam is confined between the layers having the highest dielectric constants. In order to obtain a photodetector device, the doping of the third layer (n-InGaAs) is designed to

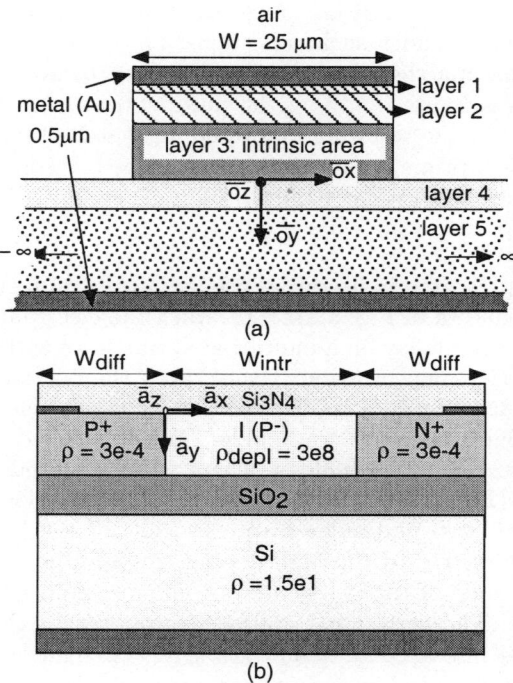

Fig. 5.21 *Travelling wave photodetectors topologies (a) mesa p-i-n GaAs structure (layer 1: p+ InGaAs, $\varepsilon_r = 15$, $H = 0.1$ μm, $\sigma = 13.4$ S/mm; layer 2: p+ InP, $\varepsilon_r = 14.5$, $H = 1$ μm, $\sigma = 6.4$ S/mm; layer 3: n- InGaAs, $\varepsilon_r = 14.5$, $H = 2$ μm, $\sigma = 0.0$ S/mm; layer 4: n+ InP, $\varepsilon_r = 15$, $H = 2$ μm, $\sigma = 128.0$ S/mm; layer 5: InP, $\varepsilon_r = 14.5$, $H = 500$ μm, $\sigma = 0.0$ S/mm); (b) coplanar p-i-n SOI structure*

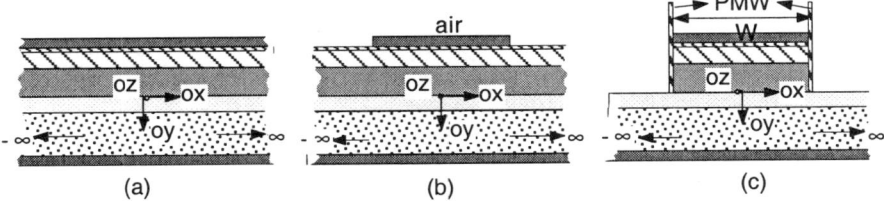

Fig. 5.22 *Approximate topologies for mesa p-i-n (a) parallel plate; (b) microstrip; (c) hybrid model*

absorb the light. The highest doping level occurs in the fourth layer which is neither a good conductor nor a good dielectric.

5.4.2.1 Simulation using a parallel plate model

The geometry of Figure 5.21a is first simplified into a parallel plate waveguide (Fig. 5.22a), by Jäger [5.67] and Guckel *et al.* [5.57]. For this TEM structure, however, a specific numerical algorithm is required because the propagation constant is the complex root of a determinantal equation associated with the corresponding eigenvalue problem. Tedious numerical iterations are necessary because of the multilayered character and the high conductivity of some layers. Curves marked □ in Figure 5.23a,b represent the results obtained using this model. A slow wave phenomenon is clearly predicted, which illustrates the importance of a good model for designing broadband TWPDs.

5.4.2.2 Measured transmission line parameters

We have measured the complex propagation constant of the structure shown in Figure 5.21a, applying a two-line calibration technique [5.68] for planar lines, described in the next section. It yields the complex propagation constant without using the inverse Fourier transforms proposed by Giboney *et al.* [5.65], because the measurement is carried out in the frequency domain. Measured results are reported in Figure 5.23a,b (solid lines). The results obtained using the parallel plate model (□) of Figure 5.22a do not agree with the measurements. The parallel plate model indeed does not take into account the air area above the strip and the finite width of the strip.

5.4.2.3 Simulation using a microstrip model

Hence, we introduce the effect of the strip by using the variational approach (4.21) applied to the microstrip configuration shown in Figure 5.22b: the various layers are infinite in the x-direction, and a strip conductor of finite width lies between the air and first layer. Each layer i is characterized by its complex permittivity. The imaginary part of the complex permittivity is derived from the layer conductivity σ_i using formula (5.14a). The application of the variational principle (4.21) is similar to that performed for the microstrip line

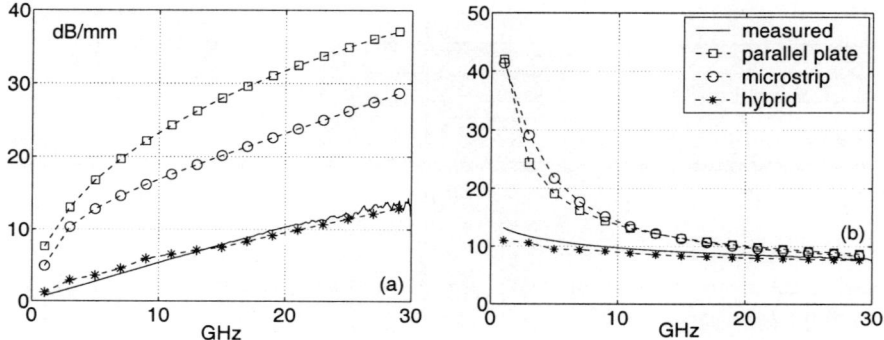

Fig. 5.23 *Comparison between measurement and simulations (a) losses; (b) effective refractive index*

on low-resistivity silicon, except for the number of layers and strip geometry. Simulated results using the microstrip structure of Figure 5.22b are reported on Figure 5.23 (curves marked o). The simulated slow wave phenomenon is still more important than the measured effect: the microstrip quasi-TEM structure does not take into account the finite width of some layers.

5.4.2.4 Simulation using extended formulation

Keeping in mind the high dielectric constant in the layers (values between 14 and 15 in Fig. 5.21a), we assume that Perfect Magnetic Walls (PMW) limit the layers of finite extent at planes $x = W/2$, to find trials in those layers (Fig. 5.22c) [5.69]. According to the spectral domain technique, applying this PMW-boundary condition means that the spatial field may be described by a summation of periodic functions of period $n\pi/W$. As a consequence, only the values of the spectral fields for discrete values k_{xn} of the k_x-variable are added up:

$$\tilde{\bar{e}}_i(k_x, y) = \sum_{n=-\infty}^{\infty} \tilde{\bar{e}}_i(k_{xn}, y)\delta(k_x - k_{xn}) \tag{5.31a}$$

and

$$\bar{e}_i(x, y) = \sum_{n=-\infty}^{\infty} \tilde{\bar{e}}_i(k_{xn}, y)e^{-jk_{xn}x} \tag{5.31b}$$

where

$$k_{xn} = \frac{n\pi}{W} \tag{5.31c}$$

where W is the width of layers of finite extent. The choice of n is related to the symmetry of trial fields describing the particular mode under investigation, and to the nature of the shielding. For the dominant mode of a shielded stripline with PMW as considered here, n is taken as even. Since all the other boundary conditions imposed on the fields in a layer of finite extent are identical to those applied in the same layer of infinite extent along the x-axis, the k_x- and y-dependencies of function $\tilde{\bar{e}}_i(k_x, y)$ are taken to be identical for a layer of finite or infinite extent. Adequate trial functions $\tilde{\bar{e}}_i(k_x, y)$ for each layer are thus calculated as in the previous section, following the method detailed in Chapter 4. Simply, the whole k_x-spectrum is considered for infinite layers, while only a set of discrete values (5.31c) is taken in layers of finite extent along the x-axis.

As a result, applying Parseval's theorem together with the spectral form (5.31a) yields expressions for the spectral coefficients of equation (4.21), which are summations of discrete values of their integrands, evaluated at discrete values of the infinite set (5.31c). For example, coefficient $\tilde{A}_i$ for layer i of finite extent becomes

$$\tilde{A}_i(\tilde{\bar{e}}) = \frac{1}{2\pi} \sum_{n=-\infty}^{\infty} \int_{y_i} \left[\bar{a}_z \times \tilde{\bar{e}}_i(k_{xn}, y)^* \right] \cdot \left[\bar{a}_z \times \tilde{\bar{e}}_i(k_{xn}, y) \right] k_{x1} \, dy \qquad (5.32)$$

A similar expression is easily deduced for coefficients B to E.

To summarize, we determine first trial spectral fields in each layer for the structure having all layers of infinite extent along the x-axis (identical to those found in the previous section). Next, we impose a PMW-boundary condition at the left- and right-hand sides of the layers of finite extent. We express this condition by extending the formulation of the spectral coefficients for each layer i in equation (4.21): for infinite layers of the mesa structure in Figure 5.22c, expressions (4.19b) to (4.19e) are used, while for layers of finite extent, only discrete values (5.31a) of the spectral trial field are considered, leading to the series expression (5.32).

The Spectral Index method was introduced in 1989 for finding the guided modes of semiconductor rib waveguides [5.70]. The transcendental equations developed by the authors are variational in nature. The method posseses the useful variational property that each value of "beta" is a lower bound. It improves the calculation by compensating the penetration into the air by slightly displacing the boundaries, taking advantage of the fact that guided waves in semiconductors have a low penetration depth into air. The concept of effective depth is used to accomodate the large jump in dielectric constant at the air/semiconductor interface.

Figure 5.23 shows that this extended variational formulation significantly improves the agreement between theory (curves *) and experiment (solid). We demonstrate with this analysis the possibility of using variational formulation (4.21) for structures combining layers of various conductivity and of finite and infinite extent, while considering simple trial fields. It has to be mentioned that simulation (*) is obtained on-line on a regular PC.

With our approach, we totally neglect the influence of the air regions to the left and right of the layers of finite extent. To validate this assumption, the method of lines [5.71][5.72] was used for computing the shape of the transverse electric field in the various layers of Figure 5.21a. Results presented in [5.71] show that in the three top layers, and in particular in the intrinsic area, the field is confined in the layers and vanishes rapidly in air, so that neglecting the fields in the air areas $x < W/2$ and $x > W/2$ outside of the layers of finite extent is relevant. On the other hand, in the fourth layer having a high conductivity (n+InP), the longitudinal current density is spread outside the area under the strip, so that fields and currents cannot be considered confined in area $|x| \leq W/2$. Those effects are observed at relatively low (15 GHz) and high frequencies (50 GHz).

5.4.3   Modeling coplanar p-i-n structures

The layout of the SOI p-i-n photodetector is shown in Figure 5.21b. Values of resistivity are in Ωm. Figure 5.24a shows the basic p-i-n structure, which consists of highly doped P$^+$ and N$^+$ zones, separated by an intrinsic zone. Ohmic contacts are considered to exist between the P$^+$ and N$^+$ zones and the two metal electrodes. The structure is reverse biased with an adequate saturating voltage, so that the intrinsic zone is depleted. The transmission line parameters along the z-axis are obtained by generalizing the previous approach, which has already successfully been applied to mesa InP/GaAs p-i-n photodetectors. The structures of Figure 5.21b are divided into n layers (perpendicular to y-axis). In each layer i, $j(i)$ sub-layers are defined, with their number depending on the variation of carrier concentrations along the x-axis. For instance, the p-i-n layers of Fig. 5.24a and of Fig. 5.21b contain $j(2) = 3$ sub-layers. Each sub-layer is characterized by its complex permittivity. The imaginary part of the complex permittivity of the sub-layer is derived from its resistivity, while the oxide layers have of course a purely real permittivity:

$$\varepsilon_{i,j(i)} = \varepsilon_{ri}\left[1 - \frac{j}{2\omega\varepsilon_{ri}\rho_{i,j(i)}}\right] \tag{5.33a}$$

$$\varepsilon_{SiO_2} = \varepsilon_{rSiO_2} \tag{5.33b}$$

$$\varepsilon_{Si_4N_3} = \varepsilon_{rSi_4N_3} \tag{5.33c}$$

where $j(i)$ is the number of sub-layers in layer i. The present case consists of a new application of the method, since for the first time planar layers having a strong variation of resistivity along the x-axis are treated. In the p-i-n layer, we have

$$\rho_{pin,1} = \rho_{pin,3} = 3\ 10^{-4}\ \Omega\text{m}$$

$$\rho_{pin,2} = 3\ 10^{-8}\ \Omega\text{m}$$

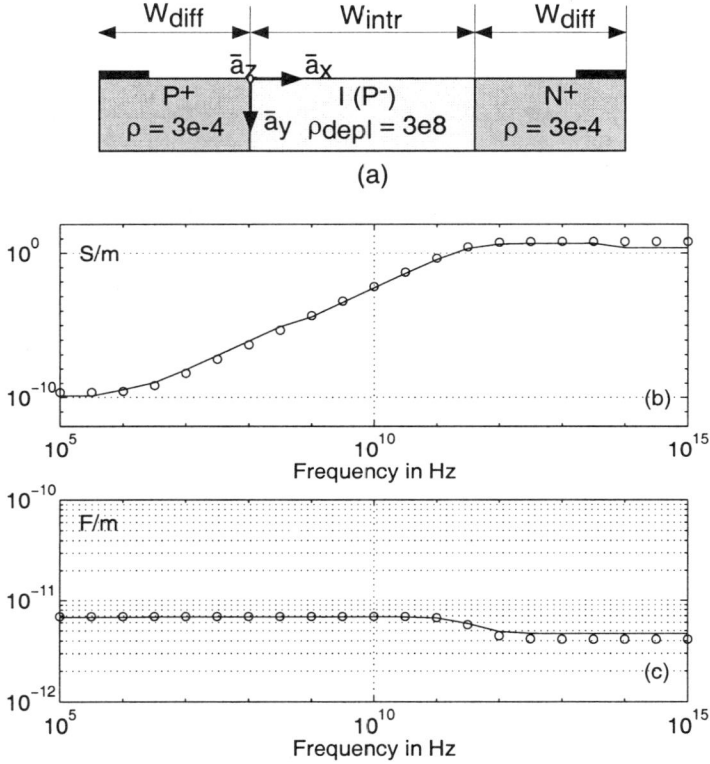

Fig. 5.24 *CPW SOI p-i-n structure (a) simplified p-i-n junction; (b) modeling p-i-n conductance; (c) modeling p-i-n capacitance; solid curves are for variational principles and symbols o for MEDICI simulations*

Hence, the variational principle yielding the complex propagation constant γ of the structure along the z-axis is rewritten as

$$\gamma^2 \sum_{i,j(i)} \tilde{A}_{i,j(i)}(\tilde{e}) + \gamma \sum_{i,j(i)} \tilde{C}_{i,j(i)}(\tilde{e}) - \sum_{i,j(i)} \tilde{B}_{i,j(i)}(\tilde{e})$$
$$+ \omega^2 \varepsilon_0 \mu_0 \sum_{i,j(i)} \varepsilon_{i,j(i)} \tilde{E}_{i,j(i)}(\tilde{e}) = 0 \tag{5.34}$$

where the trial field (5.31b) can be used in each sub-layer. For the characteristic impedance, a power-current definition (5.12) is used:

$$Z_{cPI} \triangleq \frac{P}{I^2} = \frac{-1}{2j\omega I^2} \left(\sum_{i,j(i)} \tilde{C}_{i,j(i)}(\tilde{e}) - \gamma \sum_{i,j(i)} \tilde{A}_{i,j(i)}(\tilde{e}) \right)^* \tag{5.35}$$

where I is the magnitude of the current flowing on one of the electrodes, and P is the power flowing through the cross-section of the p-i-n structure.

Results are compared with simulations using MEDICI software [5.66]. Figure 5.24b,c shows the frequency response of the conductance and capacitance per unit length of the p-i-n structure, computed using MEDICI software (curve o) and our variational approach (curve —). From the complex propagation constant and characteristic impedance, the equivalent conductance G and capacitance C per unit line length are derived as

$$G = \text{Re}\left(\frac{\gamma}{Z_c}\right) \tag{5.36a}$$

$$C = \text{Im}\left(\frac{\gamma}{\omega Z_c}\right) \tag{5.36b}$$

The agreement is excellent up to optical frequencies. The advantage of the variational approach is that results can be obtained for the whole frequency range (about 21 frequencies) within a few seconds, while several minutes are necessary when using MEDICI Version 4.0 with the same number of frequency points. Also, our model yields electromagnetic parameters such as propagation constant and characteristic impedance, while MEDICI solves equations involving electric charges, potentials and fields, with no access to magnetic parameters and propagation characteristics. Hence, the efficiency of the variational approach yields an accurate model of TWPD equivalent transmission line parameters for both RF and optical frequency ranges. Based on this formalism, an extensive comparative study of the bandwidth of SOI p-i-n TWPD structures has been presented, in terms of load matching, velocity matching, type of semiconductor material used, and a helpful scattering matrix formalism has been established [5.73].

5.5   Measurement methods using distributed circuits

In [5.74] an accurate method is presented for determining the complex permittivity of planar substrates, at frequencies up to 40 GHz. It is based on the comparison between calculated and measured values of the effective dielectric constant and losses on two complementary geometries, namely a microstrip and a slot-line. The complex permittivity is determined with an accuracy of 1 %, over the whole frequency range. A major advantage of the method is that, as it is based upon transmission lines and not on resonant structures, it offers a wideband characterization with a reduced number of test boards. The method requires an accurate measurement of the complex propagation constant of transmission lines. The measurement procedure is described in Subsections 5.5.1 to 5.5.3, while Subsections 5.5.4 and 5.5.5 discuss dielectrometric standards and present a new standard based on the measurement procedure, for a wide variety of substances.

5.5.1 L-R-L calibration method

The measurement method for the propagation constant is derived from a simplification of the Line-Return-Line (L-R-L) calibration method. Measurements on planar lines can indeed be achieved with the classical L-R-L calibration method described extensively in [5.75]-[5.80]. It requires two identical lines of different length and a reflective device (short or open circuit) (Fig. 5.25a). The magnitude and phase of the scattering parameters of these elements are measured, using a Vector Network Analyzer (VNA). This method allows the movement of the reference plane after calibration to the middle plane of the shortest line (plane PP', Fig. 5.25b). This is of significant interest for characterizing a transmission line which does not have simple and good transitions to the coaxial connectors of the measurement setup, such as a slot-line, for example. A simple explanation of the method can be found when looking at the schematic representation of the device to be measured and of its two-ports A' and B', with the following definitions:
- X: two-port device to be measured
- A' and B': two-ports characterizing the transition between the two-port and coaxial ports of VNA
- A and B: two-port networks characterizing the transition between reference plane PP' and the coaxial ports of VNA, for each element to be measured.

The method assumes that two-port networks A' and B' are the same for all the measurements illustrated in Figure 5.25a,b, and that the two lines are identical, except for their length. This is the case if the transition between coaxial ports and lines allows reproducible measurements (using the VNA's broadband test fixture) and if the etching of the planar lines can be well reproduced. Under these assumptions, the scattering matrices of two-port networks characterizing the transition between the left-hand side coaxial port of VNA and reference plane PP' are identical, as well as those of two-port networks characterizing the transition between the reference plane PP' and right-hand side coaxial port of VNA. As a consequence, the same two-ports A and B are considered for each element to be measured. Hence, only two-ports A and B have to be characterized. Considering the set of measurements illustrated in Figure 5.25b, we have as unknowns the two scattering matrices of the two-port networks $\overline{\overline{S}}^A$, $\overline{\overline{S}}^B$, and the complex terms $e^{-\gamma \Delta L}$ and Γ^{react}, yielding 10 complex unknowns. On the other hand, the measurement of the four configurations (Fig. 5.25b) provides the two scattering matrices of the short and long lines, and the two reflection coefficients of configurations 3 and 4, resulting in 10 complex quantities. The real and imaginary parts of the scattering matrices and reflection coefficients of the four configurations are expressed as a function of the real and imaginary parts of the four complex unknowns, providing a system of 20 equations for 20 real unknowns.

Hence, the scattering matrices of two-ports A and B may be determined and introduced as correcting two-ports in processing the measurement setup. Finally, the measurement of the device X (Fig. 5.25a) provides a scattering matrix which can be expressed as a function of scattering matrix $\overline{\overline{S}}^X$ of X and of scattering matrices $\overline{\overline{S}}^A$ and $\overline{\overline{S}}^B$ which have been determined from the calibration measurements, from which $\overline{\overline{S}}^X$ is extracted. This method has been used for measuring the coplanar waveguide impedance step presented in Figure 5.3.

An important comment has to be made about the L-R-L algorithm. The ΔL line is modeled as a single transmission term, which implies that the reference impedance at plane PP' for the scattering matrices is the characteristic impedance of the line. Hence, the reference impedance for $\overline{\overline{S}}^X$ is the characteristic impedance of the line, which still remains unknown after calibration because it is not needed in the calibration algorithm. As a consequence, measurements of characteristic impedance are possible after calibration if, and only if, the line calibration standards are precisely known over the whole calibrating frequency range.

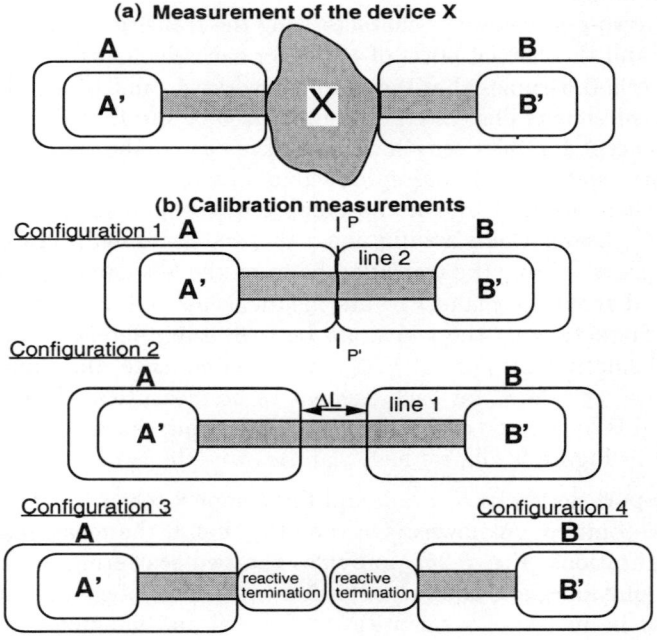

Fig. 5.25 *Measurement procedure using the Line-Return-Line calibration method (a) measurement of the unknown device; (b) calibration procedure*

5.5.2   L-L calibration method

When looking into the algorithm of the L-R-L calibration method, there is one piece of information involved in the process which is never extracted and used: the transmission line matrix corresponding to the length difference ΔL between the two calibration lines. This matrix could be used as a first step for the extraction of the propagation constant from the measurement of the two calibration lines [5.81]. The major problem, however, is the instability when the phase of the ΔL line approaches 0 or 180 degrees. Usually this phase has to be kept around 90 degrees (a difference of a quarter-wavelength) to ensure computational stability of the L-R-L calibration. Such an electrical length is not sufficient to extract insertion losses with an acceptable precision. A third line of sufficient length, typically three or four wavelengths at the lowest frequency of the operating band, must be used to obtain accurate results. Hence, *four elements* (two lines and one reflective device for calibration, and a very long line used as device X for wavelength and losses measurements) are required to characterize the propagation constant of one line with the classical L-R-L calibration method.

The method we proposed in [5.68] *reduces the number of necessary elements to only two*. It is based on the assumption that the two-ports describing the transition from each coaxial connector to the measured ΔL line are identical from mechanical and electrical points of view (Fig. 5.25b). Two-ports A and B are identical and have reciprocity properties, *i.e.*

$$S_{ii}^A = S_{ii}^B \qquad \text{for } i = 1, 2 \tag{5.37a}$$

$$S_{12}^X = S_{21}^Y \qquad \text{for } X, Y = A, B \tag{5.37b}$$

Hence only four complex variables have to be determined:

$$S_{11}, S_{22}, S_{12}(= S_{21}) \qquad \text{and} \qquad e^{-\gamma \Delta L} \tag{5.37c}$$

From the two line circuits, one can measure eight complex quantities (the four scattering parameters provided by the VNA for each circuit), but owing to the symmetry properties of these circuits, only four of them are significant. These measurements are performed after a coaxial calibration (using through-line and matched, short and open circuit loads) of the VNA, because the two-ports modeling the internal behavior of VNA from its source to its coaxial connectors do not satisfy (5.37a,b). The measured S-parameters of the two lines are expressed as a function of the 4 complex unknowns, yielding a system of 4 non-linear equations. Solving this system yields the term $e^{-\gamma \Delta L}$, hence the wavelength and losses corresponding to the length difference ΔL between the two lines. The Line-Line (L-L) method has been used for extracting the measured complex propagation constant for all transmission line results presented in Section 5.1.

It has to be mentioned that Janezic [5.82] has proposed an equivalent L-L procedure which does not assume that two-ports A and B are identical. The disadvantage of this procedure is that eight complex quantities have to be measured and manipulated for each characterization.

5.5.3   Reduction to one-line calibration method

One can achieve a calibration with a single line if the two-port networks are not only symmetric but also lossless. Lossless two-ports satisfy 4 scalar equations relating their 3 unknown S-parameters (reciprocity is assumed), so that finally one of these complex S-parameters remains unknown. Hence, only two complex quantities, provided by reflection and transmission measurements on a single line, are needed to determine one of the S-parameters of the two-port networks and the unknown term $e^{-\gamma \Delta L}$. This has been investigated by T. Kezai *et al.* [5.22] for shielded low-loss lines (finlines). They extracted the losses and propagation constant of a finline from a single line measurement. This measurement method has been used for the experimental results on finline presented in Section 5.1 (Fig. 5.4).

5.5.4   Extraction method of constitutive parameters

5.5.4.1   Overview of the existing standards

At low frequencies (up to 300 MHz), the complex permittivity of a planar substrate used in MIC structures is deduced from the measurement of the complex capacitance of concentric electrodes etched on the dielectric test sample [5.83].

At microwave frequencies, the methods are divided into two classes: metal waveguide or cavity methods, and stripline resonator methods. In the first one, the complex permittivity is derived from the measured variation of the reflection coefficient at the input of a short-circuited waveguide [5.84] or of a resonant cavity [5.85], loaded by a dielectric sample. The drawbacks of this method is the amount of time necessary to prepare, adjust and make measurements, as well as the rather extreme precision with which the position or frequency changes have to be measured.

For these reasons, the stripline resonator method seems more attractive for MICs designers : the complex permittivity of a planar substrate is derived from the measured resonant frequency and loaded-Q of four boards etched on the substrate, each of them consisting of a stripline or microstrip line resonator loosely coupled to two access lines. The stripline resonator method, however, presents some disadvantages related to the use of a measured resonant curve for determining the unknown permittivity: allowance should be made for the radiation losses, shape effects of the resonator, insertion losses of the coaxial launching connectors [5.86], and that the dimensions of the resonator become too large below 3 GHz. Finally, manufacturing and measuring four circuits is necessary at each frequency where the characterization is needed, because a linear regression is performed over the measured four resonant frequencies to eliminate the shape effect of the resonator at the frequency of interest.

The method we present in the next section extracts the complex permittivity of the planar substrate by comparing measurements carried out on two lines with very different topologies etched on the plate and calculations of the effective dielectric constant and the losses of the lines, based on our variational model. It is valid up to 40 GHz. The main advantage is that it offers a wideband characterization, because transmission lines are used instead of resonant elements and discontinuities and radiating sources are removed by the two-line calibration method. Only three circuits are needed to obtain a characterization valid over the frequency range 0 to 40 GHz, because we make use of the broadband two-line calibration method for measuring losses and effective dielectric constants.

5.5.4.2 Description of L-L planar dielectrometer

Previous sections have demonstrated the efficiency of the variational model developed for planar lines on dielectric or gyrotropic lossy substrates. It takes into account the geometrical parameters of the line and the substrate, together with the electrical parameters of the structure: conductivity of the metallization, and complex permittivity and/or permeability of the substrate. The complex propagation constant γ of the dominant mode of a planar line is obtained on-line as a result of the variational formulation: the real part of the propagation constant provides the losses per unit length, while the imaginary part yields the effective dielectric constant expressed as $(\mathrm{Im}(\gamma)c_0/\omega)^2$. It should be emphasized again that equation (4.21) has not been proven to be variational with respect to the constitutive parameters of the layers. The sensitivity of the value of the propagation constant to a variation of these parameters is preserved.

Upon performing the experimental validation of the variational formulation for planar lines on dielectric substrates, a significant difference was observed between the measured (Fig. 5.26a, c curve —) and calculated (Fig. 5.26 a,c curve - -) values of the effective dielectric constant and losses of a slot-line when using the value of the complex permittivity given by the manufacturer. A significant difference was also observed between the measured (Fig. 5.26b,d curve —) and calculated (Fig. 5.26b,d curve - -) values of the effective dielectric constant and losses of a microstrip etched on the same substrate when using the same manufacturer's value. It is observed on Figure 5.26 that the relative difference between curves - - and — is of the same order for the microstrip (b,d) and the slot-line (a,c). The two topologies are electromagnetically complementary of each other, which ensures that the observed discrepancy between modeling and measurement in the two cases is mainly due to a difference on a parameter which is common to the two topologies, *i.e.* the complex permittivity of the substrate. Furthermore it is well known that manufacturer's values have to be used with care: at X-band and higher one cannot guarantee an accuracy better than 10 % for permittivity obtained with the existing standards. Hence, adjusting the value of the permittivity to eliminate the difference between calculations and measurements seems a pertinent way to obtain an improved value. When increasing the real part of the complex permittivity by 5 % above the manufacturer's value and the dielectric loss tangent from 0.0028 to 0.0038, the calculated curves (Fig. 5.26

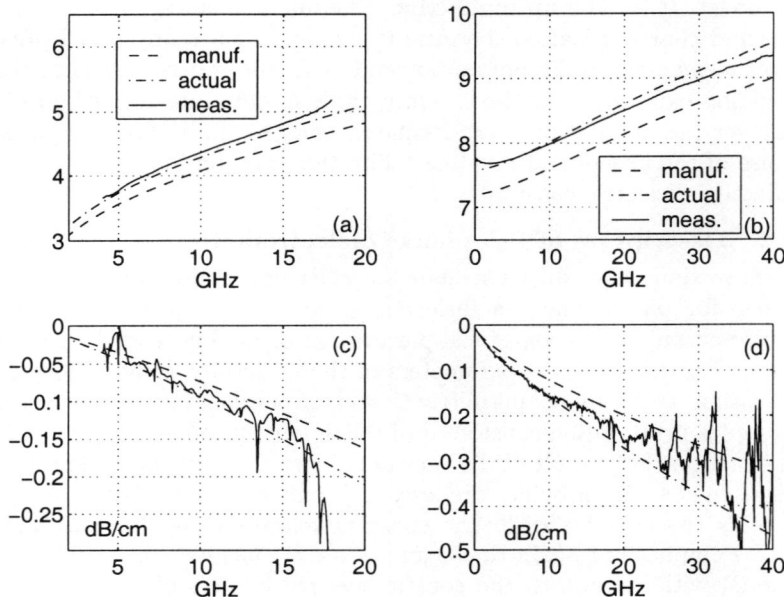

Fig. 5.26 *Comparison of measured and calculated effective dielectric constant and losses of two complementary geometries etched on a RT duroid 6010 dielectric substrate plate of thickness 0.635 mm (a),(c) slot-line (width 0.389 mm); (b),(d) microstrip line (width 0.580 mm): curve - -: calculated with manufacturer's complex relative permittivity 10.8(1 − j0.0028); curve —: measured; curve ·-·: calculated with modified complex relative permittivity 11.4(1 − j0.0038)*

curve ·-·) yield excellent agreement with the measured values over the wide frequency range for the two lines. The resulting uncertainty on the permittivity is less than 1 %, yielding 11.4(1 − j0.0038) over the frequency range 0-40 GHz, instead of 10.8(1 − j0.0028) given by the manufacturer at X-band.

5.5.4.3   Proposed new standard

The relative permittivity of a planar substrate is obtained as the complex value which, when using accurate wideband models, provides the best agreement over the 0-40 GHz band between calculated and measured effective dielectric constant and losses of two lines with very different topologies (one microstrip line and one slot-line), etched on the same substrate.

The two-line calibration method is used to measure $e^{\gamma \Delta L}$, i.e. the effective dielectric constant and the losses of the lines. Hence, only two lines of different lengths are needed for each of the two topologies considered. Three topologies are in competition to serve as a standard: microstrip, slot-line and coplanar waveguide. The microstrip-slot-line combination is the best choice for the following reason: the coplanar waveguide is too similar to the microstrip, with a quasi-TEM propagation on both topologies. This reduces its efficiency with respect to the microstrip-slot-line combination.

Hence, *three circuits* have to be etched, respectively two slot-lines, with their transitions to microstrip lines, and one microstrip line. Using this arrangement, the complex permittivity of a layer is determined with only three circuits over the whole band, and not only at discrete frequencies.

The proposed method combines rapidity and accuracy of both theoretical and experimental characterization of any planar line. Because planar lines are used instead of resonators, a number of spurious limiting effects are avoided. Also, the method removes the effects of the connectors and of the transitions between the coaxial plane of measurement of the VNA and each of the access planes of the ΔL line, provided that two-port networks A and B (Fig. 5.25a,b) are identical (reproducible), for each line and from line to line. Hence, a high accuracy is obtained over a wide frequency range with a reduced number of measurements. Such results have been obtained for a wide variety of planar substrates, with a dielectric permittivity range from 2.33 to 10.8 and a substrate thickness from 0.254 mm to 2 mm.

5.5.5 Other dielectrometric applications of the L-L method

5.5.5.1 Soils

The L-L dielectrometric method is also applicable to the electromagnetic characterization of soils. We have extensively developed the method in the frequency range 0-18 GHz in order to extract the complex permittivity, from which the equivalent conductivity is obtained. This is necessary to evaluate the performances of microwave technologies in landmine detection for humanitarian purposes [5.87], like ground penetrating radars and radiometers.

Measurements of various soils, both sandy and silty, have been carried out, using the L-L measurement method with two rectangular waveguides of width a, differing only by their length, and homogeneously filled with the soil to be characterized. Assuming, as previously, that the mechanical and electrical identity of the transitions between the VNA coaxial output and the lines, and an identical material distribution in the two guides, the propagation constant related to length ΔL of filled waveguide is extracted. Then, instead of using a variational principle as for planar lines, the complex dielectric constant extracted from measured propagation constant γ is given by

$$\varepsilon_r' = \mathrm{Re}\left\{ \frac{c_0^2}{\omega_2}\left(-\gamma^2 + (\pi/a)^2 \right) \right\} \tag{5.38a}$$

$$\varepsilon_r'' = \mathrm{Im}\left\{ \frac{c_0^2}{\omega_2}\left(-\gamma^2 + (\pi/a)^2 \right) \right\} \tag{5.38b}$$

Measurements were carried out at room temperature on four soils with different textural compositions: two pure sands and two silty soils. For each soil, measurements were done for different water contents, up to 20 %. From the extracted permittivities of those soils, a model has been elaborated by Storme *etal*. [5.88], yielding the expression of the complex permittivity as a function of easily available soil parameters such as textural composition and density. Using volume fractions and permittivities of bound water, free water, air and dry soil, the formula obtained for the dielectric constant of a water-soil mixture is

$$\varepsilon_m = \frac{3\varepsilon_s + 2V_{fw}(\varepsilon_{fw} - \varepsilon_s) + V_{bw}(\varepsilon_{bw} - \varepsilon_s) + 2V_a(\varepsilon_a - \varepsilon_s)}{3 + V_{fw}\left(\frac{\varepsilon_s}{\varepsilon_{fw}} - 1\right) + V_{bw}\left(\frac{\varepsilon_s}{\varepsilon_{bw}} - 1\right) + V_a\left(\frac{\varepsilon_s}{\varepsilon_a} - 1\right)} \qquad (5.38c)$$

where subscripts bw, fw, a, and s refer to bound water, free water, air and dry soil, respectively, and the following hypotheses have been made:

1. $V_{bw} = 0.06774 - 0.00064 \times \text{SAND} + 0.00478 \times \text{CLAY}$
   where CLAY and SAND are clay and sand contents, respectively, in percent of weight of dry soil
2. Permittivity of bound water is put equal to $35 - j15$
3. Contribution of bulk water takes into account the dielectric constant of pure water, Debye model [5.89] at room temperature
4. Volume of free water V_{fw} is taken as the difference between total water volume and V_{bw}.

5.5.5.2   Bioliquids

A good knowledge of the complex permittivity of biological media is necessary for adequately determining their response to electromagnetic fields, both for the study of biological effects as well as for medical applications. Based on the L-L method, we have set up a new procedure for measuring the permittivity of liquids [5.89]. The measurement procedure uses waveguide spacers of different thickness, placed between the two waveguide ports of VNA. A synthetic film (Parafilm), is placed at both ends between waveguides and spacer, containing the liquid in a known volume. Because of the L-L method, the film has no effect on the measurements, provided that the transition is reproducible when using both spacers. A preliminary measurement on dioxane, having a known constant permittivity, is used to validate the measurement set-up. Measurements have been carried out for the complex permittivity of biological and organic liquids at frequencies above 20 GHz up to 110 GHz, on methanol, axoplasm, and beef blood. The values obtained have been compared with Debye's law. We observed, as presented in [5.90], that for biological liquids the first-order Debye model using only one relaxation time is not sufficient. Several relaxation phenomena occur, requiring the recalculation of higher-order relaxation terms. Finally, we found that the Cole-Cole diagram [5.91], representing ε'' as a function of ε' with frequency as a parameter, is very efficient for improving models for dielectric relaxation.

5.6 Summary

Chapter 5 is devoted to a number of applications of the variational principle to calculate transmission line parameters. First, expressions for the parameters are recalled. Then, they are calculated for simple line configurations on dielectric substrates: microstrip, slot-line, coplanar waveguide, and finline. Lossy semiconducting substrates are considered for microstrip and coplanar waveguide. A recent application has been described: the calculation of parameters of a microstrip on a magnetic nanostructured substrate, consisting of an array of magnetic metallic nanowires embedded in a thin porous dielectric substrate. Another class of examples were junctions: microstrip-to-slot-line junction and planar T-junction. Gyrotropic devices were also considered, such as YIG-tuned planar MSW devices, undercoupled as well as overcoupled, for one-port and two-port configurations. Optomicrowave devices have also been analyzed, using a variational principle, illustrating how to model mesa p-i-n structures and how important the differences can be when using different topology models, and coplanar p-i-n structures. The chapter ends with the description of a measurement method based on a variational principle to determine the complex permittivity of a planar substrate with an excellent accuracy. The measurement method is then illustrated with examples for characterizing soils for demining operations, and bioliquids for medical purposes (blood and axoplasm).

5.7 References

[5.1] A. Vander Vorst, *Transmission, propagation et rayonnement*. Brussels: De Boeck-Wesmael, 1995.

[5.2] A. Vander Vorst, *Electromagnétisme*. Brussels: De Boeck-Wesmael, 1994, Ch. 5.

[5.3] T. Rozzi, F. Moglie, A. Morini, E. Marchionna, and M. Politi, "Hybrid modes, substrate leakage and losses of slot-lines at millimetre wave frequencies", *IEEE Trans. Microwave Theory Tech.*, vol. MTT-38, no. 8, pp. 1069-1078, Aug. 1990.

[5.4] I. Huynen, D. Vanhoenacker, A. Vander Vorst, "Spectral Domain Form of New Variational Expression for Very Fast Calculation of Multilayered Lossy Planar Line Parameters", *IEEE Trans. Microwave Theory Tech.*, vol. 42, no. 11, pp. 2099-2106, Nov. 1994.

[5.5] R. Faraji-Dana, Y. Leonard Chow, "The Current Distribution and AC Resistance of a Microstrip Structure", *IEEE Trans. Microwave Theory Tech.*, vol. MTT-38, no. 9, pp. 1268-1277, Sep. 1990.

[5.6] E. Yamashita, "Variational method for the analysis of microstrip-like transmission lines", *IEEE Trans. Microwave Theory Tech.*, vol. MTT-16, no. 8, pp. 529-535, Aug. 1968.

[5.7] T. Kitazawa, T. Itoh, "Asymmetrical Coplanar Waveguide with Finite Metallization Thickness Containing Anisotropic Media", *IEEE Trans. Microwave Theory Tech.*, Vol. MTT-39, no. 8, pp. 1426-1433, Aug. 1991.

[5.8] L. Zhu, E. Yamashita, "New method for the Analysis of Dispersion Characteristics of Various Planar Transmission Lines with Finite Metallization Thickness", *IEEE Microwave Guided Wave Letters*, vol. 1, no. 7, pp. 164-166, July 1991.

[5.9] W.H. Haydl, J. Braunstein, T. Kitazawa, M. Schlechtweg, P. Tasker, L.F. Eastman, "Attenuation of Millimeter Wave Coplanar Lines on Gallium Arsenide and Indium Phosphide over the range 1-60 GHz", *IEEE MTT-S Int. Microwave Symp. Dig.*, pp. 349-352, June 1992.

[5.10] W.H. Haydl, "Experimentally Observed Frequency Variation of the Attenuation of Millimeter-Wave Coplanar Transmission Lines with Thin Metallization", *IEEE Microwave Guided Wave Letters*, vol. 2, no. 8, pp. 164-166, August 1992.

[5.11] T. Itoh, *Numerical methods for Microwave and Millimeter Wave Passive structures*. New York: John Wiley&Sons, 1989, Ch. 5.

[5.12] A. Vander Vorst, D. Vanhoenacker-Janvier, *Bases de l'ingénierie microonde*. Brussels: De Boeck&Larcier, 1996, Ch. 2.

[5.13] J.B. Knorr, P. Shayda, "Millimeter wave fin line characteristics", *IEEE Trans. Microwave Theory Tech.*, vol. MTT-28, no. 7, pp. 737-743, July 1980.

[5.14] L.P. Schmidt, T. Itoh, "Spectral domain analysis of dominant and higher order modes in fin lines", *IEEE Trans. Microwave Theory Tech.*, vol. MTT-28, no. 9, pp. 981-985, Sept. 1980.

[5.15] J.B. Knorr, K.D. Kuchler, "Analysis of coupled slots and coplanar strips on dielectric substrate", *IEEE Trans. Microwave Theory Tech.*, vol. MTT-23, no. 7, pp. 541-548, July 1975.

[5.16] J.B. Knorr, A. Tufekcioglu, "Spectral-Domain Calculation of Microstrip Characteristic Impedance", *IEEE Trans. Microwave Theory Tech.*, vol. MTT-23, no. 9, pp. 725-728, Sept. 1975.

[5.17] R.H. Jansen, M. Kirschning, "Arguments and an Accurate Model for the Power-Current Formulation of Microstrip Characteristic Impedance", *Arch. Elek. Übertragung.*, vol. AEÜ-37, no. 3-4, pp. 108-112, March-April 1982.

[5.18] D.F. Williams, "Characteristic impedance, power and causality", *IEEE Microwave Guided Wave Lett.*, vol 9, no. 5, pp. 181-182, May 1999.

[5.19] R. Gillon, J.P. Raskin, D. Vanhoenacker, J.P. Colinge, "Determining the reference impedance of on-wafer TRL calibrations on lossy substrates", *Proc. 26th Eur. Microwave Conf.*, Prague, pp. 170-176, Sept. 1996.

[5.20] R.B. Marks, D.F. Williams, "Characteristic Impedance Determination Using Propagation Constant Measurement", *IEEE Microwave Guided Wave Lett.*, vol. 1, no. 6, pp. 141-143, June 1991.

[5.21] I. Huynen, A. Vander Vorst, "Variational non quasi-static formulations for the impedance of planar transmission lines", *IEEE Trans. Microwave Theory Tech.*, vol. 47, no. 7, pp. 995-1003, July 1999.

[5.22] T. Kezai, R. Sciuto, A. Vander Vorst, "Méthode d'étalonnage et de mesure pour des lignes à ailettes", *Journées Nationales Microondes (JNM 93)*, Proc., pp. D6.6-7, Brest, May 1993.

[5.23] M. Serres, *Optical control of microwave planar devices on semiconductor substrate*, Ph.D. dissertation. Louvain-la-Neuve, Belgium: Microwaves UCL, 1999.

[5.24] I. Huynen, G. Goglio, D. Vanhoenacker, A. Vander Vorst, "A novel nanostructured microstrip device for tunable stopband filtering applications at microwaves", *IEEE Microwave Guided Wave Lett.*, vol. 9, no. 10, pp. 401-403, Oct. 1999.

[5.25] I. Huynen, L. Piraux, D. Vanhoenacker, A. Vander Vorst, "Ferromagnetic resonance in metallic nanowires for tunable microwave planar circuits", *Proc. Progr. Electromagnetic Res. Symp. (PIERS 2000)*, Cambridge, pp. 320, July 2000.

[5.26] S.B. Cohn, "Slot Line on a dielectric substrate", *IEEE Trans. Microwave Theory Tech.*, vol. MTT-30, no. 10, pp. 259-269, Oct. 1969.

[5.27] E.A. Mariani, C.P. Heinzman, J.P. Agrios, S.B. Cohn, "Slot Line characteristics", *IEEE Trans. Microwave Theory Tech.*, vol. MTT-17, no. 12, pp. 1091-1096, Dec. 1969.

[5.28] G.H. Robinson, J.L. Allen, "Slot Line Application to Miniature Ferrite Devices", *IEEE Trans. Microwave Theory Tech.*, vol. MTT-16, no. 12, pp. 1097-1101, Dec. 1969.

[5.29] J. Lee, "Slot-line Impedance", *IEEE Trans. Microwave Theory Tech.*, vol. MTT-39, no. 4, pp. 666-672, April 1991.

[5.30] J.B. Knorr, "Slot-Line transitions", *IEEE Trans. Microwave Theory Tech.*, vol. MTT-22, no. 5, pp. 548-554, May 1972.

[5.31] C.P. Wen, "Coplanar Waveguide : A Surface Strip Transmission Line Suitable for Nonreciprocal Gyromagnetic Device Application", *IEEE Trans. Microwave Theory Tech.*, vol. MTT-16, no. 12, pp. 1087-1090, Dec. 1969.

[5.32] R.A. Pucel, D.J. Massé, "Microstrip Propagation on Magnetic Substrates - Part I : Design Theory", *IEEE Trans. Microwave Theory Tech.*, vol. MTT-20, no. 5, pp. 304-308, May 1972.

[5.33] R.A. Pucel, D.J. Massé, "Microstrip Propagation on Magnetic Substrates - Part II : Experiment", *IEEE Trans. Microwave Theory Tech.*, vol. MTT-20, no. 5, pp. 309-313, May 1972.

[5.34] T. Kitazawa, "Variational Method for Planar Transmission Lines with Anisotropic Magnetic Media", *IEEE Trans. Microwave Theory Tech.*, Vol. MTT-37, no. 11, pp. 1749-1754, Nov. 1989.

[5.35] M. Horno, F.L. Mesa, F. Medina, R. Marques, "Quasi-TEM Analysis of Multilayered Multiconductor Coplanar Structures with Dielectric and Magnetic Anisotropy Including Substrate Losses", *IEEE Trans. Microwave Theory Tech.*, vol. MTT-40, no. 3, pp. 524-531, March 1992.

[5.36] E.B. El-Sharawy, R.W. Jackson, "Coplanar Waveguide and Slot Line on Magnetic Substrates: Analysis and Experiment", *IEEE Trans. Microwave Theory Tech.*, vol. MTT-36, no. 6, pp. 1071-1079, June 1988.

[5.37] F.L. Mesa, M. Horno, "Quasi-TEM and full wave approach for coplanar multistrip lines including gyromagnetic media longitudinally magnetized", *Microwave Opt. Technol. Lett.*, vol. 4, no. 12, pp. 531-534, Nov. 1991.

[5.38] M. Tsutsumi, T. Asahara, "Microstrip Lines Using Yttrium Iron Garnet Films", *IEEE Trans. Microwave Theory Tech.*, vol. MTT-38, no. 10, pp. 1461-1470, Oct. 1990.

[5.39] I. Huynen, B. Stockbroeckx, A. Vander Vorst, "Variational Principles are efficient CAD tools for planar tunable MICs involving lossy gyrotropic layers", *International Journal of Numerical Modelling : Electronic networks, devices and fields*, vol. 12, no. 5, pp. 417-440, Sept.-Oct. 1999.

[5.40] S.N. Bajpai, "Insertion losses of electronically variable magnetostatic wave delay lines", *IEEE Trans. Microwave Theory Tech.*, vol. MTT-37, no. 10, pp. 1529-1535, Oct. 1989.

[5.41] A.K. Ganguly, D.C. Webb, "Microstrip excitation of Magnetostatic Surface Waves : Theory and Experiment", *IEEE Trans. Microwave Theory Tech.*, vol. MTT-23, no. 12, pp. 998-1006, Dec. 1975.

[5.42] P.R. Emtage, "Interaction of magnetostatic waves with a current", *J. Appl. Physics*, vol. 49, no. 8, pp. 4475-4484, Aug. 1978.

[5.43] J.D. Adam, S.N. Bajpai, "Magnetostatic Forward Volume Wave Propagation in YIG strips", *IEEE Trans. Magnetics*, vol. MAG-18, no. 6, pp. 1598-1600, Nov. 1982.

[5.44] I.J. Weinberg, "Insertion Loss for Magnetostatic Volume Waves", *IEEE Trans. Magnetics*, vol. MAG-18, no. 6, pp. 1607-1609, Nov. 1982.

[5.45] A.D. Berk, "Variational principles for electromagnetic resonators and waveguides", *IEEE Trans. Antennas Propagat.*, vol. AP-4, no. 2, pp. 104-111, Apr. 1956.

[5.46] I. Huynen, G. Verstraeten, A. Vander Vorst, "Theoretical and experimental evidence of non-reciprocal effects on magnetostatic forward volume wave resonators", *IEEE Microwave Guided Waves Lett.*, vol. 5, no. 6, pp. 195-197, June 1995.

[5.47] J.P. Parekh, K.W. Chang, H.S. Tuan, "Propagation characteristics of magnetostatic waves", *Circuits, Systems and Signal Proceedings*, vol. 4, pp. 9-38, 1985.

[5.48] P. Kabos and V.S. Stalmachov, *Magnetostatic waves and their applications*. London: Chapman&Hall, 1994.

[5.49] L. Kajfez , P. Guillon, *Dielectric resonators*. London: Artech House, 1986.

[5.50] I. Huynen, B. Stockbroeckx, G. Verstraeten, D. Vanhoenacker, A. Vander Vorst, "Planar gyrotropic devices analyzed by using a variational principle", *Proc. 24th Eur. Microw. Conf.*, Cannes, pp. 309-314, Sept. 1994.

[5.51] I. Huynen, A. Vander Vorst, "A new variational formulation, applicable to shielded and open multilayered transmission lines with gyrotropic non-hermitian lossy media and lossless conductors", *IEEE Trans. Microwave Theory Tech.*, vol. 42, no. 11, pp. 2107-2111, Nov. 1994.

[5.52] D.W. Van Der Weide, "Thin-film oscillators with low phase noise and high spectral purity", *IEEE MTT-S Int. Microwave Symp. Dig.*, vol. 1, pp. 313-316, 1992.

[5.53] W. Kunz, K.W. Chang, W. Ishak, "MSW-SER based tunable oscillators", *Proc. Ultrasonics Symposium*, pp. 187-190, 1986.

[5.54] I. Huynen, B. Stockbroeckx, G. Verstraeten, "An efficient energetic variational principle for modelling one-port lossy gyrotropic YIG Straight Edge Resonators", *IEEE Trans. Microwave Theory Tech.*, vol. 46, no. 7, pp. 932-939, July 1998.

[5.55] J.B. Davies in T. Itoh, *Numerical methods for Microwave and Millimeter Wave Passive structures*, Chapter 2. New York: John Wiley&Sons, 1989.

[5.56] R.E. Collin, *Field theory of guided waves (second edition)*. New York: IEEE Press, 1991.

[5.57] H. Guckel, P. Brennan I. Palocz, "A parallel-plate waveguide approach to microminiaturized planar transmission lines for integrated circuits", *IEEE Trans. Microwave Theory Tech.*, vol. MTT-11, no. 8, pp. 468-476, Aug. 1967.

[5.58] D. Jäger, "Slow-wave Propagation along variable Schottky-Contact Microstrip Line", *IEEE Trans. Microwave Theory Tech.*, vol. MTT-24, no. 9, pp. 566-573, Sept. 1976.

[5.59] K. Wu, R. Vahldiek, "Hybrid-Mode Analysis of Homogeneously and Inhomogeneously Doped Low-Loss Slow-Wave Coplanar Transmission Lines", *IEEE Trans. Microwave Theory Tech.*, vol. MTT-39, no. 8, pp. 1348-1360, Aug. 1991.

[5.60] J.K. Liou, K.M. Lau, "Analysis of Slow Wave Transmission Lines on Multi-Layered Semiconductor Structures Including Conductor Losses", *IEEE Trans. Microwave Theory Tech.*, vol. MTT-41, no. 5, pp. 824-829, May 1993.

[5.61] K.S. Giboney, M.J.W. Rodwell, J.E. Bowers, "Travelling-Wave Photodetector Theory", *IEEE Trans. Microwave Theory Tech.*, vol. 45, no. 8, pp. 1310-1319, Aug. 1997.

[5.62] V. M. Hielata, G.A. Vawter, T.M. Brennan, B.E. Hammons, "Traveling-Wave Photodetectors for High-Power, Large-Bandwidth Applications", *IEEE Trans. Microwave Theory Tech.*, vol 43, no. 9, pp. 2291-2298, Sept. 1995.

[5.63] L.Y Lin, M.C. Wu, T. Itoh, T.A. Vang, R.E. Muller, D.L. Sivco, A.Y. Cho, "High-Power High-Speed Photodetectors - Design, Analysis, and Experimental Demonstration", *IEEE Trans. Microwave Theory Tech.*, vol. 45, no. 8, pp. 1320-1331, Aug. 1997.

[5.64] K. Giboney, J. Bowers, M. Rodwell, "Travelling-wave photodetectors", *1995 IEEE MTT-S Symp. Dig.*, pp. 159-162, 1995.

[5.65] K. Giboney, R. Nagarajan, T. Reynolds, S. Allen, R. Mirin, M. Rodwell "Travelling-wave photodetectors with 172-GHz bandwidth and 76-GHz bandwidth-efficiency product", *IEEE Photon. Technol. Let.*, vol. 7, no. 4, pp. 412-414, Apr. 1995.

[5.66] I. Huynen, A. Salamone, M. Serres, "A traveling wave model for optimizing the bandwidth of p-i-n photodetectors in Silicon-On-Insulator technology", *IEEE J. Sel. Top. Quantum Electronics*, vol. 4, no. 6, pp. 953-963, Nov.-Dec. 1998.

[5.67] D. Jäger, R. Kremer, A. Stohr, "Travelling-wave optoelectronic devices for microwave applications", *1995 IEEE MTT-S Symp. Dig.*, pp. 163-166, 1995.

[5.68] M. Fossion, I. Huynen, D. Vanhoenacker, A. Vander Vorst, "A new and simple calibration method for measuring planar lines parameters up to 40 GHz", *Proc. 22nd Eur. Microw. Conf.*, Helsinki, pp. 180-185, Sept. 1992.

[5.69] Z. Zhu, I. Huynen, A. Vander Vorst, "An efficient microwave characterization of PIN photodiodes", *Proc. 26th European Microwave Conference*, Prague, vol. 2, pp. 1010-1014, Sept. 1996.

[5.70] W.A. McIlroy, M.S. Stern, P.C. Kendall, "Fast and accurate method for calculation of polarised modes in semiconductor rib waveguides", *Electronics Lett.*, vol. 25, no. 23, pp. 1586-1587, 9th Nov. 1989.

[5.71] Z. Zhu, A. Vander Vorst, "Microwave Propagation in p-i-n Transmission Lines", *IEEE Microwave Guided Wave Lett.*, vol. 7, no. 6, pp. 159-161, June 1997.

[5.72] R. Pregla, W. Pascher, "The method of lines", in T. Itoh, *Numerical methods for Microwave and Millimeter Wave Passive structures*. New York: John Wiley&Sons, 1989, Chap. 6.

[5.73] I. Huynen, A. Vander Vorst, "A 4-port scattering matrix formalism for p-i-n traveling wave photodetectors", *IEEE Trans. Microwave Theory Tech.*, vol. 48, no. 6, pp. 1007-1016, May 2000.

[5.74] I. Huynen, D. Vanhoenacker, A. Vander Vorst, "Nouvelle méthode de détermination des paramètres électriques ou magnétiques de substrats planaires (théorie et expérimentation)", *Journées de Caractérisation des Matériaux Microondes* (JCMM 92), Arcachon, Actes, pp. II.010.1-6, Oct. 1992.

[5.75] N.E. Buris, "Lecture note on TRL calibration", University of Massachusetts, ECE 697B, Spring 1990.

[5.76] G.F. Engen, C.A. Hoer, "Thru-reflect-line: An improved technique for calibrating the dual six-port automatic Network Analyser", *IEEE Trans. Microwave Theory Tech.*, vol. MTT-27, no. 12, pp. 987-993, Dec. 1979.

[5.77] N.R. Franzen, R.A. Speciale, "A new procedure for system calibration and error removal in automated S-parameter measurement", *Proc. 5th Eur. Microw. Conf.*, Hamburg, pp. 69-73, 1975.

[5.78] Hewlett-Packard, "Applying the HP8510B TRL calibration for non-coaxial measurements", Product note 8510-8.

[5.79] R.R. Pantoja, M.J. Howes, J.R. Richardson, R.D. Pollard, "Improved calibration and measurement of the scattering parameters of microwave integrated circuits", *IEEE Trans. Microwave Theory Tech.*, vol. MTT-37, no. 11, pp. 1675-1680, Nov. 1989.

[5.80] D. Rubin, "De-embedding mm-wave MICs with TRL", *Microwave J.*, vol. 33, no. 6, pp. 141-150, June 1990.

[5.81] H.J. Eul, B. Schiek, "A Generalized Theory and New Calibration Procedures for Network Analyzer Self-Calibration ", *IEEE Trans. Microwave Theory Tech.*, vol. MTT-39, no. 4, pp. 724-731, Apr. 1991.

[5.82] M.D. Janezic, J.A. Jargon, "Complex permittivity determination from propagation constant measurements", *IEEE Microwave Guided Wave Lett.*, vol. 9, no. 2, pp. 76-78, Feb. 1999.

[5.83] International Electrotechnical Commission, *"Recommended Methods for the determination of the permittivity and dielectric dissipation factor of electrical insulating materials at power, audio and radio frequencies including metre wavelengths"*, Publication 250, 1969.

[5.84] G. Kent, "A dielectrometer for the measurement of substrate permittivity", *Microwave J.*, vol. 34, pp. 72-82, Dec. 1991.

[5.85] M. Olyphant, Jr., "Practical dielectric measurement of CLAD microwave substrates", *Cu-Tips Note no. 8, 3M Co. Publication EL-CT/8(70.6)R*.

[5.86] Appendix 1, MIL-P-13949G, Military specification, Plastic Sheet, Laminated, Metal-Clad, U.S. Army Electronics Command, Fort Montmouth, N.J., Feb. 1987.

[5.87] C. Steukers, M. Storme, I. Huynen, "Modelisation of electrical properties of soils", in Chapter 4 of the book "Sustainable Humanitarian Demining: Trends, Techniques, and Technologies", edited by Humanitarian Demining Information Center, James Madison University (Co-Editors: D. Barlow, C. Bowness, A. Craib, G. Gately, J.-D. Nicoud, J. Trevelyan). Verona, Virginia, USA: Mid Valley Press, pp. 305-313, 1998.

[5.88] M. Storme, I. Huynen, A. Vander Vorst, "Characterisation of wet soils in the 2-18 GHz frequency range", *Microwave Opt. Technol. Lett.*, vol. 21, no. 5, pp. 333-335, June 1999.

[5.89] F. Duhamel, I. Huynen, A. Vander Vorst, "Measurements of complex permittivity of biological and organic liquids up to 110 GHz", *1997 IEEE MTT-S Symp. Dig.*, Denver, pp. 107-110, June 1997.

[5.90] H.P. Schwan, K.R. Forster, "RF-field with biological systems: electrical properties and biophysical mechanisms", *Proc. of the IEEE*, vol. 68, no. 1, pp. 104-113, Jan. 1981.

[5.91] V. Daniel, *Dielectric relaxation*. London: Academic Press, 1967, Ch. 7.

Green's formalism

A.1 Definition and physical interpretation

Green's functions have been used for many years in electromagnetics and physics. A review of the most useful applications of Green's functions is given in Morse and Feshbach [A.1]. Only the fundamentals are summarized in this appendix, referring first to an analogy with ordinary differential equations.

A linear system is described by an ordinary differential equation with constant coefficients. As an example, the general form of the equation for a second-order system is

$$A\frac{d^2 f(t)}{dt^2} + B\frac{df(t)}{dt} + Cf(t) = x(t) \tag{A.1}$$

where the unknown response $f(t)$ is, for instance, a voltage or a current, while the excitation $x(t)$ comes from independent sources. The equation is often solved using a transformation. Using the Fourier transformation (Appendix C) yields

$$(j\omega)^2 AF(\omega) + j\omega BF(\omega) + CF(\omega) = X(\omega) \tag{A.2}$$

from which is obtained

$$F(\omega) = H(\omega)X(\omega) \tag{A.3}$$

with

$$H(\omega) = \left[(j\omega)^2 A + j\omega B + C\right]^{-1} \tag{A.4}$$

The function $H(\omega)$ is called the *transfer function* of the system. It is obviously the response of the system to an excitation with uniform spectrum:

$$F(\omega) = H(\omega) \quad \text{if } X(\omega) = 1 \tag{A.5}$$

Noting that the inverse transform of the function with uniform spectrum is the Dirac impulse $\delta(t)$, the inverse transform $h(t)$ of the transfer function $H(\omega)$ is obviously the impulse response of the system. Hence, the solution of (A.1) is obtained by calculating the inverse transform of the product $H(\omega)X(\omega)$. It is well known that this transform is the convolution integral

$$f(t) = \int_{-\infty}^{+\infty} x(t - \tau)h(\tau)d\tau \tag{A.6}$$

This physically means that the response $f(t)$ to an excitation $x(t)$ is the superposition of a continuous sequence of weighted impulse responses. In practice, however, the integral can only be calculated in a rather small number of cases.

What has been presented for ordinary differential equations can be extended to equations involving partial derivatives, such as Maxwell's equations in electromagnetics, with their boundary conditions and constitutive relations, and where sources are in general charge and current densities. In this case, scalar and vector potentials satisfy an equation such as

$$\nabla^2\overline{\mathcal{F}} + a\frac{\partial^2\overline{\mathcal{F}}}{\partial t^2} + b\frac{\partial\overline{\mathcal{F}}}{\partial t} = \overline{\mathcal{X}} \tag{A.7}$$

which is the case for Poisson's equation, the wave equation, and the diffusion equation for instance. The multi-dimensional Fourier transformation (C.7) is applied to this vector equation, transforming the space domain into the spectral domain:

$$\overline{F}(\overline{k}) = \int_V \overline{\mathcal{F}}(\overline{r})e^{-j\overline{k}\cdot\overline{r}}\, dV(\overline{r}) \tag{A.8a}$$

$$\overline{\mathcal{F}}(\overline{r}) = \frac{1}{(2\pi)^3}\int_V \overline{F}(\overline{k})e^{j\overline{k}\cdot\overline{r}}\, dV(\overline{k}) \tag{A.8b}$$

When the problem is time-variant, the corresponding equations have, of course, to be extended to 4 dimensions:

$$\overline{F}(\overline{k},\omega) = \int_{V(\overline{r})}\int_t \overline{\mathcal{F}}(\overline{r},t)e^{-j(\overline{k}\cdot\overline{r}+\omega t)}\, dV(\overline{r})dt \tag{A.9a}$$

$$\overline{\mathcal{F}}(\overline{r},t) = \frac{1}{(2\pi)^4}\int_{V(\overline{k})}\int_\omega \overline{F}(\overline{k},\omega)e^{j(\overline{k}\cdot\overline{r}+\omega t)}\, dV(\overline{k})d\omega \tag{A.9b}$$

We can now extend to partial derivative equations what has been shown to be valid for ordinary differential equations. To simplify the notation, we assume that there is no time dependency. By doing so, we shall illustrate the correspondence from the time-to-frequency transformation in problems described by ordinary differential equations, and the space-to-state space (spectral domain) transformation in problems described by partial derivative equations. There should be no difficulty in writing this transformation in 4 dimensions, transforming the equation from a space-time volume into a state space-frequency volume. Assuming that the equation is

$$\mathcal{L}(\bar{r})\overline{\mathcal{F}}(\bar{r}) = \mathcal{X}(\bar{r}) \tag{A.10}$$

where $\mathcal{L}$ is a differential operator and $\mathcal{F}$ the system response to the excitation $\mathcal{X}$, its 3-D Fourier transform is

$$L(\bar{k})F(\bar{k}) = X(\bar{k}) \tag{A.11}$$

where L, F, and X are the transforms of operator, response, and excitation, respectively. The solution is

$$F(\bar{k}) = G(\bar{k})X(\bar{k}) \qquad \text{with } G(\bar{k}) = 1/L(\bar{k}) \tag{A.12}$$

where $G(\bar{k})$ is the "transfer function", *i.e.* the Fourier-transform of the "impulse response" of the system. It is indeed the response of the system submitted to an excitation with "uniform spectrum":

$$F(\bar{k}) = G(\bar{k}) \qquad \text{if } X(\bar{k}) = 1 \tag{A.13}$$

The inverse transform of the uniform spectrum function is the generalization of the Dirac impulse

$$\int_V \delta(\bar{r})e^{-j\bar{k}\cdot\bar{r}}\,dV(\bar{r}) = 1 \tag{A.14}$$

Hence, the solution of (A.11) is the inverse transform of product (A.12), which is obviously the generalization of the convolution integral:

$$\overline{\mathcal{L}}(\bar{r}) = \int_V G(\bar{r} - \bar{s})p(\bar{s})\,dV(\bar{s}) \tag{A.15}$$

The function $G(\bar{r} - \bar{s})$, the inverse transform of the "transfer function", is the impulse response of the system, so that

$$\mathcal{L}(\bar{r})G(\bar{r}) = \delta(\bar{r}) \tag{A.16}$$

It is the response at observation point $\bar{r}$ of a point source located at the center of coordinates. It is called the Green's function.

Extending this 3-D transformation result into the corresponding 4-D transformation result shows that Green's functions describe basically the space and time response of a physical system at point $\bar{r}$ at time t when a point source of excitation is applied to the system at point $\bar{r}'$ at time t'. By response, one means the space and time-description of a particular scalar or vector quantity $\overline{P}$ characterizing the system. A general excitation $\overline{X}$ can be either a scalar or a vector. There are eight variables involved in the description of the physical system: the six space coordinates x, y, z, x', y', z' represented by the vectors

$\bar{r}$ and $\bar{r}'$, and the time variables t, t'. The response of the system to a given excitation $\overline{X}$ is totally determined once Green's dyadic $\overline{\overline{G}}$ has been found for each pair of points $(\bar{r}, \bar{r}')$ of domain V_x. Defining V_x and T_x as the space and time domains of the excitation, respectively, Green's dyadic notation yields response

$$\overline{P}(\bar{r}, t) = \int_{V_x} \int_{T_x} \overline{\overline{G}}(\bar{r}, t | \bar{r}', t') \cdot \overline{X}(\bar{r}', t') dr' dt' \tag{A.17a}$$

Intuitively, the response of the system to excitation $\overline{X}$ at any observation point is obtained by superimposing the contributions at the observation point of infinitesimal unit point sources of excitation located at the various points $(\bar{r}', t')$ of domain (V_x, T_x) and having the magnitude and orientation of $\overline{X}$ at these points.

From definition (A.17a), the Green's function can also be viewed as a linear integral dyadic operator applied to excitation $\overline{X}$ with the view of finding the corresponding distribution of the parameter $\overline{P}$. It is denoted by $\overline{\overline{\mathcal{G}}}$:

$$\overline{P}(\bar{r}, t) = \overline{\overline{\mathcal{G}}}\{\overline{X}(\bar{r}', t')\} \tag{A.17b}$$

with

$$P_u = \sum_v \left[\overline{\overline{\mathcal{G}}}\{\overline{X}(\bar{r}', t')\} \right]_{uv} \tag{A.17c}$$

$$= \sum_v \int_{V_x} \int_{T_x} G_{uv}(\bar{r}, t | \bar{r}', t') X_v(\bar{r}', t') dr' dt' \tag{A.17d}$$

For a unit point source oriented along the u-axis and applied at $(\bar{r}_0, t_0)$

$$\overline{X}(\bar{r}, t) = \delta(\bar{r} - \bar{r}_0)\delta(t - t_0)\bar{a}_u (u = x, y, z) \tag{A.18a}$$

the three components of the u-th column of Green's dyadic are obtained as the three components of vector when an excitation $\delta(\bar{r}' - \bar{r}_0)\delta(t' - t_0)\bar{a}_u$ is applied:

$$\overline{P}(\bar{r}, t) = \int_{V_x} \int_{T_x} \overline{\overline{G}}(\bar{r}, t | \bar{r}', t') \cdot \{\delta(\bar{r}' - \bar{r}_0)\delta(t' - t_0)\bar{a}_u\} dr' dt' \tag{A.18b}$$

$$= G_{xu}(\bar{r}, t | \bar{r}_0, t_0)\bar{a}_x + G_{yu}(\bar{r}, t | \bar{r}_0, t_0)\bar{a}_y + G_{zu}(\bar{r}, t | \bar{r}_0, t_0)\bar{a}_z \tag{A.18c}$$

In expressions (A.18a,b), the notation $\delta(\bar{r}' - \bar{r}_0)$ represents for the triple product

$$\delta(x' - x_0)\delta(y' - y_0)\delta(z' - z_0) \tag{A.18d}$$

expressing the presence of a point source with unit magnitude, located at point $\bar{r}_0$ at time t_0. The same formalism can also be used for applying specific boundary conditions to a system, that is searching for the space and time distribution of the response $\overline{P}$ associated with the system which satisfies specific boundary conditions on a surface Ω_x. A suitable expression for the Green's dyadic $\overline{\overline{G}}_s$ is obtained by calculating $\overline{P}$ at the points of V_x which are not on the boundary Ω_x while imposing a homogeneous zero boundary condition at every point located on Ω_x, except for one point $\overline{r}_0$ on Ω_x where a unit boundary condition is imposed. The boundary condition is imposed on $\overline{P}$ (Dirichlet condition) or on its gradient (Neumann condition). Hence, the required solution is obtained when summing over Ω_x the partial solutions corresponding to a point boundary condition imposed on an infinitesimal portion of Ω_x around $\overline{r}'$:

$$\overline{P}(\overline{r}, t) = \int_{\Omega_x} \int_{T_x} \overline{\overline{G}}_s(\overline{r}, t | \overline{r}', t') \cdot \overline{X}(\overline{r}', t') dr' dt' \tag{A.19}$$

An important feature of the Green's formalism is that the Green's function $\overline{\overline{G}}$, providing for a given system the distribution $\overline{P}$ of corresponding to any source distribution (A.17), is essentially the same as Green's function $\overline{\overline{G}}_s$ (A.19) yielding specific boundary conditions on a surface for the same system:

1. for Neumann boundary conditions

$$\overline{\overline{G}}_s = \overline{\overline{G}} \quad \text{and} \quad \overline{X}(\overline{r}', t') = \left. \frac{\partial \overline{P}}{\partial \overline{n}} \right|_{\Omega_x} \tag{A.20a}$$

yielding

$$\overline{P}(\overline{r}, t) = \int_{\Omega_x} \int_{T_x} \overline{\overline{G}}(\overline{r}, t | \overline{r}', t') \cdot \left. \frac{\partial \overline{P}(\overline{r}', t')}{\partial \overline{n}} \right|_{\Omega_x} dr' dt' \tag{A.20b}$$

2. for Dirichlet boundary conditions

$$\overline{\overline{G}}_s = \frac{\partial \overline{\overline{G}}}{\partial \overline{n}} \quad \text{and} \quad \overline{X}(\overline{r}', t') = \left. \overline{P} \right|_{\Omega_x} \tag{A.20c}$$

yielding

$$\overline{P}(x, y, z, t) = \int_{\Omega_x} \int_{T_x} \frac{\partial \overline{\overline{G}}(\overline{r}, t | \overline{r}', t')}{\partial \overline{n}} \cdot \left. \overline{P}(\overline{r}', t') \right|_{\Omega_x} dr' dt' \tag{A.20d}$$

where $\overline{n}$ is the normal at the surface [A.1]. The Green's function is in fact the solution for a situation which is homogeneous everywhere except at one point. The solution of a homogeneous equation satisfying inhomogeneous boundary conditions is equivalent, using the Green's formalism, to the solution of an inhomogeneous equation satisfying homogeneous boundary conditions, with the inhomogeneous part being a surface layer of charge. More complicated situations are then handled as linear combinations of the solutions of a homogeneous problem with inhomogeneous boundary conditions, and of an inhomogeneous problem with homogeneous boundary conditions, which is equivalent to inhomogeneous problems with homogeneous boundary conditions.

A.2   Simplified notations

Generally speaking, the notation $\overline{\overline{G}}(\overline{r}, t|\overline{r}', t')$ holds for a 3×3 tensor whose elements are functions of the variables $\overline{r}$, t, $\overline{r}'$, t'. Simpler Green's functions may be defined when the geometry of the problem yields a constant or a a priori-known dependence of the problem along a particular direction, or when its time-dependence is well known or equal to a constant (steady-state systems). For instance, 2-D Green's formulations similar to (A.18b) are defined as

$$\overline{P}(x, y, t) = \int \overline{\overline{G}}(x, y, t|x', y', t')\delta(x' - x_0)\delta(y' - y_0)\delta(t' - t_0)dr'dt'$$

$$= \overline{\overline{G}}(x, y, t|x_0, y_0, t_0)$$

$$\text{(A.21a)}$$

when no variation of the quantities of interest occurs along the z-axis. The 2-D Green's function is then denoted by

$$\overline{\overline{G}}(\overline{r}, t|\overline{r}_0, t_0) = \overline{\overline{G}}(x, y, t|x_0, y_0, t_0) \tag{A.21b}$$

with

$$\overline{r} = x\overline{a}_x + y\overline{a}_y \tag{A.21c}$$

$$\overline{r}' = x'\overline{a}_x + y'\overline{a}_y \tag{A.21d}$$

$$\overline{r}_0 = x_0\overline{a}_x + y_0\overline{a}_y \tag{A.21e}$$

The one-dimensional Green's function is similarly noted

$$\overline{\overline{G}}(x, t|x_0, t_0) \tag{A.21f}$$

When the excitation $\overline{X}$ is described by a scalar, the dyadic reduces to a vector Green's function of the eight variables $(\overline{r}, \overline{r}', t, t')$, and denoted by $\overline{G}(\overline{r}, t|\overline{r}', t')$. When both $\overline{P}$ and $\overline{X}$ are scalar, the dyadic reduces to an algebraic Green's function of the eight variables $(\overline{r}, \overline{r}', t, t')$, noted $G(\overline{r}, t|\overline{r}', t')$. For vector and algebraic Green's formulations, 1- and 2-D forms are readily deduced by substituting respectively the adequate scalar coordinate or $\overline{r}$ vector to the 3-D vector in the previous notations.

A.3   Green's functions and partial differential equations

The Green's formalism has been developed in connection with problems which can be described by inhomogeneous partial differential equations and suitable boundary conditions. It offers an elegant way of handling these problems, because instead of solving the partial differential equation for the quantity $\overline{P}$ in the presence of excitation $\overline{X}$ or for specific boundary conditions, the corresponding equation for the Green's function is solved in the presence of the unit point source of excitation, and (A.17) or (A.20) is then applied. It is expected that solving the equation for $\overline{\overline{G}}$ is much easier than solving it for $\overline{P}$ in the presence of $\overline{X}$ and/or specific boundary conditions.

Denoting $\mathcal{L}$ the differential operator associated with the partial differential equation, the problem becomes

$$\mathcal{L}\{\overline{\overline{G}}(\overline{r}, t|\overline{r}', t')\} \cdot \overline{a}_u = \delta(\overline{r} - \overline{r}')\delta(t - t')\overline{a}_u \tag{A.22a}$$

instead of

$$\mathcal{L}\{\overline{P}(\overline{r}, t)\} = \overline{X}(\overline{r}, t) \tag{A.22b}$$

In electromagnetics, five equations are of main interest.

1. Poisson's equation:

$$\nabla^2 \Phi(\overline{r}) = -\rho(\overline{r}) \qquad \text{where } L_P = \nabla^2 \tag{A.23}$$

The electrostatic field distribution is obtained as the gradient of electrostatic potential Φ.

2. Homogeneous Helmholtz equation:

$$\nabla^2 \Phi(\overline{r}) + k^2 \Phi(\overline{r}) = 0 \qquad \text{where } L_H = \nabla^2 + k^2 \tag{A.24}$$

The harmonic TE and TM fields are obtained as particular combinations of potential Φ solution of the Helmholtz equation and its partial derivatives. A time dependence $e^{j\omega t}$ is assumed for both potentials and fields, so that one has

$$k^2 = \omega^2 \varepsilon \mu \tag{A.25}$$

3. Scalar wave equation:

$$\nabla^2 \Phi(\overline{r}) - (1/c)^2 \frac{\partial^2 \Phi(\overline{r})}{\partial t^2} = -\rho(\overline{r}, t) \qquad \text{where } L_0 = \nabla^2 - (1/c)^2 \frac{\partial^2}{\partial t^2} \tag{A.26}$$

4. Vector wave equation:

$$\nabla^2 \overline{A}(\overline{r}) + \omega^2 \varepsilon \mu \overline{A}(\overline{r}) = -\mu \overline{J}(\overline{r}) \qquad \text{where } L_d = \nabla^2 + \omega^2 \varepsilon \mu \tag{A.27}$$

5. Diffusion equation:

$$\nabla^2 \Phi(\overline{r}) - a^2 \frac{\partial \Phi(\overline{r})}{\partial t} = 0 \qquad \text{where } L_d = \nabla^2 - a^2 \frac{\partial}{\partial t} \tag{A.28}$$

For a vector wave equation, a Green's function can also be derived, which relates the vector field to the vector source. Usually, those Green's functions are derived from the Green's functions relating the source to a scalar or vector potential.

A.4   Green's formulation in the spectral domain

The Green's formulation in the spectral domain (Appendix C) is obtained by taking the Fourier transform along the three space variables, yielding the spectral domain, and along the time variable, yielding the harmonic domain. When the medium is multilayered, particular solutions must be taken into account, so that the boundary conditions at each interface are satisfied. This has been considered in Subsection 3.3.1. The differential operator ∇ is transformed into $j\overline{k}$ and the partial time derivative into $j\omega t$, which yields

$$\tilde{\overline{P}}(k_x, k_y, k_z, \omega) = \frac{1}{(2\pi)^4} \iiiint \overline{P}(x, y, z, t) e^{j\omega t} e^{j\overline{k}\cdot\overline{r}} dk d\omega \tag{A.29}$$

and the corresponding forms of the various equations in the spectral domain, for homogeneous layers. We obtain for Poisson's equation:

$$-|k|^2 \tilde{\Phi}(k_x, k_y, k_z) = -\tilde{\rho}(k_x, k_y, k_z) \qquad \text{where } \tilde{L}_d = -|k|^2 \tag{A.30}$$

The corresponding spectral scalar Green's function $\tilde{G}_P$ satisfies

$$-|k|^2 \tilde{G}_P(k_x, k_y, k_z) = 1 \tag{A.31}$$

and the general form of the spectral Green's function associated with Poisson's equation is

$$\tilde{G}_P(k_x, k_y, k_z) = -\frac{1}{|k|^2} \tag{A.32}$$

Equation (A.30) combined with (A.32) provides the spectral form of the potential

$$\tilde{\Phi}(k_x, k_y, k_z) = \tilde{G}_P(k_x, k_y, k_z)\tilde{\rho}(k_x, k_y, k_z) \tag{A.33}$$

Its inverse Fourier-transform is

$$\Phi(\overline{r}, t) = \int_{V_x} \int_{T_x} G_P(\overline{r} - \overline{r}', t - t')\rho(\overline{r}', t') dr' dt' \tag{A.34}$$

The Green's formulation in the spectral domain has the advantage that the differential or integral operators are replaced by simple algebraic expressions, so that the derivation of the spectral Green's function is very simple (A.31). Also, the solution is obtained by taking the product of the spectral Green's function by the Fourier-transform of the excitation (A.33), instead of using the Green's function as a linear integral operator for elaborating the solution. The user has the choice either to invert the spectral Green's function and apply (A.17) in the space domain, or to invert the obtained spectral solution (A.33). A Green's function may of course be deduced for relating the electrostatic field gradient of the potential to the excitation distribution $\rho(\overline{r})$.

Green's functions offer an elegant way of characterizing a physical system, for any source or boundary conditions, since the knowledge of the Green's function is sufficient to calculate the response of the system to any excitation. Hence, only the Green's function is usually mentioned for the five categories of equations introduced above. For example, we have:

a. for the Helmholtz equation:

$$\tilde{G}_H(k_x, k_y, k_z, \omega) = \frac{1}{-|k|^2 + \omega^2 \varepsilon \mu} \tag{A.35}$$

b. for the diffusion equation:

$$\tilde{G}_D(k_x, k_y, k_z, \omega) = \frac{-1}{|k|^2 + j\omega a^2} \tag{A.36}$$

A.5 Reciprocity

Green's functions are subject to the fundamental theorems of electromagnetics. The most important for the present purpose is the reciprocity theorem: for steady-state systems in an unbounded free-space, the effect at the observation point $\bar{r}$ of a source located at point $\bar{r}'$ is identical to the effect observed at point $\bar{r}'$ when the same source is located at point $\bar{r}$:

$$\overline{\overline{G}}(\bar{r}|\bar{r}_0) = \overline{\overline{G}}(\bar{r}_0|\bar{r}) \tag{A.37a}$$

with

$$\overline{\overline{G}} = \overline{\overline{G}}^T \tag{A.37b}$$

This property is satisfied by the scalar Green's function associated to Poisson's equation for quasi-static electromagnetic problems and by the Green's dyadic associated with the vector Helmoltz equation solved in free-space [A.2]. Reciprocity in a free-space unbounded medium is thus expressed by a symmetry property of the corresponding Green function in the position vectors $\bar{r}_0$ and $\bar{r}$. Property (A.37a,b) is valid for most electromagnetic problems, provided they are described in a unbounded free-space medium. For more practical situations (inhomogeneous and/or bounded media), Green's functions satisfy some reciprocity properties, however they are no longer symmetrical dyadics.

When time is introduced, identity (A.37a) is no longer valid. It has to be replaced by

$$\overline{\overline{G}}(\bar{r}, t|\bar{r}_0, t_0) = \overline{\overline{G}}(\bar{r}_0, -t_0|\bar{r}, -t) \tag{A.38}$$

The sign inversion for the time variable ensures the causality of the physical system described by the Green's function.

A.6   Distributed systems

This review of the main characteristics of the Green's formalism is now adapted to distributed systems, where propagation occurs along a particular direction, taken as the z-axis. Assuming k_{zo} as the propagation constant, the source distribution can be written

$$\overline{X}(\overline{r}', t') = \overline{x}(x', y', t') e^{-jk_{z0}z'} \tag{A.39}$$

where the lower case $\overline{x}$ holds for transverse dependence of fields for which dependence along the longitudinal direction may be separated. Introducing this expression into general definition (A.18) yields the relationship between the response $\overline{P}$ and the transverse dependence of the excitation:

$$\overline{P}(\overline{r}, t) = \int_{V_x} \int_{T_x} \overline{\overline{G}}(\overline{r}, t | \overline{r}', t') \cdot \overline{x}(x', y', t') e^{-jk_{z0}z'} dr' dt' \tag{A.40a}$$

$$= \int_{V_x} \int_{T_x} \overline{\overline{G}}(\overline{r}, t | \overline{r}', t') \cdot \overline{x}(x', y', t') e^{-jk_{z0}(z'-z)}$$
$$e^{-jk_{z0}z} dx' dy' d(z' - z) dt' \tag{A.40b}$$

$$= e^{-jk_{z0}z} \int_{V_x} \int_{T_x} \overline{\overline{G}}(x, y, t | x', y', t'; k_{z0}) \cdot \overline{x}(x', y', t') dx' dy' dt' \tag{A.40c}$$

which is a Green's formulation with six variables.

A.7   References

[A.1]   P. M. Morse and H. Feshbach, *Methods of Theoretical Physics*. New York: McGraw-Hill, 1953, Ch. 7.

[A.2]   R. E. Collin, *Field theory of guided waves*, 2nd edn. New York: IEEE Press, 1991, Ch. 2.

APPENDIX B

Green's identities and theorem

B.1   Scalar identities and theorem

Let V be a closed region of space bounded by a regular surface S, and let a and b be two scalar functions of position which together with their first and second derivative are continuous throughout V and on surface S. (It is mentioned by Stratton that this condition is more stringent than is necessary, and that the second derivative of one function b need not be continuous [B.1]). Then the divergence theorem applied to the vector $b\nabla a$ gives

$$\int_V \nabla \cdot (b\nabla a)\, dV = \int_S (b\nabla a) \cdot \bar{n}\, dS \tag{B.1}$$

Expanding the divergence to

$$\nabla \cdot (b\nabla a) = \nabla b \cdot \nabla a + b\nabla \cdot \nabla a = \nabla b \cdot \nabla a + b\nabla^2 a \tag{B.2}$$

and noting that

$$\nabla a \cdot \bar{n} = \frac{\partial a}{\partial n} \tag{B.3}$$

where $\partial a/\partial n$ is the derivative in the direction of the positive normal, yields what is known as *Green's first scalar identity* [B.1][B.2]:

$$\int_V \nabla b \cdot \nabla a\, dV + \int_V b\nabla^2 a\, dV = \int_S b\frac{\partial a}{\partial n}\, dS \tag{B.4}$$

If, in particular, we place $b = a$ and let a be a solution of Laplace's equation, (B.4) reduces to

$$\int_V (\nabla a)^2\, dV = \int_S a\frac{\partial a}{\partial n}\, dS \tag{B.5}$$

If the roles of the functions a and b are interchanged, *i.e.* applying the divergence theorem to the vector $a\nabla b$, one obtains

$$\int_V \nabla a \cdot \nabla b\, dV + \int_V a\nabla^2 b\, dV = \int_S a\frac{\partial b}{\partial n}\, dS \tag{B.6}$$

Upon subtracting (B.6) from (B.4) a relation between a volume integral and a surface integral is obtained of the form

$$\int_V \left(b\nabla^2 a - a\nabla^2 b\right) dV = \int_S \left(b\frac{\partial a}{\partial n} - a\frac{\partial b}{\partial n}\right) dS \tag{B.7}$$

known as *Green's second scalar identity* or also frequently as *Green's theorem* [B.1].

B.2   Vector identities

It is possible to establish a set of vector identities wholly analogous to those of Green for scalar functions. Let $\overline{P}$ and $\overline{Q}$ be two functions of position which together with their first and second derivatives are continuous throughout V and on the surface S. Then, if the divergence theorem is applied to the vector $\overline{P} \times \nabla \times \overline{Q}$, one has

$$\int_V \nabla \cdot (\overline{P} \times \nabla \times \overline{Q})\, dV = \int_S (\overline{P} \times \nabla \times \overline{Q}) \cdot \overline{n}\, dS \tag{B.8}$$

Upon expanding the integrand of the volume integral one obtains the vector analogue of Green's first identity

$$\int_V (\nabla \times \overline{P} \cdot \nabla \times \overline{Q} - \overline{P} \cdot \nabla \times \nabla \times \overline{Q})\, dV = \int_S (\overline{P} \times \nabla \times \overline{Q}) \cdot \overline{n}\, dS \tag{B.9}$$

The analogue of Green's vector second identity is obtained by an interchange of the roles of $\overline{P}$ and $\overline{Q}$ in (B.9) followed by subtracting from (B.9). As a result

$$\int_V (\overline{Q} \cdot \nabla \times \nabla \times \overline{P} - \overline{P} \cdot \nabla \times \nabla \times \overline{Q})\, dV$$
$$= \int_S (\overline{P} \times \nabla \times \overline{Q} - \overline{Q} \times \nabla \times \overline{P}) \cdot \overline{n}\, dS \tag{B.10}$$

B.3   References

[B.1]   J.A. Stratton, *Electromagnetic Theory*. New York: McGraw-Hill, 1941, Chs. 3 and 4.

[B.2]   E. Roubine, *Mathematics applied to physics*, 1970. Berlin: Springer-Verlag, Ch. V.

Fourier transformation and Parseval's theorem

C.1 Fourier transformation into spectral domain

A number of mathematical texts have been published on the Fourier transformation. Some others, like [C.1], are intended for those who are concerned with applying Fourier transforms to physical situations rather than with furthering the mathematical subject as such. The purpose of this appendix is not to review or summarize the subject. Quite specifically, it is intended to show how to extend the classical Fourier transform, relating time and frequency domain, to the relation between space and spectral domain, for scalar as for vector quantities. Most usually, the Fourier transformation is applied to a function $f(t)$, described in the *time domain*. It is defined by

$$F(\omega) = \int_{-\infty}^{+\infty} f(t)e^{-j\omega t}\, dt \tag{C.1}$$

which is a function described in what is called the *frequency domain*. Using a similar transformation, which is called the inverse transformation, yields

$$f(t) = \frac{1}{2\pi} \int_{-\infty}^{+\infty} F(\omega)e^{j\omega t}\, d\omega \tag{C.2}$$

Calculating the Fourier transform of an ordinary differential equation offers the enormous advantage of transforming the operators contained in the equation into terms of a polynomial expression. It must be observed that the second transformation (C.2) is not exactly the same as the first.

On the other hand, the Fourier transformation can be applied to any pair of quantities, relating the first quantity to its *spectral* transform. Regarding (x, s) as such a pair, and applying the Fourier transformation twice yields

$$F(s) = \int_{-\infty}^{+\infty} f(x)e^{-j2\pi xs}\, dx \tag{C.3}$$

and

$$f(w) = \int_{-\infty}^{+\infty} F(s)e^{-j2\pi ws}\, ds \tag{C.4}$$

This time, the same transformation has been applied twice. When $f(x)$ is an even function of x, that is when $f(x) = f(-x)$, the repeated transformation yields the initial function. This is the *cyclical* property of the Fourier transformation, and since the cycle is in two steps, the *reciprocal* property is implied: if $F(s)$ is the Fourier transform of $f(x)$, then $f(x)$ is the Fourier transform of $F(s)$. The cyclical and reciprocal properties are imperfect, however, because when the initial function is odd, the repeated transformation yields $f(-w)$. When $f(x)$ is neither even nor odd, the imperfection is still more pronounced. The customary formulas exhibiting the reversibility of the Fourier transformation are

$$F(s) = \int_{-\infty}^{+\infty} f(x)e^{-j2\pi xs}\, dx \qquad (C.5)$$

$$f(x) = \int_{-\infty}^{+\infty} F(s)e^{j2\pi xs}\, ds \qquad (C.6)$$

When variable x is time t, then variable $2\pi s$ is frequency ω, which yields expressions (C.1) and (C.2). Factor 2π is often taken out of the integral in the expressions.

Fourier transformation can easily be applied to functions with several variables. In electromagnetics, the pairs of variables will often be the space variables (x, y, z) on one side and the spectral variables (k_x, k_y, k_z) on the other, to obtain

$$F(k_x, k_y, k_z) = \int_{-\infty}^{+\infty} \int_{-\infty}^{+\infty} \int_{-\infty}^{+\infty} f(x,y,z)e^{-j(xk_x+yk_y+zk_z)}\, dx\, dy\, dz \quad (C.7)$$

$$f(x, y, z) = \frac{1}{(2\pi)^3} \int_{-\infty}^{+\infty} \int_{-\infty}^{+\infty} \int_{-\infty}^{+\infty} F(k_x, k_y, k_z)e^{j(xk_x+yk_y+zk_z)}\, dk_x\, dk_y\, dk_z$$
$$(C.8)$$

Similarly, the Fourier transformation can be used to transform the four variables of space-time domain into the four spectral variables k_x, k_y, k_z and ω. The extension to vector fields is immediate: the component of the transform is the transform of the corresponding component, and the transform of a vector field is also a vector field. Transforming, for instance, from the time domain into the frequency domain for an electric field, yields

$$\overline{E}(\overline{r}, \omega) \triangleq \int_{-\infty}^{+\infty} \overline{\mathcal{E}}(\overline{r}, t)e^{-j\omega t}\, dt \qquad (C.9)$$

$$\overline{\mathcal{E}}(\overline{r}, t) \triangleq \frac{1}{2\pi} \int_{-\infty}^{+\infty} \overline{E}(\overline{r}, \omega)e^{j\omega t}\, d\omega \qquad (C.10)$$

while transforming from the space domain into the spectral domain yields expressions which are identical to (C.7)(C.8) except for replacing $F(k_x, k_y, k_z)$ and $f(x, y, z)$ by the correspondent vector fields.

It is interesting to observe that the sum $xk_x + yk_y + zk_z$ is the scalar product of the vectors position in space domain and in spectral domain, respectively. It is rather customary to express this sum as a dot product in the integrals, which renders the expressions more compact. A further improvement is to express $dx\,dy\,dz$ and $dk_x\,dk_y\,dk_z$ as unit volumes $dV(r)$ and $dV(k)$ in space domain and spectral domain, respectively.

C.2   Parseval's theorem

There is some ambiguity about what is called Parseval's theorem. Rigorously speaking, Parseval's theorem is related to Fourier series, while its equivalent for Fourier transforms is sometimes called Rayleigh's theorem [C.1].

Calling $F(u, v)$ the two-dimensional Fourier transform of the two-dimensional function $f(x, y)$, Parseval's theorem for Fourier series is written [C.1]

$$\int_{-\infty}^{+\infty} \int_{-\infty}^{+\infty} |f(x,y)|^2 = \sum \sum a_{mn}^2 \tag{C.11}$$

where

$$\left| F(u,v) \right|^2 = \sum \sum a_{mn}^2 [\delta(u - m, v - n)]$$

When related to Fourier transforms, Parseval's theorem is sometimes called Rayleigh's theorem. The reason, reported in [C.1], is that the theorem was first used by Rayleigh in his study of black-body radiation, published in 1899. It states that the integral of the squared modulus of a function is equal to the integral of the squared modulus of its spectrum; that is

$$\int_{-\infty}^{+\infty} |f(x)|^2 \, dx = \frac{1}{2\pi} \int_{-\infty}^{+\infty} |F(s)|^2 \, ds \tag{C.12}$$

or

$$
\begin{aligned}
\int_{-\infty}^{+\infty} f(x) f^*(x) \, dx &= \int_{-\infty}^{+\infty} f(x) f^*(x) e^{-j2\pi x s'} \, dx & s' = 0 \\
&= F(s') * F^*(-s') & s' = 0 \\
&= \int_{-\infty}^{+\infty} F(s) F^*(s - s') \, ds & s' = 0 \\
&= \int_{-\infty}^{+\infty} F(s) F^*(s) \, ds
\end{aligned}
\tag{C.13}
$$

Each integral represents the amount of energy in a system, one integral being taken over all values of a coordinate, the other over all spectral components.

The theorem is true if one of the integrals exists. Form (C.13) is used in Chapters 3 and 4 of this book.

It has become customary to replace continuous Fourier transforms by an equivalent discrete Fourier transform (DFT) and then evaluate the DFT using the discrete data. Parseval's theorem is then usually written as [C.2]:

$$\sum_{n=0}^{N-1} f^2(n) = \frac{1}{N} \sum_{k=0}^{N-1} |F(k)|^2 \tag{C.14}$$

where

$$|F(k)|^2 = F(k)F^*(k)$$

C.3 References

[C.1] R.N. Bracewell, *The Fourier Transform and Its Applications*. New York: McGraw-Hill, 1965.

[C.2] R.A. Dorf, *The Electrical Engineering Handbook*. Boca Raton: CRC Press, 1993.

APPENDIX D

Ferrites and Magnetostatic Waves

D.1 Permeability tensor of a Ferrite YIG-film

Assuming a DC field along the y-axis, the magnetic permeability tensor of a YIG-film is defined as [D.1]

$$\overline{\overline{\mu}} = \begin{bmatrix} \mu_{11} & 0 & \mu_{21} \\ 0 & 1 & 0 \\ \mu_{12} & 0 & \mu_{11} \end{bmatrix} \tag{D.1a}$$

where

$$\mu_{11} = \mu_0 \left\{ 1 + \frac{(f/\omega_0)(\omega_h g M_s + jfa)}{(\omega_h g M_s + jfa)^2 - f^2} \right\} \tag{D.1b}$$

$$\mu_{12} = \mu_0 \left\{ j \frac{f^2/\omega_0}{(\omega_h g M_s + jfa)^2 - f^2} \right\} \tag{D.1c}$$

$$= -\mu_{21} \tag{D.1d}$$

with

a	$= \Delta H/(2H_{int})$
ω_h	$= H_{int}/M_s$
ω_0	$= f/(gM_s)$
H	thickness of film [m]
H_{int}	DC-magnetic field inside YIG-film [Oe]
	when a field H_0 is created by magnetic poles
M_s	saturation magnetization of the film [G]
	(usually noted $4\pi M_s$ in data sheets)
ΔH	linewidth of YIG-film [G]
g	gyromagnetic ratio $= 2.8$ [MHz/Oe]

D.2 Demagnetizing effect

The YIG-film is placed in a uniform DC-magnetic field applied along the y-axis (Fig. 4.2). The demagnetizing effect exists because a YIG-sample placed into a uniform DC-magnetic field H_0, generates an internal component of field necessary to ensure the boundary conditions at the interfaces between YIG and air. The result is that the total internal magnetic field inside the YIG-sample, denoted by H_{int}, is lower than the external applied field, hence a demagnetizing effect occurs. It is usually expressed via a demagnetizing factor N_y:

$$H_{int} = H_0 - N_y M_s \tag{D.2}$$

For ellipsoidal bodies, the demagnetizing effect is easy to calculate [D.2]. For a YIG-film infinite along the x- and z-axis, one has

$$N_y = 1 \tag{D.3a}$$

so that

$$H_{int} = H_0 - M_s \tag{D.3b}$$

For a YIG-film of finite width, the exact calculation of the demagnetizing factor is difficult. One solution proposed by Joseph and Schlomann [D.3] develops the total internal magnetic field as a series of ascending powers in (H_0/M_s). The first-order term is used in our models and is sufficient to predict the internal magnetic field. The main characteristics of the demagnetizing effect obtained when using those expressions is that the resulting internal field is nonuniform over the sample width. The demagnetizing factor is about 1 at the center of the film, while it decreases to $1/2$ at its edges.

Hence, the permeability tensor to be considered in the film varies with the x-position in the sample, since the internal field varies. The edge values of the permeability tensor are calculated by using the value of H_{int} at the edges of the film. The position-dependent inverted permeability tensor required in (4.2) is calculated for each x-position and the integral of each term of (4.2) over the film width is performed numerically (simple trapezoidal method).

D.3   Magnetostatic Wave Theory

D.3.1   Magnetostatic assumption

Magnetostatic waves are derived from the solution of Maxwell's equations for magnetically ordered matter, completed with the constitutive equations:

$$\begin{aligned} \nabla \times \overline{E} = -j\omega\overline{B} \qquad & \qquad \nabla \cdot \overline{B} = 0 \\ \nabla \times \overline{H} = \overline{J}_\nu + j\omega\overline{D} \qquad & \qquad \nabla \cdot \overline{D} = \rho_\nu \end{aligned} \tag{D.4a}$$

$$\overline{D} = \overline{\overline{\varepsilon}} \cdot \overline{E} \qquad \overline{B} = \overline{\overline{\mu}} \cdot \overline{H} \qquad \overline{J}_\nu = \sigma\overline{E} \tag{D.4b}$$

In this appendix we only deal with magnetic dielectrics (ferrites) in which free charge carriers do not exist and the current density is negligible. In addition, we assume that this matter is electrically isotropic, so that the permittivity tensor reduces to the scalar ε. If we separate the quantities in (D.4a) into DC and RF components:

$$\overline{H} = \overline{H}_0 + \overline{h}\, e^{j\omega t} \qquad\qquad \overline{E} = \overline{E}_0 + \overline{e}\, e^{j\omega t}$$
$$\overline{B} = \overline{B}_0 + \overline{b}\, e^{j\omega t} \qquad\qquad \overline{D} = \overline{D}_0 + \overline{d}\, e^{j\omega t}$$

we obtain

$$\nabla \times \overline{e} = -j\omega \overline{b} \qquad\qquad \nabla \cdot \overline{b} = 0$$
$$\nabla \times \overline{h} = j\omega \overline{d} \qquad\qquad \nabla \cdot \overline{d} = 0 \tag{D.5a}$$

$$\overline{d} = \varepsilon \overline{e} \qquad \overline{b} = \overline{\overline{\mu}} \cdot \overline{h} \tag{D.5b}$$

We now simplify system (D.5a) [D.4], taking advantage of the fact that we are only interested in slow electromagnetic waves in ferrimagnetic media (*i.e.* $k >> k_0 \sqrt{\varepsilon/\varepsilon_0}$), so that the following condition is met:

$$\frac{2\pi}{\lambda} >> \frac{\omega}{c} \qquad \text{or} \qquad \omega << \frac{2\pi c}{\lambda} \tag{D.6}$$

where λ is the wavelength of the slow electromagnetic wave in the medium and c is the light velocity in a medium of permittivity ε. Form (D.6) is valid when ω is small (quasi-stationary approximation) or when the wavelength of the excitation in medium λ is much smaller than the wavelength in vacuum, *i.e.* $\lambda << \frac{2\pi c}{\omega}$ (quasi-static approximation). Taking this into account, then system (D.5a) can be split into two independent systems for electric and magnetic fields:

$$\nabla \times \overline{e} = 0 \qquad\qquad \nabla \cdot (\overline{\overline{\mu}} \cdot \overline{h}) = 0$$
$$\nabla \times \overline{h} = 0 \qquad\qquad \nabla \cdot (\varepsilon \overline{e}) = 0 \tag{D.7}$$

System (D.7) is similar to the equations of statics, the only difference being that the permeability is frequency-dependent. For this reason slow electromagnetic waves in magnetically ordered matter are called magnetostatic waves. There is, however, a comment to be made. System (D.4a) can be solved for special cases, yielding expressions for the components of high-frequency electric and magnetic fields [D.5]. If condition (D.6) is applied to those solutions, *i.e.* quasi-static approximation, we see that the magnetic fields obtained satisfy the magnetostatic equations ($\nabla \times \overline{h} = 0$) while the electric fields do not satisfy equation $\nabla \times \overline{e} = 0$, except in the limit when the high-frequency electric field vanishes. This requirement is not realized since magnetostatic waves, although slow, are electromagnetic waves: they have to contain both magnetic and electric components of the high-frequency field. That is why system (D.7) for magnetostatic waves has no physical basis and has to be replaced by system (D.4a). Maxwell's equations can then be written for magnetostatic approximation:

$$\nabla \times \overline{e} = -j\omega \overline{\overline{\mu}} \cdot \overline{h} \qquad\qquad \nabla \cdot (\overline{\overline{\mu}} \cdot \overline{h}) = 0$$
$$\nabla \times \overline{h} = 0 \qquad\qquad \nabla \cdot (\varepsilon \overline{e}) = 0 \tag{D.8}$$

D.3.2　Types of Magnetostatic Waves

The Damon-Eshbach study [D.6] establishes that one can distinguish between three MSW-types: magnetostatic surface waves (MSSW), magnetostatic forward volume waves (MSFVW) and magnetostatic backward volume waves (MSBVW), with the orientation of the internal bias field relative to the YIG-film and the propagation direction determining which particular wave-type can exist. We assume that the propagation occurs along the z-axis. A wave having its phase and group velocities in the same direction propagates in a perpendicularly magnetized chain of magnetic dipoles. This wave is analogous to a slow electromagnetic wave in perpendiculary magnetized ferrite slabs (H_{dc} y-directed). It is a MSFVW. A wave having its phase and group velocities in opposite directions propagates in a longitudinally magnetized chain of magnetic dipoles. This wave is analogous to a slow electromagnetic wave in the longitudinally magnetized ferrite slabs (H_{dc} z-directed). It is a MSBVW. Magnetostatic surface waves are the most common and well investigated magnetostatic waves. They propagate in ferromagnetic materials magnetized in the layer plane perpendicularly to the direction of the magnetic field (H_{dc} x-directed).

D.4　Perfect Magnetic Wall Assumption

The physical phenomenon giving the boundary condition at the film edges is described as follows by O'Keeffe and Patterson [D.7]: "In most experimental situations the sample edges are defined either by slicing or chemical etching. The combination of the edge roughness produced by the sample preparation and demagnetizing fields is sufficient to pin the spins at the sample edges. We shall approximate the pinning condition by assuming the microwave fields [tangential] $B_x = B_y = 0$, at $z = 0$, w." A number of authors refer to this explanation.

D.5　References

[D.1]　A. Vander Vorst, D. Vanhoenacker, *Bases de l'ingénierie micro-onde*, Ch. 2. Brussels: De Boeck-Wesmael, 1996.

[D.2]　B. Lax, K.J. Button, *Microwave Ferrites and Ferrimagnetics*. New York: McGraw Hill, 1962.

[D.3]　R.I. Joseph, E. Schlomann, "Demagnitizing field in nonellipsoidal bodies", *J. Appl. Phys.*, vol. 36, no. 5, pp. 1579-1593, May 1965.

[D.4]　P. Kabos, V.S. Stalmachov, *Magnetostatic waves and their applications*. London: Chapman&Hall, 1994.

[D.5]　A.V. Vashkovsky, V.S. Stalmachov, Yu. N. Sharaevsky, "Magnetostatic waves in high-frequency electronics", *SGU, USSR, 1991* (in Russian), in P. Kabos, V.S. Stalmachov, *Magnetostatic waves and their applications*. London: Chapman&Hall, 1994.

[D.6] R.W. Damon, J.R. Eshbach, "Magnetostatic modes of a ferromagnetic slab", *J. Phys. Chem. Solids*, vol. 19, pp. 308-320, in J.P. Parekh, K.W. Chang, H.S. Tuan, "Propagation characteristics of magnetostatic waves", *Circuits, Systems and Signal Proc.*, vol. 4, no. 1-2, pp. 9-38, 1985.

[D.7] T.W. O'Keeffe and R.W. Patterson, "Magnetostatic surface-wave propagation examples", *J. Appl. Phys.*, vol. 49, no. 9, pp. 4886-4895, Sept. 1978.

APPENDIX E

Floquet's theorem and Mathieu functions

E.1 Floquet's theorem

Wave propagation in a periodic medium has always received much interest, largely because of the importance of cristalline media [E.1] and periodically loaded structures, such as waveguides for example [E.2]. The resolution of those problems is based on Floquet's theorem [E.3], applicable to linear differential equations with periodic coefficients. When applied to wave propagation in periodic structures its main result is that, for a certain oscillation mode of the structure at a given frequency, the wave function, *i.e.* the electromagnetic field, is multiplied by a complex constant when moving one period along the structure. Floquet's theorem applies to equations of the form

$$y'' + p(x)\, y' + q(x)\, y = 0 \tag{E.1}$$

where $p(x)$ and $q(x)$ are periodical, with a period D, which are parametric linear equations. Hill's and Mathieu's equations are particular cases of (E.1), with Mathieu's equation being a particular case of Hill's equation. We describe briefly the main results [E.4]. If $y(x)$ is a solution of (E.1) then $y(x+D)$ is of course also a solution, as well as the expressions

$$y_1(x + D) = a\, y_1(x) + b\, y_2(x) \tag{E.2}$$

$$y_2(x + D) = c\, y_1(x) + d\, y_2(x) \tag{E.3}$$

Writing these expressions in matrix form yields

$$\begin{vmatrix} y_1(x + D) \\ y_2(x + D) \end{vmatrix} = \overline{\overline{A}} \begin{vmatrix} y_1(x) \\ y_2(x) \end{vmatrix} \tag{E.4}$$

with

$$\overline{\overline{A}} = \begin{vmatrix} a & b \\ c & d \end{vmatrix} \tag{E.5}$$

where $\overline{\overline{A}}$ is independent of x. It appears that for most cases in electromagnetics

$$\det(\overline{\overline{A}}) = 1 \tag{E.6}$$

From solution y_1 and y_2 of equation (E.4) one can derive other solutions, Y_1 and Y_2, for instance. These have also to satisfy (E.4). It can be shown that it is possible to diagonalize the relation between $Y_i(x + D)$ and Y_i, yielding

$$Y_1(x + D) = a\, Y_1(x)$$
$$Y_2(x + D) = b\, Y_2(x)$$
(E.7)

The diagonalization leads to $ab = 1$, with as general solution

$$a = e^{-\mu} \quad ; \quad b = e^{\mu}$$
(E.8)

where μ is complex. Relations (E.6) can then be written

$$Y_1(x + D) = e^{-\mu}\, Y_1(x)$$
$$Y_2(x + D) = e^{\mu}\, Y_2(x)$$
(E.9)

which form Floquet's theorem.

E.2   Mathieu's equation and functions

When written in elliptical and hyperbolic coordinates, Laplace's equation becomes

$$\frac{d^2y}{dz^2} + (a - 2q\cos 2z)y = 0$$

which is called Mathieu's differential equation [E.4]-[E.7]. It is obviously a particular case of (E.1). Here, we limit ourselves to the situation where variable z and parameters a and q are real. When the observation point is able to move freely along a complete ellipse, the solutions of (E.1) are periodic of period 2π; they are the *Mathieu functions* of integer order. They exist only when parameters a and q are linked by a rather intricate relationship. If q goes to zero, the equation becomes

$$\frac{d^2y}{dz^2} + a(0)y = 0$$
(E.10)

with periodic solutions when $a(0)$ is the square of an integer m. These solutions are $\cos mz$ and $\sin mz$. The functions of period 2π becoming $\cos mz$ and $\sin mz$ when q goes to zero are called Mathieu functions of index m. Mathieu functions are usually described in an elliptical system. (They can also be transformed into *modified Mathieu functions*, to be described in an hyperbolic system.) By denoting ce and se the elliptical cosine and sine respectively, those functions are designated respectively by

$$\mathrm{ce}_m(z, q) \quad \text{and} \quad \mathrm{se}_m(z, q)$$
(E.11)

When $m = 0$, the limit of ce_0 is unity. The relation between a and q is called the *characteristic equation*. It can be written

$$F_m(a, q) = 0 \quad \text{with} \quad F_m(m^2, 0) = 0 \quad \text{for} \quad \mathrm{ce}_m$$
$$\Phi_m(a, q) = 0 \quad \text{with} \quad \Phi_m(m^2, 0) = 0 \quad \text{for} \quad \mathrm{se}_m$$
(E.12)

Equation (E.11) shows that Mathieu functions of integer order are functions of the two variables z and q which depend on the index m, with a 2π period for variable z. The functions (E.12) are represented in the space (a, q) [E.5] which yields a set of curves characterizing Mathieu functions of integer order. These curves have no point in common, except on the a-axis: it is impossible to find a pair a, q satisfying simultaneously both equations (E.12). Hence solutions ce_m and se_m cannot co-exist, except when $a = 0$. So, the general solution of (E.10) cannot be a linear combination of two Mathieu functions of integer order. This leads to the definition of *Mathieu functions of second kind*, solution linearly independent of ce_m and se_m.

Mathieu functions have properties similar to those of the other eigenfunctions of Laplace's equation, described in other coordinate systems. They are *orthogonal*, in the sense that the integrals over 2π of the products $ce_m\ ce_p$, $se_m\ se_p$, and $ce_m\ se_p$ are zero if m differs from p. They can be decomposed into Fourier series and into Bessel functions. In addition, Mathieu functions can be obtained for any a and q.

E.3 References

[E.1] G. Floquet, "Sur les équations différentielles linéaires à coefficients périodiques ", *Ann. Ecole Norm. Sup.*, vol. 12, no. 47, pp. 47-88, 1883.

[E.2] L. Brillouin, *Wave Propagation in Periodic Structures*. New York: Dover, 1953.

[E.3] J.C. Slater, *Microwave Electronics*. New York: Van Nostrand, 1950.

[E.4] A. Vander Vorst, D. Vanhoenacker-Janvier, *Bases de l'ingénierie micro-onde*. Brussels: De Boeck&Larcier, 1996, App. C.

[E.5] A. Angot, *Compléments de mathématiques*. Paris: Revue d'Optique, 1961, Ch. 7.

[E.6] I.M. Abramowitz, I.A. Stegun, *Handbook of mathematical functions*. New York: Dover, 1965, 20.

[E.7] E. Jahnke, F. Emde, *Tables of functions*. New York: Dover, 1945, XI.

Index